PROCESSING AND PRESERVATION

PROCESSING AND PRESERVATION

Editor

Rodrigo Affini

scitus academics

Processing and Preservation

Edited by **Rodrigo Affini**

This Book is Printed on Acid-Free Paper. ♾

ISBN: 9781681179407
Library of Congress Control Number: 2017949079

SCITUS Academics LLC,
616, Corporate Way, Suite 2, 4766,
Valley Cottage, NY 10989

www.scitusacademics.com

Notice

Printed in United States of America.

Preface

Horticultural crops such as fruits and vegetables are very important in human nutrition because when consumed in adequate amounts, they are an important source of vitamins, mineral nutrients and fiber. Processing of fruits and vegetables is very important to preserve quality. The magnitude of postharvest losses of fresh fruits and vegetables is estimated to be 5 to 25% in developed countries and 20 to 50% in developing countries depending upon commodity. In spite of having varied agro-climatic conditions, abundance of natural resources like sunlight and water, sufficient labor availability and abundant produce, our country is trailing behind in productivity, export and processing of horticultural produce as compared with other horticultural advanced countries. Further, due to lack of adequate post-harvest handling, processing and infrastructure facilities, post-harvest losses caused by spoilage are very high. It is estimated that post-harvest losses of horticultural produce range between 8-37 per cent. Generally losses occur during pre-harvesting, harvesting, transportation, storage, and processing, packing, marketing and distribution stages. When conditions are not suitable for storage or immediate marketing of fresh produce, many horticultural crops can be processed using simple technologies. There are some processing methods that can be used by small-scale handlers, including drying, fermenting, canning, freezing, preserving and juicing. Fruits, vegetables and flowers can all be dried and stored for use or sale in the future. Fruits and vegetables can be canned or frozen, and fruits are often preserved in sugar or juiced. The main objective of fruit and vegetable processing is to supply wholesome, safe, nutritious and acceptable food to consumers at affordable prices throughout the year. Successful fruit and vegetable processing enterprises require a lot of skill and provision of a consistently high quality product. Processors must therefore employ quality control measures to ensure uniformity and consistency in their products.

The book covers the processing of a wide range of fruit, vegetable and root produce. These foods are the cornerstone of good health and contain many different vitamins and minerals. This book presents several pre- and postharvest strategies that have been developed to modify these physiological activities, resulting in increased shelf life. This will be of invaluable for students, professional horticulturalists and readers with interest in the subject.

Table of Contents

CHAPTER 1

Value-Added Fruit Processing for Human Health

H.P. Vasantha Rupasinghe[1] and Li Juan Yu[1]

[1]*Faculty of Agriculture, Dalhousie University, Truro, Nova Scotia, Canada*

1. INTRODUCTION

Fruits are staple food in human diet. There has been a growing interest in the connection of fruit and vegetable consumption and improved health. Research have shown that biologically active components in plant-based foods, particularly phytochemicals such as polyphenolics and carotenoids, have important role in reducing the risks of chronic diseases, including cancer, cardiovascular disease, diabetes and Alzheimer's disease, among others. The first part of the chapter provides a brief update of the links between fruit-based antioxidants and other biologically active compounds and potential health benefits.

Fruit production is increasing globally. Despite the increasing fruit production at the global level, a significant amount of fruit produced is lost or wasted due to poor post-harvest management. The second part of the chapter provides information on current status of post-harvest losses in selected fruits and methods to prevent these losses. Therefore, processing fruits into value-added products is one of the strategies to reduce post-harvest losses and promote consumption of fruits.

Fresh-cut fruits, also called minimally processed fruits, are products that are partially prepared, maintain a fresh-like state and ready for use and eating. Recently, fresh-cut fruits have become popular because they meet the consumer demand for convenient ready-to-eat foods with fresh-like quality. However, fresh-cut fruits are more perishable than whole fruits.

The third part of the chapter covers some recently developed approaches for the value addition of fresh-cut fruits with respect to the use of natural antimicrobials, anti-browning agents, edible coating, modified atmosphere packaging (MAP), 1-methylcyclopropene (1-MCP) application and vacuum impregnation (VI).

2. FRUITS AND HUMAN HEALTH

Consumption of fruits and vegetables is increasing because of strong evidence that many beneficial effects for human health are associated with the dietary intake of fruits and vegetables (Kaur & Kapoor 2001, Rupasinghe et al. 2012). As suggested by epidemiological studies, the consumption of fruit and vegetables may lead to prevention of many chronic diseases, including cardiovascular disease (Weichselbaum 2010, Al-Dosari et al. 2011; Thilakarathna and Rupasinghe 2012), type II diabetes (Johnston et al. 2002, Yu et al. 2012b) and some cancers (De Mejía & Prisecaru 2005, Lala et al. 2006, Sun & Liu 2008, Lippi & Targher 2011). These disease prevention effects of fruits could be due to the presence of health promoting phytochemicals such as carotenoids (Chichili et al. 2006), flavonoids (Yu et al. 2012a), other phenolic compounds (Masibo & He 2008) and vitamins (Lippi & Targher 2011, Gutierrez 2008). Furthermore, the health-protective effects may be rather produced by complex mixtures of interacting natural chemicals than a single component in these plant-derived foods (Lila 2007). Table 1 gives a summary of selected fruit-based antioxidants and other health promoting compounds for disease prevention.

3. FRUIT PRODUCTION AND POST-HARVEST LOSS

3.1. Fruit Production

Fruit production is increasing dramatically worldwide. According to the FAO, the total world fruit production in 2008 was 572.4 million tons, and the number climbed to 609.2 million tons in 2010 (FAO 2010). Among these fruits, thirty percent of which were tropical fruits, with water melon occupied of 59.2%, mongo and guavas of 20.5% and pineapple of 11.4% (Rawson et al. 2011).

Despite the increasing food production at the global level, about one-third of the food produced in the world is lost or wasted (Prusky 2011), among which, post-harvest stage losses and marketing stage losses are major losses.

TABLE 1. Fruit-based health promoting compounds and postulated disease prevention

Source	Active component	Prevention mechanism	Disease	References
Grape	Anthocyanins	Anti-proliferative	Cancer	Lala et al. 2006
	Flavonoids	Inhibition of HNR-adduct formation	Macular degeneration and cataract	Yu et al. 2012a
	Resveratrol	Antioxidant, anti-inflammatory, activation of SIRT1	Alzheimer's	Sun et al. 2010
	Resveratrol	Normalize iron and Ca^{2+}, increase SOD activity	Cardiotoxicity	Mokni et al. 2012
	Resveratrol	Enhance insulin secretion	Diabetes	Yu et al. 2012b
Apple	Polyphenols	Antioxidant, cell cycle modulation	Cancer	Sun & Liu 2008
	Polyphenols	Antioxidant, multiple mechanisms	Cardiovascular	Weichselbaum 2010
	Phloridzin	Anti-inflammatory, bone resorption	Bone protection	Puel et al. 2005
	Polyphenols	Reduce amyloid-β formation	Alzheimer's	Chan & Shea 2009
	Phloretin-2'-O-Glucoside	Delay glucose absorption	Diabetes	Johnston et al. 2002
Banana	Lectins (Bioactive protein)	Cell cycle arrest, apoptosis	Cancer	De Mejía & Prisecaru 2005
	Polyphenols	Antioxidant, reduce LDL modification	cardiovascular	Yin et al. 2008
	Polyphenols	Antioxidant	Alzheimer's	Heo et al. 2008
Pineapple	Bromelain	Proteolytic enzyme regulation	Anti-inflammatory	Hale et al. 2010
Mango	Phenolic compounds	Antioxidant, multiple mechanisms	Degenerative diseases	Masibo & He 2008

3.2. Post-Harvest Loss of Fruits

Despite of food production is increasing globally, a significant amount of the food for human consumption is lost or wasted, especially perishable foods such as fruits and vegetables (Prusky 2011). The amount of food lost each year is equivalent to more than half of the world's annual cereals production (2.3 billion tonnes in 2009/2010) (Gustavsson et al. 2011).

It is hard to give precise information on the amount of fruit losses generated globally, because fruit losses vary greatly among varieties, countries, and climatic regions, and there is no universally applied method for measuring losses. As a consequence, the food loss data during post-harvest are mostly estimated and the variations are from 10% to 40% (Prusky 2011). Table 2 lists some examples of post-harvest losses of selected fruits in India, Egypt and United States.

TABLE 2. Post-harvest losses in selected fruits

Source	Post-harvest loss %				References
	Farm	Wholesale	Retail	Total	
Grape	15.1	6.9	6.0	28.0	Kader 2010
Grape	7.3	4.2	2.9	14.4	Murthy et al. 2009
Grape	N/A	N/A	7.6	N/A	Buzby et al. 2009
Mango	15.6	8.9	5.3	29.7	Murthy et al. 2009
Mango	N/A	N/A	14.5	N/A	Buzby et al. 2009
Banana	5.5	6.7	16.7	28.8	Murthy et al. 2009
Banana	N/A	N/A	8.0	N/A	Buzby et al. 2009
Papaya	N/A	N/A	54.9	N/A	Buzby et al. 2009
Pineapple	N/A	N/A	14.6	N/A	Buzby et al. 2009
Kiwi	N/A	N/A	12.7	N/A	Buzby et al. 2009
Apple	N/A	N/A	8.6	N/A	Buzby et al. 2009
Avocado	N/A	N/A	9.3	N/A	Buzby et al. 2009
Tomato	9.0	17.9	16.3	43.2	Kader 2010

3.3. Prevention and Reduction of Post-Harvest Loss

Methods of preventing losses of fruits and vegetables could be found from papers and fact sheet written by Singh and Goswami (2006), Sonkar et al. (2008), Prusky (2011) and DeEII and Murr (2009). These methods include selection of new cultivars with firm fruits and longer postharvest life, minimizing physical damage during harvesting and postharvest handling, control and monitoring of temperature and relative humidity, use of controlled or modified atmosphere storage, use of pre- and post-harvest fungicides (hydrogen peroxide) before and after harvest and use of physical treatment such as ozonation technology. Table 3 gives examples of use of controlled atmosphere storage of selected fruits.

TABLE 3. Controlled atmosphere storage conditions of selected fruits

Source	Temp (°C)	RH (%)	O_2 (%)	CO_2 (%)	Storage life	References
Grape	0-5	90-95	5-10	15-20	> two weeks	Singh and Goswami 2006
Mango	10-15	90	3-7	5-8	> two weeks	Singh & Goswami 2006
Banana	12-16	90	2-5	2-5	> two weeks	Singh & Goswami 2006
Papaya	10-15	90	2-5	5-8	> two weeks	Singh & Goswami 2006
Pineapple	8-13	90	2-5	5-10	> two weeks	Singh & Goswami 2006
Kiwi	0-5	90	1-2	3-5	> two weeks	Singh & Goswami 2006
Avocado	5-13	90	2-5	3-10	> two weeks	Singh & Goswami 2006
Apple (Empire)	1-2	N/A	1.5-2.5	1.5-2.0	5-8 months	DeEll & Murr 2009
Apple (Gala)	0	N/A	1.5-2.5	1.5-2.5	5-8 months	DeEll & Murr 2009
Apple (Golden Delicious)	0	N/A	1.5-2.5	1.5-2.5	5-8 months	DeEll & Murr 2009
Apple (McIntosh)	3	N/A	1.0-2.5	0.5-2.5	5-8 months	DeEll & Murr 2009

* Temp: temperature; RH: Relative humidity

4. FRUIT PROCESSING AND PRESERVATION

Processed fruit products generally include minimally processed fruit products such as fresh-cut fruit, fermented fruit products such as cider, wine and vinegar, traditional thermally processed fruit products such as jam, jelly, juice and beverage, novel non-thermal processed fruit products such as juice and beverage, etc. A comprehensive review has been given by the same authors on novel non-thermal processed fruit product preservation including juices and beverages (Rupasinghe & Yu 2012). At the same time, fresh-cut fruit stands out to be a promising food that meets the demand of consumers for convenient and ready-to-eat fruits with a fresh-like quality. In this case, this part of the chapter would give emphasis on fresh-cut fruit processing and preservation.

4.1. Fresh-Cut Fruit Processing

The sales of fresh-cut produce have grown from approximately $5 billion in 1994 to $10–12 billion in 2005, which is about 10% of total produce sales (Rupasinghe et al. 2005). Fresh-cut fruits and vegetables are products that are partially prepared, maintain a fresh-like state and no additional preparation is necessary for use and eating (Watada & Qi 1999). Figure

1 shows the flowchart of fresh-cut fruit processing. It generally includes washing, and/or peeling, cutting, and/or slicing or wedging and packaging. Dipping solutions or edible coating materials could be applied during dipping or coating process.

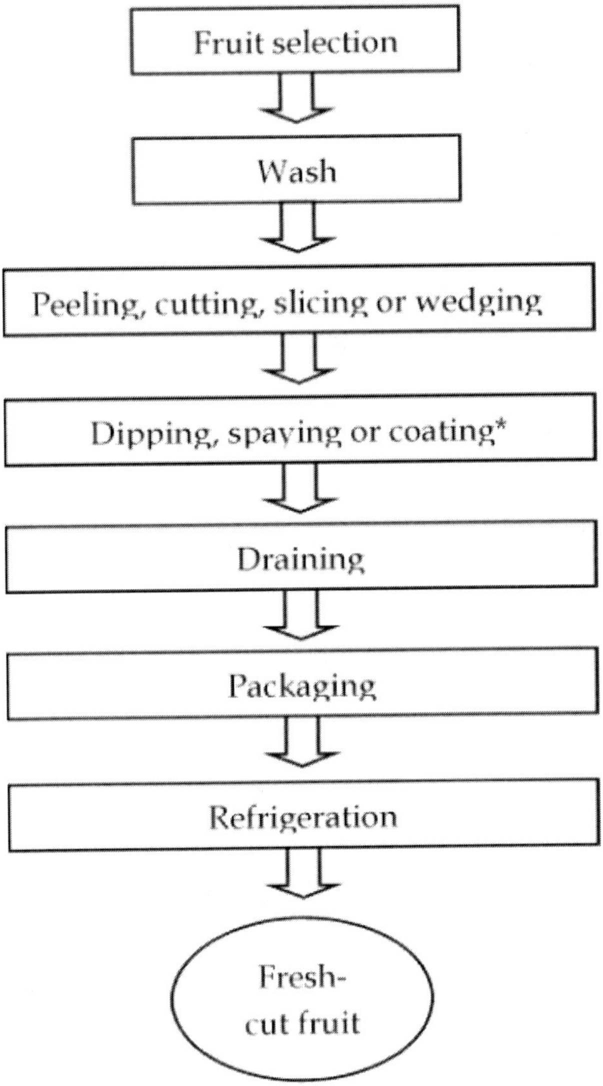

FIGURE 1. Major steps for fresh-cut fruit processing (revised from Corbo et al. 2010) * During this process, natural preservatives or edible coating materials could be applied

4.2. Fresh-Cut Fruit Preservation

Fresh-cut fruits are more perishable than whole fruits, because the tissue integrity of fruits is more easily altered during processing. Post-cut quality of fresh-cut fruits suffers from wound induced biochemical and physiological changes such as water loss, accelerated respiration and cut-surface browning as well as microbiological spoilage (Kader 2002, Chiabrando & Giacalone 2012). Therefore, preservation of fresh-cut fruits needs combinative efforts of antimicrobial agents, anti-browning substances as well as packaging strategies. A detailed review was given by Oms-Oliu et al. (2010) about recent approaches for preserving quality of fresh-cut fruits.

4.2.1. Antimicrobial Agent

During the preparatory steps of fresh-cut fruit processing, the natural protection of fruit is removed and chances of contamination may increase. Damage of tissues allows the growth and fermentation of some species of yeasts such as *Saccharomyces cerevisiae* and the attack by pathogenic microorganisms such as *Listeria monocytogenes*, *Salmonella* spp., *Staphylococcus aureus* and *Escherichia coli* O157:H7 (Martin-Belloso et al. 2006). Therefore, the searching for methods to retard microbial growth is of great interest to researchers and fresh-cut industry.

Traditionally, the most commonly used antimicrobials are potassium sorbate and sodium benzoate. However, consumer demand for natural origin, safe and environmental friendly food preservatives is increasing. Natural antimicrobials such as organic acids, herb leaves extracts and oils, chitozan and *bacteriocins* have shown feasibility for use in some food products including fresh-cut fruits (Gould 2001, Corbo et al. 2009). Some of them have been considered as Generally Recognized As Safe (GRAS) additives in foods. Selected natural antimicrobials and their status for GRAS additives are listed in Table 4.

Cinnamon as an antimicrobial agent has been investigated in fresh-cut apple slices (Muthuswamy et al. 2008). Ethanol extract of cinnamon bark (1% to 2% w/v) and cinnamic aldehyde (2 mM) could reduce *E. coli* O157:H7 and *L. innocua in vitro*. Ethanol extract of cinnamon bark (1% w/v) reduced significantly the aerobic growth of bacteria inoculated in fresh-cut apples during storage at 6°C up to 12 days. It was also found that cinnamic aldehyde has greater antimicrobial activity than potassium sorbate (Muthuswamy et al. 2008).

TABLE 4. Selected natural antimicrobial agents and their status for GRAS additives*

Name	Origin	GRAS status
Rosemary	Plant	Yes
Cinnamon	Plant	Yes
Cinnamic acid	Plant	Yes
Clove	Plant	Yes
Lactoperoxidase	Animal	No
Lemon (peel, balm, grass)	Plant	Yes
Lime	Plant	Yes
Nisin	Microorganism	Yes
Chitozan	Animal	No
Carvacrol	Plant	Yes
Citric acid	Plant	Yes
Ascorbic acid	Plant	Yes
Vanillin	Plant	Yes

* Revised from USFDA (2006). Food Additive Status List

Carvacrol and cinnamic acid could delay microbial spoilage of fresh-cut melon and kiwifruit. Dipping of fresh-cut kiwifruit in carvacrol solutions at up to 15 mM reduced total viable counts from 6.6 to less than 2 *log* CFU/g for 21 days of storage at 4°C. Also, treatment with 1 mM of carvacrol or cinnamic acid reduced viable counts on kiwifruit by 4 and 1.5 *log* CFU/g for 5 days of storage at 4°C and 8°C, respectively (Roller & Seedhar 2002).

Vanillin was also proved to be a practical preservative for processing fresh-cut mango and apples under refrigerated conditions. Fresh-cut mango slices were dipped for 1 min in solutions containing 80 mM vanillin before being packaged. Results indicated that treatment with 80 mM vanillin significantly delayed ($P < 0.05$) the development of total aerobic bacteria and yeast and mold populations of fresh-cut mangoes stored at 5 and 10 ◦C for up to 14 and 7 d, respectively (Ngarmsak et al. 2006). Also, a dip of 12 mM vanillin incorporated with a commercial anti-browning dipping solution (calcium ascorbate, NatureSeal™) inhibited the total aerobic microbial growth by 37% and 66% in fresh-cut 'Empire' and 'Crispin' apples, respectively, during storage at 4 °C for 19 days. Furthermore, vanillin (12 mM) did not influence the control of enzymatic browning and softening by NatureSeal (Rupasinghe et al. 2006).

4.2.2. Anti-Browning Agents

Enzymatic browning is also a major concern on the extension of shelf-life of fresh-cut fruit (Oms-Oliu et al. 2010). It is caused by the enzymatic oxidation of phenols to quinones by enzymes, typically polyphenoloxidases, in the presence of oxygen. Quinones are then subjected to further reactions, leading to the formation of browning pigments (Ozoglu & Bayindirli 2002, Jeon & Zhao 2005). Traditionally, sulfites have been used for browning prevention. However, their use on fresh-cut fruit and vegetables was banned in 1986 by the FDA owing to their potential hazards to health (Buta et al. 1999). Therefore, various alternative substances, such as honey, citric acid, ascorbic acid, calcium chloride, calcium lactate and calcium ascorbate, among others, have being used to retard browning in fresh-cut fruit (Jeon & Zhao 2005, Oms-Oliu et al. 2010). These anti-browning products are not often used alone because it is difficult to achieve efficient browning inhibition, and combination of them would give preferable results. Table 5 gives examples of anti-browning treatment on fresh-cut Apples. Examples of anti-browning treatment on other fresh-cut fruits including banana, kiwifruits, mango, among others could be found in Table 3 in Oms-Oliu et al. (2010)'s paper.

TABLE 5. Examples of anti-browning treatments assessed on fresh-cut fruits

Cultivar of apple	Anti-browning agent	Storage conditions	References
Gala	10% honey solution with vacuum impregnating	3°C for 14 days	Jeon & Zhao (2005)
Granny Smith	1% w/v of citric acid/CaCl2 and 1% of ascorbic acid/CaCl2	4°C for 5 days	Chiabrando & Giacalone (2012)
Granny Smith	0.05% w/v of sodium chlorite and 1% of calcium propionate	10°C for 14 days	Guan & Fan (2010)
Golden Delicious	1% w/v of citric acid/CaCl2 and 1% of ascorbic acid/CaCl2	4°C for 5 days	Chiabrando & Giacalone (2012)
Golden Delicious	80 mg/L acidic electrolyzed water (AEW) followed by 5%calcium ascorbate	4°C for 11 days	Wang et al (2007)
Granny Smith	1% w/v of citric acid/CaCl2 and 1% of ascorbic acid/CaCl2	4°C for 5 days	Chiabrando & Giacalone (2012)
Red Delicious	300 mg /L sodium chlorite (SC) and 300 mg /L citric acid	5°C for 14 days	Luo et al. (2011)
Scarlet Spur	1% w/v of citric acid/CaCl2 and 1% of ascorbic acid/CaCl2	4°C for 5 days	Chiabrando & Giacalone (2012)

4.2.3. Edible Coating

The incorporation of antimicrobial and anti-browning agents to fresh-cut fruits could be done by dipping, spaying or edible coating treatment. Dipping or spraying aqueous solutions to fruit pieces containing antimicrobial agents, antioxidants, calcium salts or functional ingredients such as minerals and vitamins are widely used to improve quality of fresh-cut fruit. However, the effectiveness of these compounds could be better improved with their incorporation into edible coatings. The application of edible coatings to deliver active ingredients is one of the recent progresses made for shelf-life extension of fresh-cut fruits. Detailed information on edible coating for fresh-cut fruits could be found in review papers from Vargas et al. (2008), Rojas-Graü et al. (2009) and Valencia-Chamorro et al. (2011).

Edible coatings may be defined as a thin layer of material that covers the surface of the food and can be eaten as a part of the whole product. Therefore, the composition of edible coatings has to be food grade or GRAS. Furthermore, the coating materials need to be transparent, odourless, permeable for water vapour and selectively permeable to gases and volatile compounds (Kester & Fennema 1986).

Ingredients that can be used to form edible coatings include polysaccharides such as cellulose, starch, alginate, chitosan, pectin, carrageenan, gum Arabic, guar gum and xanthan gum, proteins such as zein, gluten, soy, whey protein, lipids such as beeswax, lecithin, cocoa butter and fatty acids (Vargas et al. 2008). Examples of edible coating treatment on fresh-cut apples are listed in Table 6.

4.2.4. Modified Atmosphere Packaging (MAP) And 1-Methylcyclopropene (1-MCP)

The respiration rate of fresh-cut fruits is greater than that of intact fruits (Kader 1986). The increased respiration rate can induce the ethylene synthesis, increase enzymatic activity, promote oxidation of phenolic compounds and microbial growth, and therefore contributes to quality losses such as color and firmness. In this case, the control of respiration is essential for maintaining quality and prolonging the shelf life of fresh-cut fruits (Rocha & Morais 2003).

Modified atmosphere packaging (MAP) is a technology which offers the optimum gas conditions around the product by adjusting the barrier properties of the packaging film (Simpson and Carevi 2004). Various approaches to prolong the shelf life of fresh-cut products, such as edible coatings and refrigeration could be applied in combination with MAP (Rupasinghe 2005).

TABLE 6. Examples of edible coating treatment on fresh-cut apples

Cultivar of apple	Functional ingredients	Concentration (%)	Coating materials	References
Gala	N/A	N/A	Cassava starch, glycerol, carnauba wax, stearic acid	Chiumarelli & Hubinger 2012
Gala	N/A	N/A	Chitosan	Wu et al. 2005
Fuji	Oregano oil	0.1 – 0.5 (v/v)	Apple puree, alginate	Rojas-Graü et al. 2007
Fuji	Lemongrass	1.0 – 1.5 (v/v)	Apple puree, alginate	Rojas-Graü et al. 2007
Fuji	Vanillin	0.3 – 0.6 (v/v)	Apple puree, alginate	Rojas-Graü et al. 2007
Fuji	Cinnamon	0.7 (v/v)	Alginate	Raybaudi-Massilia et al. 2008
Fuji	Clove	0.7 (v/v)	Alginate	Raybaudi-Massilia et al. 2008
Fuji	Lemongrass	0.7 (v/v)	Alginate	Raybaudi-Massilia et al. 2008
Fuji	Cinnamaldehyde	0.5 (v/v)	Alginate	Raybaudi-Massilia et al. 2008
Fuji	Citral	0.5 (v/v)	Alginate	Raybaudi-Massilia et al. 2008
Fuji	Ascorbic acid, $CaCl_2$	1.0 (w/v)	Carrageenan	Lee et al. 2003
Fuji	Ascorbic acid, $CaCl_2$	1.0 (w/v)	Whey protein concentrate	Lee et al. 2003
Fuji	Ascorbic acid, $CaCl_2$	1.0 (w/v)	Whey protein concentrate	Lee et al. 2003
Golden Delicious	Ascorbic acid	0.5-1.0 (w/v)	Whey protein concentrate, beeswax	Perez-Gago et al. 2006
Golden Delicious	Cysteine	0.1-0.5 (w/v)	Whey protein concentrate, beeswax	Perez-Gago et al. 2006
Granny Smith	Ascorbic acid, citric acid	0.5 (w/v)	Pectin, apple purée	McHugh & Senesi 2000

1-Methylcyclopropene (1-MCP) may retard or inhibit the generation of ethylene, the natural ripening hormone which is undesirable in terms of storage of certain fruits. Therefore, 1-MCP is becoming a commercial tool (SmartFresh, AgroFresh Inc., Philadelphia) for extending the shelf-life and quality of certain fruits and plant products (Rupasinghe et al. 2005). 1-MCP can be applied immediately after harvest (Aguayo et al. 2006; Mao & Fei 2007), just before fresh-cut processing or at both steps (Calderón-López et al. 2005; Vilas-Boas & Kader 2007). However, treatment of intact fruit with

1-MCP before fresh-cut processing is easier and more convenient than after processing. Moreover, the increase in ethylene production promoted by peeling, slicing or wedging could be prevented by the pre-use of 1-MCP (Rupasinghe et al., 2005).

4.2.5. Vacuum Impregnation

Osmotic treatments have been traditionally used as a pre-treatment step in freezing, canning and frying to improve the quality of the final produce (Alzamora et al., 2000). Among developments in osmotic treatments of fruit products, vacuum impregnation (VI) may be the latest (Zhao & Xie 2004). The VI technique is performed by applying a vacuum pressure in a tank or oven containing the immersed product for a short time and then restoring the atmospheric pressure with the product remains immersed (Martínez-Monzó et al., 1998). The process of VI is a hydrodynamic mass transfer process based on an exchange between internal gas or liquid and an external liquid phase (Zhao & Xie, 2004). VI technique could be used to develop novel minimally processed fruit products with value-addition since nutritional and bioactive ingredients could be incorporated into the fruit based products during VI process (Xie & Zhao, 2003; Guillemin et al., 2008, Rößle 2011) and which gives a bright future for VI application in fresh-cut fruits. Table 7 gives examples of VI treatment on fresh-cut fruits.

TABLE 7. Examples of VI treatment conditions on fresh-cut fruits or value-added products

Type of fruit	VI treatment conditions				References
	VI solution	VI pressure (mmHg)	VI time (min)	Restoration Time (min)	
Apple (Gala)	20% (w/w) of HFCS, Ca, Zn	50	15	30	Xie & Zhao 2003
Apple (Gala)	10% (w/w) of honey	75	15	30	Jeon & Zhao 2005
Strawberry	8°Brix of glucose solution	37.5	5	5	Castelló et al. 2006
Apple (Empire)	15°Brix of grape juice, 1.6% of $CaCl_2$ (w/v), 0.05% of NaCl (w/v), 0.1% of vitamin E (v/v)	152.4	10	22	Joshi et al. 2010
Apple (Empire)	20-40 % (v/v) of maple syrup, 1.6% of $CaCl_2$ (w/v), 0.05% of NaCl (w/v)	152	10	22	Joshi et al. 2011
Apple (Granny Smith)	50% (v/v) of honey	525	10	10	Rößle et al. 2011

HFCS: High fructose corn syrup

5. CONCLUSION

Fruits are not only consumed as stable food but also provide desirable health benefits beyond their basic nutrition. However, the quantitative and qualitative losses of fruits are significant during post-harvest, marketing, processing and storage. Prevention of these losses during post-harvest management could be done by multiple steps and methods such as controlled or modified atmosphere packaging and application of ozonation technology.

On the other hand, promotion of minimally processed fruit products such as fresh-cut fruit into the commercial market is a practical, economical, and consumer and environmental friendly approach compared with traditional processing methods. However, fresh-cut fruits are more perishable than whole fruits in terms of biochemical and physiological changes such as water loss, accelerated respiration and cut-surface browning as well as microbiological spoilage. Therefore, preservation of fresh-cut fruits needs combinative efforts of antimicrobial agents, anti-browning substances as well as packaging strategies.

Natural or GRAS additives have been the popular ingredients used as antimicrobial agents and anti-browning agents, or bioactive ingredients. The incorporation of antimicrobial and anti-browning agents to fresh-cut fruits could be done by dipping, spaying or edible coating treatment. The application of edible coatings to deliver active ingredients is one of the recent progresses made for shelf-life extension of fresh-cut fruits. It could be used in combination with modified atmosphere packaging (MAP), 1-methylcyclopropene (1-MCP) and refrigeration for better results.

In addition for edible coating, vacuum impregnation (VI) may be another practical approach for incorporation of health promoting natural ingredients into fresh-cut fruits. VI technique could be used to develop novel minimally processed fruit products with value-addition through incorporation of nutritional and bioactive ingredients.

REFERENCES

1. Aguayo, E., Jansasithorn, R. & Kader, A. A. (2006). Combined effects of 1-Methylcyclopropene, calcium chloride dip, and/or atmospheric modification on quality changes in fresh-cut strawberries, *Postharvest Biology and Technology* 40(3): 269-278.
2. Alzamora SM, Tapia MS, Leunda A, Guerrero SN, Parada-Arias E. 2000. Minimal Presser vation of fruits: a cited project, *in*: Lozano, J.E., Anon, C., Parada-Arias, E. & Borabosa-Canovas GV (ed.), *Trend in Food Engineering*, Technomic Publishing Company,Pennsylvania, USA, pp. 205-225.

3. Buta, J. G., Moline, H. E., Spaulding, D. W. & Wang, C. Y. (1999). Extending storage life of fresh-cut apples using natural products and their derivatives, *Journal of Agricultural and Food Chemistry* 47(1): 1-6.
4. Buzby, J. C.,Wells, H. F., Axtman, B. & Mickey, J. (2009). Supermarket loss estimates for fresh fruit, vegetables, meat, poultry, and seafood and their use in the ERS loss-adjusted food availability data, *United States Department of Agriculture*, Economic Research Service, Economic Information Bulletin, 44: 20.
5. Calderon-Lopez, B., Bartsch, J. A., Lee, C. Y. & Watkins, C. B. (2005). Cultivar effects on quality of fresh cut apple slices from 1-methylcyclopropene (1-MCP)-treated apple fruit, *Journal of Food Science* 70(3): S221-227.
6. Castelló, M. L., Fito, P. J. & Chiralt, A. (2006). Effect of osmotic dehydration and vacuum impregnation on respiration rate of cut strawberries, *LWT - Food Science and Technology* 39(10): 1171-1179.
7. Chan, A. & Shea, T. B. (2009). Dietary supplementation with apple juice decreases endo- genous amyloid-β levels in murine brain, *Journal of Alzheimer's Disease* 16(1): 167- 171.
8. Chiabrando, V. & Giacalone, G. (2012). Effect of anti-browning agents on color and related enzymes in fresh-cut apples during cold storage, *Journal of Food Processing and Preservation* 36(2): 133-140.
9. Chiumarelli, M. & Hubinger, M. D. (2012). Stability, solubility, mechanical and barrier properties of cassava starch - carnauba wax edible coatings to preserve fresh-cut apples, *Food Hydrocolloids* 28(1): 59-67.
10. Corbo, M. R., Bevilacqua, A., Campaniello, D., D'Amato, D., Speranza, B. & Sinigaglia, M. (2009). Prolonging microbial shelf life of foods through the use of natural compounds and non-thermal approaches - A review, *International Journal of Food Science and Technology* 44(2): 223-241.
11. Corbo, M.R., Speranza, B., Campaniello, D., D'Amato, D. & Sinigaglia, M. (2010). Fresh-cut fruits preservation: current status and emerging technologies, *in* Mendez-Vilas, A (Ed.), *Current research, Technology and Education Topics in Applied Microbiology and Microbial Biotechnology*, Formatex Research Center, Badajoz, Spain, pp. 1143-1154.
12. DeEII J. R. & Murr, D. P. (2009). CA storage guidelines and recommendations for apples, *http://www.omafra.gov.on.ca/english/crops/facts/03-073.htm*
13. De Mejía EG & Prisecaru VI. (2005). Lectins as bioactive plant proteins: a potential in cancer Treatment, *Critical Reviews in Food Science and Nutrition* 45 (6): 425-445.
14. FAO (2010). World fruit production in metric tonnes. *http://faostat.fao.org/default.aspx*.
15. Gould, G. W. (2001). Symposium on 'nutritional effects of new processing technologies' – new processing technologies: An overview, *Proceedings of the Nutrition Society* 60(4): 463-474.
16. Guan, W. & Fan, X. (2010). Combination of sodium chlorite and calcium propionate reduces enzymatic browning and microbial population of fresh-cut "granny smith" apples, *Journal of Food Science* 75920, M72-M77
17. Guillemin, A., Guillon, F., Degraeve, P. et al. (2008). Firming of fruit tissues by vacuum- infusion of pectin methylesterase: visualisation of enzyme reaction, *Food Chemistry* 109, 368–378.
18. Gustavsson, J., Cederberg, C., Sonesson, U., van Otterdijk, R. & Meybeck, A. (2011). Global food losses and food waste: extent causes and prevention, Food and Agriculture Organization of the United Nations. International congress Save Food!, *International packaging industry fair Interpack 2011*, Dusseldorf, Germany.
19. Hale, L. P., Chichlowski, M., Trinh, C. T. & Greer, P. K. (2010). Dietary supplementation with fresh pineapple juice decreases inflammation and colonic neoplasia in IL-10- deficient mice with colitis, *Inflammatory Bowel Diseases* 16(12): 2012-2021.

20. Heo, H.J., Choi, S.J., Choi, S-G, Lee. J.M. & Lee, C.Y. (2008). Effects of banana, orange, and apple on oxidative stress-induced neurotoxicity in PC12 cells, *Journal of Food Science* 73(2), H28-H32.
21. Jeon, M. & Zhao, Y. (2005). Honey in combination with vacuum impregnation to prevent enzymatic browning of fresh-cut apples, *International Journal of Food Sciences and Nutrition* 56(3): 165-176.
22. Johnston, K. L., Clifford, M. N. & Morgan, L. M. (2002). Possible role for apple juice phenolic compounds in the acute modification of glucose tolerance and gastrointestinal hormone secretion in humans, *Journal of the Science of Food and Agriculture* 82(15): 1800-1805.
23. Joshi, A. P. K., Rupasinghe, H. P. V. & Pitts, N. L. (2010). Sensory and nutritional quality of the apple snacks prepared by vacuum impregnation process, *Journal of Food Quality* 33(6): 758-767.
24. Joshi, A. P. K., Rupasinghe, H. P. V. & Pitts, N. L. (2011). Comparison of non-fried apple snacks with commercially available fried snacks, Food *Science and Technology International* 17(3): 249-255.
25. Kader, A.A., (1986). Biochemical and physiological basis for effects of controlled and Modified atmospheres on fruits and vegetables, *Food Technol.* 40: 99-104.
26. Kader, A. (2002). Quality parameters of fresh-cut fruit and vegetable products, in Lamikanra, O. (ed.), *Fresh-Cut Fruits and Vegetables, Science, Technology and Market*, CRC Press, Boca Raton, Florida, USA, pp. 11–19.
27. Kader, A. A. (2010). Handling of horticultural perishables in developing vs. developed countries, *Acta Horticulturae* 877: 121–126.
28. Kaur, C. & Kapoor, H. C. (2001). Antioxidants in fruits and vegetables - the millennium's health, *International Journal of Food Science and Technology* 36(7): 703-725.
29. Kester, J.J. & Fennema O. R. (1986). Edible films and coatings: A review, *Food Technol.* 40: 47– 49.
30. Lala, G., Malik, M., Zhao, C., He, J., Kwon, Y., Giusti, M. M. & Magnuson, B. A. (2006). Anthocyanin-rich extracts inhibit multiple biomarkers of colon cancer in rats, *Nutrition and Cancer* 54(1): 84-93.
31. Lee, J. Y., Park, H. J., Lee, C. Y. & Choi, W. Y. (2003). Extending shelf-life of minimally processed apples with edible coatings and anti-browning agents, *Lebensmittel Wissenschaft und Technologie* 36: 323-329.
32. Lila, M. A. (2007). From beans to berries and beyond: Teamwork between plant chemicals for protection of optimal human health, *Annals of the New York Academy of Sciences* 1114: 372-380.
33. Luo, Y., Lu, S., Zhou, B. & Feng, H. (2011). Dual effectiveness of sodium chlorite for enzymatic browning inhibition and microbial inactivation on fresh-cut apples, *LWT-Food Science and Technology* 44(7): 1621-1625.
34. Mao, L. & Fei, G. (2007). Application of 1-methylciclopropene prior to cutting reduces wound responses and maintains quality in cut kiwifruit, *J. Food Eng.* 78: 361–365.
35. Martin-Belloso, O., Soliva-Fortuny, R. & Oms-Oliu, G. (2006). Fresh-cut fruits, in: Hui, Y.H. (Ed.), *Handbook of Fruits and Fruit Processing*, Blackwell Publishing, Oxford, pp. 129– 144.
36. Martínez-Monzó, J., Martínez-Navarrete, N., Chiralt, A. & Fito, P. (1998). Mechanical and structural changes in apple (var. granny smith) due to vacuum impregnation with cryoprotectants, *Journal of Food Science* 63(3): 499-503.
37. Masibo, M. & He, Q (2008). Major mango polyphenols and their potential significance to human health, *Comprehensive Reviews in Food Science and Food Safety* 7(4): 309-319.
38. McHugh, T. H. & Senesi, E. (2000). Apple wraps: a novel method to improve the quality and extend the shelf life of fresh-cut apples, *Journal of Food Science* 65: 480-485.

39. Mokni, M., Hamlaoui-Guesmi, S., Amri, M., Marzouki, L., Limam, F. & Aouani, E. (2012). Grape seed and skin extract protects against acute chemotherapy toxicity induced by doxorubicin in rat heart, *Cardiovascular Toxicology* 1-8.
40. Murthy, D. S., Gajanana, T. M., Sudha, M. & Dakshinamoorthy, V. (2009). Marketing and post-harvest losses in fruits: Its implications on availability and economy, *Indian Journal of Agricultural Economics* 64(2): 259-275.
41. Muthuswamy, S., Rupasinghe, H. P. V. & Stratton, G. W. (2008). Antimicrobial effect of cinnamon bark extract on *Escherichia coli* O157:H7, *Listeria innocua* and fresh-cut apple slices, *Journal of Food Safety* 28(4): 534-549.
42. Ngarmsak, M., Delaquis, P., Toivonen, P., Ngarmsak, T., Ooraikul, B. & Mazza, G., (2006). Antimicrobial activity of vanillin against spoilage microorganisms in stored fresh- cut mangoes, *J. Food Prot.* 69: 1724–1727.
43. Oms-Oliu, G., Rojas-Graü, M. A., González, L. A., Varela, P., Soliva-Fortuny, R., Hernando, M. I. H., . . . Martín-Belloso, O. (2010). Recent approaches using chemical treatments to preserve quality of fresh-cut fruit: A review, *Postharvest Biology and Technology* 57(3): 139-148.
44. Ozoglu, H. & Bayindirli, A. (2002). Inhibition of enzymatic browning in cloudy apple juice with selected anti-browning agents, *Food Control* 13: 213-221.
45. Perez-Gago, M. B., Serra, M. & del Rio, M. A. (2006). Color change of fresh-cut apples coated with whey protein concentrate-based edible coatings, *Postharvest Biology and Technology* 39: 84-92.
46. Prusky, D. (2011). Reduction of the incidence of postharvest quality losses, and future prospects, *Food Security* 3(4): 463-474.
47. Puel, C., Quintin, A., Mathey, J., Obled, C., Davicco, M. J., Lebecque, PCoxam, V. (2005). Prevention of bone loss by phloridzin, an apple polyphenol, in ovariectomized rats under inflammation conditions, *Calcified Tissue International* 7(5): 311-318.
48. Rawson, A., Patras, A., Tiwari, B. K., Noci, F., Koutchma, T. & Brunton, N. (2011). Effect of thermal and non-thermal processing technologies on the bioactive content of exotic fruits and their products: Review of recent advances, *Food Research International*, 4(7): 1875-1887.
49. Raybaudi-Massilia, R. M., Rojas-Grau, M. A., Mosqueda-Melgar, J. & Martín -Belloso, O. (2008). Comparative study on essential oils incorporated into an alginate-based edible coating to assure the safety and quality of fresh-cut Fuji apples, *J. Food Prot.* 71: 1150–1161.
50. Rößle, C., Brunton, N., Gormley, T. R. & Butler, F. (2011). Quality and antioxidant capacity of fresh-cut apple wedges enriched with honey by vacuum impregnation, *International Journal of Food Science and Technology* 46(3): 626-634.
51. Rocha, A. M. C. N. & Morais, A. M. M. B. (2003). Shelf life of minimally processed apple (cv. Jon gold) determined by colour changes, *Food Control*14(1): 13-20.
52. Rojas-Graü, M. A., Raybaudi-Massilia, R. M., Soliva-Fortuny, R. C., Avena-Bustillos, R. J., McHugh, T. H. & Martín-Belloso, O. (2007). Apple puree-alginate edible coating as carrier of antimicrobial agents to prolong shelf-life of fresh-cut apples, *Postharvest Biol. Technol.* 45: 254–264.
53. Rojas-Graü, M. A., Soliva-Fortuny, R. & Martín-Belloso, O. (2009). Edible coatings to incorporate active ingredients to fresh-cut fruits: A review,*Trends in Food Science and Technology* 20(10): 438-447.
54. Roller, S& Seedhar, P. (2002). Carvacrol and cinnamic acid inhibit microbial growth in fresh- cut melon and kiwifruit at 4° and 8°C, *Letters in Applied Microbiology* 35: 390–394.
55. Rupasinghe, H.P.V., Murr, D.P., DeEll, J.R. & Odumeru, J. A. (2005). Influence of 1-methylcyclopropene (1-MCP) and NatureSealTM on the quality of fresh-cut 'Empire' and 'Crispin' apples, *J. Food Qual.* 28: 289-307.
56. Rupasinghe, H.P.V., Boulter-Bitzer, J, Ahn, T. & Odumeru, J.A. (2006). Vanillin inhibits pathogenic and spoilage microorganisms in vitro and aerobic microbial growth on fresh-cut apples, *Food Res. Intern.* 39: 575-580.

57. Rupasinghe, H. and Yu, L. J. 2012. Emerging preservation methods for fruit juices and beverages, in El-Samragy Y. (Eds.), *Food Additives*, InTech - Open Access Publisher, Rijeka, Croatia. ISBN: 978-953-51-0067-6. pp: 65-82.
58. Rupasinghe, H.P.V., Thilakarathna, S. & Nair, S. (2012). Polyphenols of Apples and Their Potential Health Benefits, in: J. Sun, K. N. Prasad, A. Ismail, B. Yang, X. You and L. Li (Eds.), *Polyphenols: Chemistry, Dietary Sources and Health Benefits,* Nova Science Publishers, Inc. Hauppauge, NY, USA, ISBN: 978-1-62081-868-8.
59. Simpson, R. & Carevic, E. (2004). Designing a modified atmosphere packaging system for foodservice portions on non-respiring foods: optimal gas mixture and food/headspace ratio, *Foodservice Research International* 14: 257-272.
60. Singh, A. K. & Goswami, T. K. (2006). Controlled atmosphere storage of fruits and vegetables: A review, *Journal of Food Science and Technology*43(1): 1-7.
61. Sonkar, R. K., Sarnaik, D. A., Dikshit, S. N., Saroj, P. L. & Huchche, A. D. (2008). Post-harvest management of citrus fruits: A review, *Journal of Food Science and Technology* 45(3): 199-208.
62. Sun, J. & Liu, R. H. (2008). Apple phytochemical extracts inhibit proliferation of estrogen-dependent and estrogen-independent human breast cancer cells through cell cycle modulation, *Journal of Agricultural and Food Chemistry* 56(24): 11661-11667.
63. Sun, A. Y., Wang, Q., Simonyi, A. & Sun, G. Y. (2010). Resveratrol as a therapeutic agent for neurodegenerative diseases, *Molecular Neurobiology*41(2-3): 375-383.
64. Thilakarathna, S.H. & H.P.V. Rupasinghe. 2012. Anti-atherosclerotic effects of fruit bioactive compounds: A review of current scientific evidence, *Can J. Plant Sc.* 92: 407-419.
65. U.S. Food and Drug Administration. (2006). Food Additive Status List. http://www.cfsan.fda.gov/~dms/opa-appa.html
66. Valencia-Chamorro, S. A., Palou, L., Delŕio, M. A. & Pérez-Gago, M. B. (2011). Antimicrobial edible films and coatings for fresh and minimally processed fruits and vegetables: A review, *Critical Reviews in Food Science and Nutrition* 51(9): 872-900.
67. Vargas, M., Pastor, C., Chiralt, A., McClements, D. J. & González-Martínez, C. (2008). Recent advances in edible coatings for fresh and minimally processed fruits, *Critical Reviews in Food Science and Nutrition* 48(6): 496-511.
68. Vilas-Boas, E. & Kader, A. (2007). Effect of 1-MCP on softening of fresh-cut kiwifruit, mango and persimmon slices, *Post. Biol. Technol.* 43: 238–244.
69. Wang, H., Feng, H. & Luo, Y. (2007). Control of browning and microbial growth on fresh-cut apples by sequential treatment of sanitizers and calcium ascorbate, *Journal of Food Science* 72(1), M1-M7.
70. Watada, A.E. & Qi, L. (1999). Quality of fresh-cut produce, *Postharvest Biol. Technol.* 15: 201–205.
71. Weichselbaum, E., Wyness, L. & Stanner, S. (2010). Apple polyphenols and cardiovascular disease - a review of the evidence, *Nutrition Bulletin*35(2): 92-101.
72. Wu, T., Zivanovic, S., Draughon, F.A., Conway, W. S. & Sams, C. E. (2005). Physicochemical properties and bioactivity of fungal chitin and chitosan, *J. Agric. Food Chem.* 53: 3888–3894.
73. Xie, J. & Zhao, Y. (2003). Nutritional enrichments of fresh apple (Royal Gala) by vacuum impregnation, *International Journal of Food Sciences and Nutrition* 54: 387–398.
74. Yin, X., Quan. J. & Kanazawa T. (2008). Banana prevents plasma oxidative stress in healthy individuals, *Plant Foods in Human Nutrition* 63 (2): 71-76.
75. Yu, C., Nandrot, E. F., Dun, Y. & Finnemann, S. C. (2012a). Dietary antioxidants prevent age-related retinal pigment epithelium actin damage and blindness in mice lacking αvβ5 integrin, *Free Radical Biology and Medicine* 52(3): 660-670.
76. Yu, W., Fu, Y. -. & Wang, W. (2012b). Cellular and molecular effects of resveratrol in health and disease, *Journal of Cellular Biochemistry* 113(3): 752-759.

77. Zhao, Y. & Xie, J. (2004). Practical applications of vacuum impregnation in fruit and vegetable processing, *Trends in Food Science & Technology*15; 434–451.

CHAPTER 2

Monitoring the Shelf-Life of Minimally Processed Fresh-Cut Apple Slices By Physical Chemical Analysis and Electronic Nose

Valeria Guarrasi[1*], Daniela Giacomazza[1], Maria Antonietta Germanà[2], Margherita Amenta[3], Pier Luigi San Biagio[1]

[1]*Istituto di Biofisica, Consiglio Nazionale delle Ricerche, Via Ugo La Malfa 153, 90146 Palermo, Italy*

[2]*SAF Department, University of Palermo, Viale delle Scienze 4, 90128 Palermo, Italy*

[3]*CRA — Centro di Ricerca per l'Agrumicoltura e le Colture Mediterranee, Corso Savoia 190, 95024 Acireale (CT), Italy*

ABSTRACT

Fresh-cut apples, in slices or in cubes, are minimally processed products, which are currently collecting a great interest by fruit marketers for their promising diffusion. Their shelf life, from a microbiological point of view, has been fixed about 2 or 3 weeks under refrigeration. However in a few days they undergo biochemical degradations with production of off-flavors and texture breakdown. In this work, the change of aromatic fingerprint of apple slices packaged in air and in a modified atmosphere (with 100% N_2) and stored at 4°C was measured, by using a commercial electronic nose. The obtained data were also compared with sensory evaluation of judge's panel. Moreover, quality parameters such as total acidity, total soluble solids and firmness were determined at different storage times (0, 4, 8 and 12 days).

The data show that the electronic nose is able to discriminate between the two different storage conditions applied: the multivariate analysis, Principal Component Analysis, presents clearly differences among the four sampling times when the apple slices are stored in air and in N2. Our results indicate that the electronic nose can be considered a valid supplementary tool to human sensory panel assessment especially in food quality safety and control and it can be a simple, objective and rapid method to control the food quality during the storage.

Our results indicate that the electronic nose **c**an be considered a valid supplementary tool to human sensory panel assessment especially in food quality safety and control and it can be a simple, objective and rapid method to control the food quality during the storage.

Keywords: Fresh-cut apples; Shelf-life; Electronic nose; **SENSORY ANALYSIS**; Food storage; Minimally processed fruit

INTRODUCTION

Recently the demand for minimally processed fruits and vegetables is increasing. The ready-to-eat fresh-cut products are one of the major growing segments in food markets. The rapid growth is due to the new lifestyles and to the health-consciousness of the consumers [1]. In fact, they request fresh-like processed products with high quality attributes (such as appearance, texture, and flavor) similar to those of the raw products [2] to satisfy the daily needs of antioxidants, minerals and dietary fibers. The ready-to-eat fresh-cut products are slightly processed products, retaining intact the positive characteristics of the fresh fruits and vegetables. Minimal processing has been defined as a combination of procedures, such as washing, peeling, slicing or chopping, not affecting the fresh-like quality and flavor of the food [3]. However, fresh-cut fruits are more challenging to obtain than other processed products for the difficulties in preserving their fresh-like properties during prolonged periods. The tissue integrity of fruits, in fact, can be easily altered during the shelf-life time [4-6].

Fresh-cut apples, in slices or in cubes, are lightly processed products and the fruit marketers have shown a great interest in their development. Although from a microbiological point of view, their shelf-life, in refrigerated conditions, has been fixed to 2 or 3 weeks [7], **BIOCHEMICAL DEGRADATIONS,** such as enzymatic browning, off-flavors release and texture breakdown, occur in a few days [8]. The addition of chemical additives (preservatives, antioxidants, edible coating, colorants, etc.), although used to prolong the shelf-life [9-11] can be associated with health problems such as allergies and more [12]. On this basis, new techniques to maintain the natural qualities of ready-to-eat fruits without using chemical preservatives were developed. The Nicoli's group first proposed the use of a modified atmosphere composed by 80% N_2 and 20% CO_2 to better preserve the apple slice quality [13]. Ever since, many attemps have been done to identify the better atmosphere composition to preserve the fresh fruit quality [14,15].

The fruit quality is not a single and well-defined attribute but comprises many properties or characteristics. In many cases, indicators of fruit maturity, such as color, total soluble solids and titratable acids, may not be sufficient to determine optimal sensory quality [16].

Aroma is one of the most important quality parameter perceived by consumers. Gas chromatography technique is usually employed to identify the fruit aroma, while sensory analysis is used to assess the intensity of the aroma descriptors [17]. Whether the chromatographic techniques require special equipment and dedicated staff, the sensory analysis also presents some drawbacks, such as the duration of the panel training and, sometimes, the dubious objectivity of the results. New technologies have been tested on fruits, aiming for fast and, in some cases, non-destructive volatile detection [18,19]. The electronic olfactory systems or electronic noses are the new technologies that can be used to predict fruit quality [20-23].

In this work we studied the changes of aromatic fingerprint of apple slices, packaged in atmospheric air and in a modified atmosphere (100% N_2), stored at 4°C, for 0, 4, 8 and 12 days by an electronic nose. A commercial type of electronic nose, EOS^{835}, based on metal oxide semiconductor sensors (MOS), to detect changes in volatile profiles due to storage conditions and during shelf life was used. Moreover, classical quality parameters, such as total **acidity,** total soluble solids, firmness and sensory profile by trained judges, were also determined.

MATERIALS AND METHODS

Apples and fruit processing

'Fuji' apples were harvested on September 2009 in Caltavuturo (Palermo, Italy) in an organic farm. Thirty fruits were selected for regular shape, uniform size and same ripeness stage. The fruits were cleaned with distilled water, dried and cut into slices of 1 cm thickness. The slices were individually put in the plastic bags (10×10 cm) and no pre-processing coating was applied. The bags were heat-sealed in atmospheric air (Air) and in nitrogen saturated (N_2) atmosphere. The plastic bags, purchased from Reber (Italy), presented a permeability to O_2 of 70.5 $cm^3/(m^2$ 24h atm) at the temperature of 23°C and 0% relative humidity and to N_2 of 10.5 $cm^3/(m^2$ 24h atm) at the same temperature and relative humidity. All the sealed bags were stored in refrigerator at 4 ± 1°C in darkness. The samples were analyzed just before packaging (0 day) and after 4, 8 and 12 storage days. At each chosen time, 10 packages for trial (Air and N_2) were randomly taken to perform the analysis.

Physical and chemical analyses

The juice soluble solid content (SSC) was determined by an optical refractometer (Atago Co., Ltd., Japan); results are expressed as °Brix.

Total acidity (TA) was measured by titration with 0.1 N NaOH. Three replicates were performed and the results expressed as g/L of citric acid.

Firmness measurements were performed by EFFEGI Texture Analyzer by determining the maximum force during the penetration of a 8 mm diameter stainless steel cylinder into the apple slice tissue. Results were expressed as kgf.

Statistical analysis was carried out using the STATSOFT 6.0 program (Vigonza, Padova, Italy). The statistical differences among different storage times were evaluated by variance analysis (ANOVA), and mean separation was performed by using the Tukey's test.

SENSORY EVALUATION

A judge's panel was engaged in the sensory analysis. The judges (10 persons) were trained in some preliminary sessions by using different apple samples, in order to develop a common vocabulary for the description of the sensorial attributes. Eight descriptors were chosen to describe the quality of apple's slices: appearance, browning, flavor, consistency, juiciness, sweetness, acidity, and pleasantness. Samples were evaluated by assigning a score between 1 (absence of the sensation) and 5 (extremely intense), except for the descriptor 'browning' where a reverse evaluation (1=maximum; 5=minimum) was adopted. Water at room temperature was used to rinse the sample before tasting. No statistical analysis of the data was carried out.

ELECTRONIC NOSE MEASUREMENTS

An EOS[835] (Sacmi, Italy) was used to detect the aromatic fingerprint at different storage times. The EOS[835] instrument, belonging to the so-called "*electronic noses*", consists of an array of metallic oxide semiconductor (MOS) resistances installed inside a patented measuring cell, the "sensors chamber". The interaction with odor molecules causes variation of the electrical conductivity. The MOS sensors are doped with different metallic oxides (Table 1) that react differently to the same odorous molecules, thus generating a set of signals (*olfactory imprint or aromatic fingerprint*) characteristic of each analyzed sample. All measurements were done in triplicate, cutting 3g of slices and placing them into 20 ml vials. Each vial was sealed and conditioned at 40°C for 10 minutes before analysis to ensure that the aroma reached the vial headspace. After equilibration, 4 ml of headspace was drawn in. The interaction between flavors and sensor layers causes a modification of the sensor resistance. A typical measurement profile is given in Figure 1. After each measurement the sensors are constantly purged with GC-grade air. The difference between the resistance value at rest, R_0, and that one induced by sample volatile compounds, R, is considered as the sensor response. After each measurement, the electronic nose chamber is flushed again with pure grade air. The measurement duration was 16 minutes (1 min of cleaning before, 2 min of measurement, 13 min of clearing after). Six values, $\{R_{S1}, ..., R_{S6}\}$, corresponding to the six different sensors, were obtained for each measurement and data pre-processing was carried out. The feature *Delta* was obtained from the sensor response curve (Figure 1). Principal Component Analysis (PCA) was used as unsupervised statistic method to reduce the dimensional space and plot the data, by a correlation matrix. PCA transforms the original variables into new uncorrelated ones called "principal components". The first component (PC1) describes the larger amount of the data. The second principal component (PC2) is orthogonal to the first one and covers almost the whole remaining variation and so on (PC3...PC6).

RESULTS AND DISCUSSION

PHYSICAL AND CHEMICAL PARAMETERS

The physical and chemical characteristics of apples processed in Air and in N_2 and stored for 0, 4, 8 and 12 days are reported in Table 2. The SSC and TA values do not show noteworthy changes as function of the storage time both in Air and in N_2.

Table I: MOS sensor array configuration of the EOS835.

Sensor No.	Sensor Code	Sensitive Layer	Temperature
1	CJ1316	SnO_2 cat SiO_2	450°C
2	SB0225	SnO_2 cat Ag	400°C
3	SD0515	SnO_2 cat Mo	400°C
4	SH0612	WO_3	375°C
5	SJ0717	SnO_2	450°C
6	WHT19	WO_3	400°C

Penetration measurements performed in both experimental condition (air and N_2) showed significant differences in the apple texture, especially in the sample packaged in Air ($p \leq 0.001$). In the case of apple slices packaged in N_2 the p value calculated for firmness modifications was $p \leq 0.05$. After 12 days the texture of slices stored in Air was softer than that one of samples stored in N_2, thus indicating that a modified atmosphere was more able to save the sample hardness [24,25]. The maintenance of the texture is considered a positive attribute in processed apples.

SENSORY EVALUATION

Results concerning sensory analysis are shown in the radar plot for the 8 chosen descriptors (Figures 2a and b). 'Sweetness', 'acidity' and 'consistency' are in agreement with analytical data results. Figure 2a illustrates that the scores of global 'appearance' for apple slices stored in Air after 8 and 12 days were the same, probably due to the presence of brown and soft parts that influenced the acceptance of the judges. Samples stored in N_2 (Figure 2b) showed the lowest 'appearance' score after 12-day packing time. The 'flavor' descriptor behavior is different for apple slices stored in Air and in N_2: the same value was attributed to samples packaged in Air for 0 - 4 days and smaller value was assigned to samples examined after 8 - 12 days. In the N_2 case, the same values were attributed to apples packaged for 0, 4, 8 days and smaller value for those saved for 12 days.

ELECTRONIC NOSE DATA

A PCA correlation matrix was carried out on the electronic nose data (Figure 3). For the apple slices packaged in air, the two first principal components, PC1 and PC2, accounted for 94.07% of the total variance (73.92% and 20.15% respectively) (Figure 3a). Samples were distributed along PC1 according to the storage time. The aroma showed changes from 0 to 12 days of storage. At 0 and 4 days, as well as at 8 and 12 days, the aromatic fingerprint is in agreement with the panel estimation.

In the case of apple slices packaged in nitrogen, the two first principal components, PC1 and PC2, accounted for 94.60% of the total variance (75.88% and 18.72% respectively) (Figure 3b). Also in this set, samples were distributed along PC1 according to the storage time. The behavior of aromatic fingerprint was different from that observed for samples packaged in air. In fact, data suggest that the aroma is better preserved in nitrogen: the aroma of samples stored for 0, 4, and 8 days in N2 was separated in a cluster independent on that obtained in the case of samples stored for 12 days. Also in this case the electronic nose response was similar to value assigned by the panel.

Furthermore, the electronic nose results confirm data previously obtained by penetration measurements: the aroma changes are consequence of firmness loss.

CONCLUSIONS

Fresh-cut apple tissues undergoing changes during storage, including production of compounds affecting the fruit flavor. Our data show that changes of the aroma fingerprint detected by electronic nose measurements are in agreement with results obtained by physical-chemical analysis and human evaluation panel.

Furthermore, product storage in modified atmosphere (N_2) is able to prolong the shelf-life of the fresh-cut apple slices up to 1 week.

The approach here used can be applied to evaluate the loss of aroma quality of other fresh-cut products, in which enzymatic and microbiological processes are responsible for the decrease of the shelf life.

ACKNOWLEDGEMENTS

The Authors would like to thank Dr. Laura Dollo for technical support and useful discussions.

REFERENCES

1. Michaelidou N, Hassa LM (2007) The role of health consciousness, food safety concern and ethical identity on attitudes and intentions towards organic food. Int J Consumer Studies 32: 163-170
2. Wong DWS, Tillin SJ, Hudson JS, Pavlath AE (1994) Gas exchange in cut apples with bilayer coatings. J Agric Food Chem 42: 2278-2285
3. Odriozola-Serrano I, Soliva-Fortuny R, Mantìn-Belloso O (2008) Effect of minimal processing on bioactive compounds and color attributes of fresh-cut tomatoes. LWT – Food Science and Technology 41: 217-226
4. Rolle RS, Chism GW (1987) Physiological consequences of minimally processed fruits and vegetables. J Food Qual 10: 157-177
5. Rosen JC, Kader AA (2008) Postharvest Physiology and Quality Maintenance of Sliced Pear and Strawberry Fruits. J Food Sci 54: 656-659
6. Catalano AE, Sapienza M, Peluso O, Spagna G (2008) Miglioramento qualitativo degli ortofrutticoli di IV Gamma: ruolo dei principali enzimi degradativi. Industrie alimentari – XLVII: 479-485
7. Garcia E, Barret DM (2002) Preservative treatments for fresh-cut fruits and vegetables. In: Lamikanra O (ed) Fresh-cut fruits and vegetables. CRC Press, Boca Raton, FL (USA)
8. Barry-Ryan C, Martìn-Diana AB, Rico D, Barat JM (2007) Extending and measuring the quality of fresh-cut fruit and vegetables: a review. Trends in Food Sci Technol 18: 373-386
9. Ahvenainen R (1996) New approaches in improving the shelf life of minimally processed fruit and vegetables. Trends in Food Science & Technology 7: 179-187
10. Soliva-Fortuny RC, Marti´n-Belloso O (2003) New advances in extending the shelf-life of fresh-cut fruits: a review. Trends in Food Science & Technology 14: 341-353
11. Chien PJ, Sheu F, Yang SH (2007) Effects of edible chitosan coating on quality and shelf life of sliced mango fruit. Journal of Food Engineering 78: 225-229
12. Halliwell B Murcia MA, Chirico S, Aruoma OI (1995) Free radicals and antioxidants in food and in vivo: what they do and how they work. See comment in PubMed Commons below Crit Rev Food Sci Nutr 35: 7-20
13. Nicoli MC, Anese M, Severini C (1994) Combined effects in preventing browning reactions in minimally processed fruit. J Food Qual 17: 221-229
14. Viskelis P, Rubinskiene M, Sasnaukas A, Bobinas C, Kvkliene N (2011) Changes in apple fruit quality during a modified atmoshere storage. Journal of Fruit and Ornamental Plant Research 19: 155-165
15. Rocha AMCN, Barreiro MG, Morais AMMB (2004) Modified atmosphere package for apple "Bravo de Esmolfe". Food Control 15: 61-64
16. Mehinagic E, Royer G, Symoneaux R, Jourjon F, Prost C (2006) Characterization of odor-active volatiles in apples: influence of cultivars and maturity stage. See comment in PubMed Commons below J Agric Food Chem 54: 2678-2687
17. Defilippi BG, San Juan W, Valdés H, Moya-León MA, Infante R, et al. (2009) The aroma development during storage of Castlebrite apricots as evaluated by gas chromatography, electronic nose, and sensory analysis. Postharvest Biology and Technology 51: 212–219
18. Nicolaï BM Defraeye T, De Ketelaere B, Herremans E, Hertog ML, et al. (2014) Nondestructive measurement of fruit and vegetable quality. See
19. Baldwin EA Bai J, Plotto A, Dea S (2011) Electronic noses and tongues: applications for the food and pharmaceutical industries. See comment in PubMed Commons below Sensors (Basel) 11: 4744-4766
20. Falasconi M, Concina I, Sberveglieri V, Pulvirenti A, Sberveglieri G (2012) Electronic nose for the microbiological quality control of food products. Int J Electrochemistry 2012: 715763

21. Lebrun M, Plotto A, Goodner K, Ducamp MN, Baldwin EA (2008) Discrimination of mango fruit maturity by volatiles using the electronic nose and gas chromatography. Postharvest Biology and Technology 48: 122–131
22. Concina I, Falasconi M, Sberveglieri V (2012) Electronic Noses as Flexible Tools to Assess Food Quality and Safety: Should we Trust Them? IEEE Sensors Journal 12: 3232-3237
23. Garcia-Breijo E, Guarrasi V, Masot Peris R, Alcañiz Fillol M, Pinatti CO (2013) Odour sampling system with modifiable parameters applied to fruit classification. Journal of Food Engineering 116: 277–285
24. Keenan DF, Valverde J, Gormley R, Butler F, Brunton NP (2012) Selecting apple cultivars for use in ready-to-eat desserts based on multivariate analyses of physico-chemical properties. LWT – Food Science and Technology 48: 308-315
25. Portela SI, Cantwell MI (1998) Quality changes of minimally processed honeydew melons stored in air or controlled atmosphere. Postharvest Biology and Technology 14: 351–357

CHAPTER 3

Application of Chitosan Based Coating in Fruit and Vegetable Preservation: A Review

Duan Jianglian and Zhang Shaoying*

College of Engineering, Shanxi Normal University, Linfen 041004, China

ABSTRACT

Edible coating is beneficial to the shelf life of postharvest fruit and vegetable. Chitosan-based coating was concerned in recent years owing to its non-toxic, biodegradable, and biocompatible properties. In view of the deficiency of single chitosan coating, there are two main methods to improve the property of chitosan-based coating at present. One method is that the chitosan were combined with organic compounds, inorganic compound, or biological control agents. The other method is that single chitosan coating was applied with non-coating remedies including heat treatment, hypobaric treatment, gas fumigation, and modified atmosphere packaging. After applying improved chitosan-based coating, the preserving effects were increased in most of the cases compared with single chitosan coating.

Keywords: Chitosan; Coating; Fruit; Vegetable; Preservation

INTRODUCTION

Postharvest fruit and vegetable are living organisms, undertaking metabolism ceaselessly. Their character such as nutrition, favor, and appearance deteriorated during the process of storage and transportation owing to water loss, browning, decay, and so on [1,2]. Thus, the commercial value also decreases and many damages are caused to producer. To extend the shelf life of postharvest fruit and vegetable, some effective measures including low temperature, modified atmosphere packaging, irradiation and coating, have been applied [3-5]. In those measures, edible coating is one of promising methods because of its particular properties, which could avoid moisture loss and aromas loss, and inhibit the oxygen penetration to the plant tissue or microbial growth. In addition, edible coating is convenient and conforms to food safety [6]. Many materials such as polysaccharides, proteins, essential oils, may economically serve as edible coatings [7-9].

Chitosan is a linear polysaccharide consisting of β-(1→4)-linked 2-amino-2-deoxy-D-glucose residues, originating from deacetylated derivative of chitin, which is the second most abundant polysaccharide in nature after cellulose. It was non-toxic, biodegradable, biofunctional, and biocompatible. Chitosan has strong antimicrobial and antifungal activities that could effectively control fruit decay [10]. It could easily form coating on fruit and vegetable, and the respiration rate of fruit and vegetable was reduced by adjusting the permeability of carbon dioxide and oxygen [11]. Considering the superior properties of chitosan, it has been successfully used in many postharvest fruits and vegetables, such as grape, berry, jujube and fresh-cut lotus root [9,12-14]. Though chitosan coating has many advantages to the preservation of postharvest fruit and vegetable, as for specific fruit or vegetable, single chitosan coating sometimes demonstrates a certain defect, which includes limited inhibition to especial microorganism that leads fruit to decay, and poor coating structure to adjust the permeability of carbon dioxide and oxygen [15]. To effectively apply the chitosan coating, the chitosan was combined with other substances. In addition, the single chitosan coating was often combined with physical methods such as short heating, short gas fumigation, modified atmosphere packaging, and so on [3,5,16].

This review attempts to summarize the general development of various chitosan-based coatings, based on the current situation of research and progress in the preservation of fruit and vegetable. We expect that this review will provide insights for researchers working in postharvest preservation.

Chitosan Based Composite Coating

Composite coating of chitosan and organic compounds

Chitosan combined with essential oil: Essential oils are natural antioxidant and antimicrobial lipidic substances extracted from plants. Most of them consist of a mixture of terpens, terpenoids and other aromatic and aliphatic constituents [17]. They are gaining increasing interest because of relatively safe status, wide acceptance by consumers, and exploitation for potential multi-purpose functional uses. Filmforming dispersions (FFD) were prepared with 1% high molecular weight chitosan and 3% lemon essential oil. Chitosan coatings did not show a significant effect in terms of the acidity, pH and soluble solid content of strawberries throughout storage. In contrast, coatings slowed down the respiration rate of samples when lemon essential oil was added to the FFD. Adding lemon essential oil enhanced the chitosan antifungal activity both in vitro tests and during cold storage in strawberries inoculated with a spore suspension of *Botrytis cinerea* [12].

Chitosan combined with organic acid: Organic acids are weak acids and do not dissociate completely in water, whereas the strong mineral acids do. Contrary to the use of mineral acids, organic acids are easy to handle, non-corrosive in nature, biodegradable and cause no harm to the coated fruit and vegetable. They have been used as substitutes for sulfite to prevent enzymatic browning, alone or in combination with antimicrobial agent [18]. The effect of chitosan coating combined with phytic acid on the preservation of fresh cut lotus root was investigated. The result showed that composite coating with 1% chitosan and 1% phytic acid could decrease the weight loss rate and malondialdehyde (MDA) content of fresh-cutting lotus root, postpone browning, restrain the activities of peroxidase (POD), polyphenol oxidase (PPO) and phenylalanine ammonia-lysae (PAL), and maintain the content of vitamin C and polyphenol at relatively high level. At day 8, the weight loss rate of coated fresh cut lotus root was one half of control, MDA content was 18.8% lower, and L value was 51.2% higher than that of control sample. Composite coating with chitosan combined phytic acid is an effective method to preserve fresh cut lotus root, superior to single chitosan coating [19]. However, aciddissolved chitosan solutions develop bitterness and astringency tastes, making chitosan coated foods less practical on the real market [20].

Chitosan combined with other organic matters: Other organic matters such as ethanol, wax, and so on, could also improve the properties of chitosan coating in fruit preservation. Artificially inoculated single berries or clusters were immersed in chitosan, ethanol, or their mixture. The tests with small clusters were carried out to simulate commercial prolonged cold storage of table grapes. The combination of reduced doses of chitosan and ethanol improved the control of gray mold of table grapes compared to their application alone, and the effect was at least additive and at times synergistic [21].

Composite coating of chitosan and inorganic compound

Chitosan combined with metalions: Calcium ions perform multiple roles in plant cell physiology. They are important intracellular messengers, mediating responses to hormones, biotic and abiotic stress signals and a variety of developmental processes [22]. They also play an essential role in the structural maintenance of membranes and cell walls. Calcium ions cross-link free carboxyl groups on adjacent polygalacturonate chains present in the middle lamella of the plant cell wall contributing to cell–cell adhesion and cohesion. Preharvest and postharvest treatments with calcium salts have been effective in controlling several physiological disorders, reducing the incidence of fungal pathogens and maintaining fruit firmness [23]. Strawberries (*Fragaria ×ananassa Duch.*) were coated with chitosan combined with calcium gluconate. Following treatment, strawberries were stored at 10°C and 70 ± 5% RH for one week. No sign of fungal decay was observed during the storage period for fruit coated with 1.5% chitosan or 1% chitosan+0.5% CaGlu. By contrast, 12.5% of the strawberries coated with 1% chitosan lacking calcium salt were infected after five days of storage. And addition of calcium to the chitosan solution increased the firmness and nutrients of the fruit [24]. In addition, zinc ion also attracted the interest of several researchers, because of its physiological functions. Zinc is one of the most important essential micro-metallic elements in the human body. It is an essential component of a number of important proteins and is indispensable for its stability and catalytic functions [25]. Cerium ion has good antibiotic capability and cleavage of the phosphodiester [26]. The complex of zinc (II) and cerium (IV) with chitosan had been applied for preserving Chinese jujube fruits to expand their shelf life and decrease residues of organophosphorus pesticides on the fruits [27].

Chitosan combined with inorganic nano-material: Emerging at the end of the 1980s, nanotechnology is currently rapidly developing and being widely used in the material, chemical, and physical fields. Different from normal-sized substances, nano-sized substances show quanta size, small size, surface, and macroscopic quanta size effects [28]. At present, nano-ZnO, nano-silicon and nano-$CaCO_3$ were used in postharvest fruit preservation. The effect of 1% chitosan film with 0.04% nano-silicon dioxide on the qualitative properties of harvested jujube under ambient temperature was investigated. After 32 day, the red index, decay incidence, weight loss, and respiration rate of the coated jujubes were lower compared with those of the control. The lower PAL activity and higher activities of scavenger antioxidant enzymes (i.e., SOD, POD, and CAT) of the coated jujubes can be attributed to the compound coating. Increased MDA in the coated jujubes was restrained. Composite coating had shown to be superior in preserving total flavonoid than chitosan coating alone. But no differences were observed in terms of vitamin C loss and total polyphenol content between composite coating and control [13,29,30].

Composite coating of chitosan and biological control agents

Biological control with antagonistic microorganisms has been considered the most promising alternative method for inhibition of postharvest diseases in recent years, especially for control of woundinvading pathogens [31]. At present there are several commercial products available in the market for postharvest use, such as Biosave (*Pseudomonas syringae*) registered in the USA and "Shemer" (*Metschnikowia fructicola*) registered in Israel [32]. The application of L. enzymogenes 3.1T8 in combination with chitosan reduced the number of diseased plants by 50-100% in four independent experiments relative to the Pythium control of preserved cucumber. However, application of chitosan or the bacterial inoculant alone was not effective [33].

Chitosan Coating Combined With Non-Coating Treatment

Integrated application of chitosan coating and heat treatment

Heat treatment is an effective non-chemical method for controlling postharvest pests and diseases if suitable combinations of temperatures and exposure times are selected to prevent quality loss. The use of a combination of such techniques in the postharvest handling of fresh products is increasing [34]. Apples (*Malus domestica Borkh. cv. Gala*) were heat-treated at 38°C for 4 days (heat treatment) before or after coated with 1% chitosan. Following treatment, apples were stored at 0°C for 8 weeks and 20°C for 7 days as shelf life. The heat treatment + chitosan treated fruit showed the lowest respiration rate, ethylene evolution, malondialdehyde and membrane leakage, and the highest firmness and consumer acceptance among the treatments. At the same time, this combined treatment could inhibit the lost of green color, titratable acidity and weight loss compared with HT alone [16].

Integrated application of chitosan coating and hypobaric treatment

Among the physical means for controlling postharvest decay of fruit and vegetables, the use of sub-atmospheric pressure treatments has been largely unexploited to date. It has been reported that hypobaric treatments, applied during all or almost all of the storage period, delay the ripening of some climateric fruits. The effectiveness of chitosan and short hypobaric treatments to control storage decay of sweet cherries was investigated over 2 years. The results indicate that the combination of hypobaric and chitosan treatments is a valid strategy for increasing the effectiveness of the treatments in controlling postharvest decay of sweet cherries [35,36].

Integrated application of chitosan coating and gas fumigation

1-methylcyclopropene (1-MCP), an inhibitor of ethylene action, acts at very low concentration for delaying fruit ripening and improving storage quality of many fruits and vegetables. It is nontoxic, odorless and highly effective and has also been commercially used to extend the storage life of horticultural products [37]. The combination of 1-MCP and chitosan coating could effectively improve the storage life extension and quality maintenance of Indian jujube fruit at room temperature storage [38].

Integrated application of chitosan coating and modified atmosphere packaging

Modified atmosphere packaging (MAP) is an important technique for modifying the in-package atmosphere using polymeric films with or without perforations to reduce quality deterioration and improve shelf-life of the packaged fruits and vegetables through water loss, metabolic and microbial activity reduction. Reuck et al found that the combination of chitosan (1.0 g L^{-1}) + MAP (control) was effective in preventing decay, browning and retaining the pericarp color in the cultivar McLean's Red compared with single MAP. Chitosan (1.0 gL^{-1}) + MAP significantly reduced PPO and POD activity, retained membrane integrity, anthocyanin content and prevented the decline of pericarp color values during storage [39].

CONCLUSION

To effectively extend the shelf life of postharvest fruit and vegetable, chitosan-based coating as a relatively convenient and safe measure, is more and more concerned in food industry in recent years. In view of the deficiency of single chitosan coating, there are two main methods to improve the property of chitosan coating. One method is that the chitosan were combined with organic compounds such as essential oil, organic acid, or inorganic compound including metal ions and inorganic nano-material, as well as biological control agents. The other method is that the single chitosan coating was applied with physical remedies containing heat treatment, hypobaric treatment, gas fumigation and modified atmosphere packaging. After applying improved chitosan-based coating, the preserving effects were increased in most of the cases compared with single chitosan coating.

ACKNOWLEDGEMENTS

This work was supported by project of the National Natural Science Foundation of China under grant no. 31101359, and was also supported by Program for the Innovative Talents of Higher Learning Institutions of Shanxi (2012).

REFERENCES

1. Gatto MA, Ippolito A, Linsalata V, Cascarano NA, Nigro F, et al. (2011) Activity of extracts from wild edible herbs against postharvest fungal diseases of fruit and vegetables. Postharvest Biol Tec 61: 72-82.
2. Terry LA, Joyce DC (2004) Elicitors of induced disease resistance in postharvest horticultural crops: a brief review. Postharvest Biol Tec 32: 1-13.
3. Xanthopoulos G, Koronaki ED, Boudouvis AG (2012) Mass transport analysis in perforation-mediated modified atmosphere packaging of strawberries. J Food Eng 111: 326-335.
4. Castagna A, Chiavaro E, Dall'asta C, Rinaldi M, Galaverna G, et al. (2013) Effect of postharvest UV-B irradiation on nutraceutical quality and physical properties of tomato fruits. Food Chem 137: 151-158.
5. Sañudo MB, Siller-Cepeda J, Muy-Rangel D, Heredia JB (2009) Extending the shelf-life of bananas with 1-methylcyclopropene and a chitosan-based edible coating. J Sci Food Agr 89: 2343-2349.
6. Mantilla N, Castell-Perez ME, Gomes C, Moreira RG (2013) Multilayered antimicrobial edible coating and its effect on quality and shelf-life of fresh-cut pineapple (Ananas comosus). Food Sci Technol 51: 37-43.
7. Vanzela ESL, Nascimento PD, Fontes EAF, Mauro MA, Kimura M (2013) Edible coatings from native and modified starches retain carotenoids in pumpkin during drying. Food Sci Technol 50: 420-425.
8. Lima MA, Cerqueira MA, Souza BWS, Santos ECM, Teixeir a JA, et al. (2010) New edible coatings composed of galactomannans and collagen blends to improve the postharvest quality of fruits -Influence on fruits gas transfer rate. J Food Eng 97: 101-109.
9. Santos NSD, Aguiar AJA, Oliveira CED, Sales CVD, Silva SDM, et al. (2012) Efficacy of the application of a coating composed of chitosan and Origanum vulgare L. essential oil to control Rhizopus stolonifer and Aspergillus niger in grapes (Vitis labrusca L.). Food Microbiol 32: 345-353.
10. Aider M (2010) Chitosan application for active bio-based films production and potential in the food industry: Review. Food Sci Technol-LEB 43: 837-842.
11. Elsabee MZ, Abdou ES (2013) Chitosan based edible films and coatings: A review. Mater Sci Eng C Mater Biol Appl 33: 1819-1841.
12. Perdones A, Sánchez-González L, Chiralt A, Vargas M (2012) Effect of chitosan-lemon essential oil coatings on storage-keeping quality of strawberry. Postharvest Biol Tec 70: 32-41.

13. Yu YW, Zhang SY, Ren YZ, Li H, Zhang XN, et al. (2012) Jujube preservation using chitosan film with nano-silicon dioxide. J Food Eng 113: 408-414.
14. Xing Y, Li X, Xu Q, Jiang YH, Yun J, et al. (2010) Effects of chitosan-based coating and modified atmosphere packaging (MAP) on browning and shelf life of fresh-cut lotus root (Nelumbo nucifera Gaerth). Innov Food Sci Emerg 11: 684-689.
15. Ravi KMNV (2000) A review of chitin and chitosan applications. React funct Polym 46: 1-27.
16. Shao XF, Tu K, Tu S, Tu J (2012) A Combination of Heat Treatment and Chitosan Coating Delays Ripening and Reduces Decay in "Gala" Apple Fruit. J Food Quality 35: 83-92.
17. Bakkali F, Averbeck S, Averbeck D, Idaomar M (2008) Biological effects of essential oils--a review. Food Chem Toxicol 46: 446-475.
18. Qi HP, Hu WZ, Jiang A, Tian MX, Li YQ (2010) Extending shelf-life of Fresh-cut 'Fuji' apples with chitosan-coatings. Innov Food Sci Emerg 12: 62-66.
19. Yu YW, Li H, Jinhua D, Ren YZ (2012) Study of Natural Film with Chitosan Combining Phytic Acids on Preservation of Fresh-cutting Lotus Root. J Chinese Institute Food Sci Tech 12: 131-136.
20. Han C, Lederer C, McDaniel M, Zhao YY (2005) Sensory Evaluation of Fresh Strawberries (Fragaria ananassa) Coated with Chitosan-based Edible Coatings. J Food Sci 70: S172-S178.
21. Romanazzi G, Karabulut OA, Smilanick JL (2007) Combination of chitosan and ethanol to control postharvest gray mold of table grapes. Postharvest Biol Tec 45: 134-140.
22. Reddy VS, Reddy AS (2004) Proteomics of calcium-signaling components in plants. Phytochemistry 65: 1745-1776.
23. Bakshi P, Masoodi FA, Chauhan GS, Shah TA (2005) Role of calcium in post-harvest life of temperate fruits: A review. J Food Sci Tech Mys 42: 1-8.
24. Muñoz PH, Almenar E, Valle VD, Velez D, Gavar R (2008) Effect of chitosan coating combined with postharvest calcium treatment on strawberry (Fragaria × ananassa) quality during refrigerated storage. J Food Chem 110: 428 - 435.
25. Wagner CC, Baran EJ (2009) Vibrational spectra of Zn(II) complexes of the amino acids with hydrophobic residues. Spectrochim Acta A Mol Biomol Spectrosc 72: 936-940.
26. Huang LF, Xuesong S (2002) The Biological Function of Rare Earth Element. Acta Medicinae Sinica 15: 693-695.
27. Wu H, Wang D, Shi J, Xue S, Gao M (2010) Effect of the complex of zincII and ceriumIV with chitosan on the preservation quality and degradation of organophosphorus pesticides in Chinese jujube (Zizyphus jujuba Mill. cv. Dongzao). J Agric Food Chem 58: 5757-5762.
28. Hu GH, Hoppe S, Feng LF, Fonteix C (2007) Nano-scale phenomena and applications in polymer processing. Chem Eng Sci 62: 3528-3537.
29. Jiang M, Ye XL, Qiu Q (2012) Study on the preservation effect of chitosan and nano-ZnO compound coating on Shatang mandarin. Sci Tech Food Industry 33: 348-351.
30. Luo ZS, Xu XL, Xu TQ, Xie J (2009) Effect of chitosan coating with nano-CaCO3 appendix on quality of fresh cut yam. Transactions of the Chinese Society for Agricultural Machinery 40: 125-128.
31. Janisiewicz WJ, Conway WS (2010) Combining biological control with physical and chemical treatments to control fruit decay after harvest. Stewart Postharvest Review 6: 1-16.
32. Droby S, Wisniewski M, Macarisin D, Wilson C (2009) Twenty years of postharvest biocontrol research: is it time for a new paradigm? Postharvest Biol Tec 52: 137-145.
33. Postma J, Stevens LH, Wiegers GL, Davelaar E, Nijhuis EH (2009) Biological control of Pythium aphanidermatum in cucumber with a combined application of Lysobacter enzymogenes strain 3.1T8 and chitosan. Biol Control 48: 301-309.

34. Zhao Y, Tu K, Su J, Tu S, Hou Y, et al. (2009) Heat treatment in combination with antagonistic yeast reduces diseases and elicits the active defense responses in harvested cherry tomato fruit. J Agric Food Chem 57: 7565-7570.
35. Romanazzi G, Nigro F, Ippolito A, Salerno M (2001) Effect of short hypobaric treatments on postharvest rots of sweet cherries, strawberries and table grapes. Postharvest Biol Tec 22: 1-6.
36. Romanazzi G, Nigro F, Ippolito A (2003) Short hypobaric treatments potentiate the effect of chitosan in reducing storage decay of sweet cherries. Postharvest Biol Tec 29: 73-80.
37. Valero D, Romero DM, Valverde JM, Guilléna F,Seranno M (2003) Quality improvement and extension of shelf life by 1-methylcyclopropene in plum as affected by ripening stage at harvest. Innov Food Sci Emerg 4: 339-348.
38. Zhong QP, Xia WS (2007) Effect of 1-methylcyclopropene and/or chitosan coating treatments on storage life and quality maintenance of Indian jujube fruit. Food Sci Technol 40: 404-411.
39. Reuck KD, Sivakumar D, Korsten L (2009) Effect of integrated application of chitosan coating and modified atmosphere packaging on overall quality retention in litchi cultivars. J Sci Food Agr 89: 915-920.

CHAPTER 4

Analysis of Challenges Facing Rice Processing in Nigeria

A. S. Ajala and A. Gana

Department of Food Science and Engineering, Ladoke Akintola University of Technology, PMB 4000, Ogbomoso, Nigeria

ABSTRACT

Rice is the most economically important food crop in many developing countries and has also become a major crop in many developed countries where its consumption has increased considerably. It has become necessary to meet the demand of the world's current population growth rate, and the least costly means for achieving this aim is to increase rice productivity, wherever possible. The main challenges encountered by rice processors in Nigeria are to find appropriate solutions for quality rice processing. Therefore this work provides basic information about the challenges of rice processing and focuses on the challenges faced by the small scale rice processors and reasons for continuous rice importation with a view to guiding decision-making to be self-sufficient in rice production, thereby making some improvement in Nigerian economy.

1. INTRODUCTION

(a) Background of Study. Rice (Oryza sativa) a cereal belonging to the Gramineae, a large monocotyledonous family of some 600 genera and around 10,000 species [1]. It is valued as the most important staple food for over half of the world population [2] and ranks third after wheat and maize in production on world basis. More than half of the world's population depends on rice as the major source of calories [3]. The amount consumed by all these people ranged from 100 kg to 240 kg per annum in the year 2000 alone [3]. Two species have emerged as our most popular cultivated rice, Oryza sativa and Oryza glaberrima; of these two species the more widely produced is Oryza sativa.

In Nigeria, rice consumption has risen tremendously at about 10% per annum due to changing consumer preferences [4]. However, [5] discovered that most Nigerians prefer to consume imported rice brands as compared to local rice varieties. The reason is that most Nigerian rice processors lack adequate technology of rice processing to meet international standard.

Rice processing involves several steps: removal of the husks, milling the shelled rice to remove the bran layer, and an additional whitening step to meet market expectations for appearance of the rice kernels. This process generated several streams of material which include the husks, the bran, and the milled rice kernel [6]. Nigeria has the potential to be self-sufficient in rice production, both for food and industrial raw material needs and for export purpose. However, a number of constraints have been identified as limiting factors to rice production. These include problems with research, pest and disease management. Addressing at least most of these problems is good first step towards attaining the target of rice self-sufficiency [7]. Therefore the aim of this paper is to review the causes of the challenges facing rice production in Nigeria and provide guide to overcome these problems.

(b) World Rice Production. The world rice production for over almost a decade (2003–2012) is as shown in Figure 1. In 2003, about 580 million tons of rice was produced worldwide, 602 million tons in 2004, 620 million tons in 2005, and 622 million tons in 2006. The production continued to grow yearly; by 2007, the production had risen to 648 million tons. The production reached the peak in 2011 with a total production 720 million tons in order to feed the increasing global population. Furthermore, the world's annual production growth rate was stagnated in 2012. The reason could be attributed to natural disaster such as storm, tornado, and unfavourable climate as reported by [8].

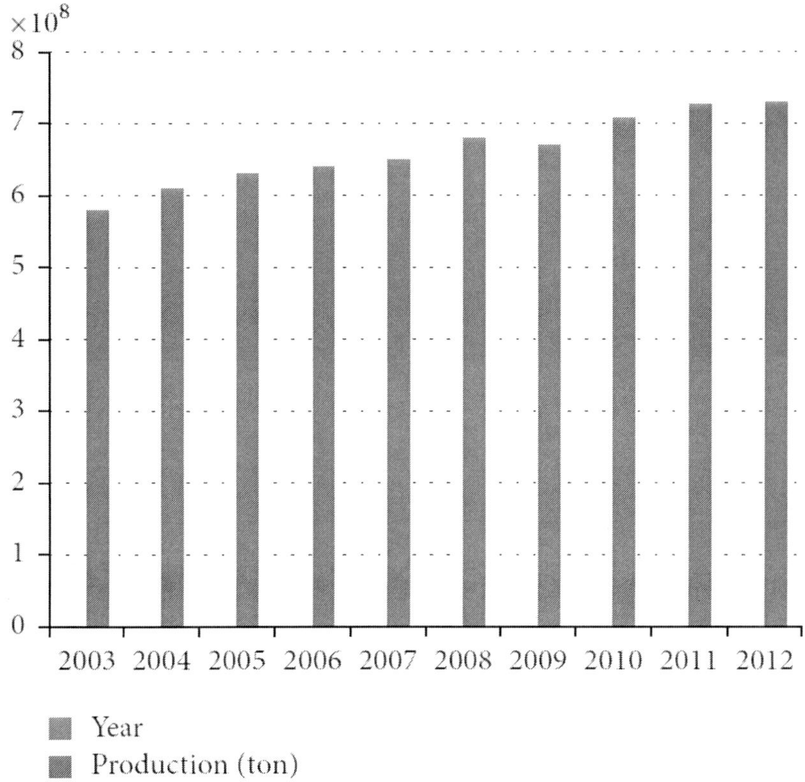

Figure 1: World rice production. Source: FAO [24].

Furthermore, [9] reported that instant low temperature below the critical point can affect seedling establishment in the early growth stage and high grain sterility in the late crop season. Unavailability of water has impeded farmers from growing rice in the southern part of California and in Southern Italy, even though these areas are more favourable in terms of climate for growing rice than the northern parts of the countries [10]. Asia accounts for 90% of the world's production and consumption of rice because of its favourable warm and humid climate, but suitable lands for increasing rice production are almost exhausted [11].

(c) Rice Production and Importation Trend in Nigeria. According to Hardcastle [12], rice production started in Nigeria in 1500 BC with the low-yielding indigenous red grain species O. glaberrima Steud and then widely grown in the Niger Delta area. WARDA [13] reported about the high-yielding white grain O. Sativa L. introduced in 1890, and by 1960 it accounted for more than 60 percent of the rice grown in the country. Also,

Figure 2 shows the trend of rice production in Nigeria since 1968 through 2008 (four-decade) study. Rice production remained at low level from 1968 to 1978 perhaps due to dietary idiosyncrasy for tubers. WARDA [13] reported that paddy rice production had risen from 13400 to 344000 tons in 1970 and area was cultivated from 156,000 to 255,000 ha. Since then, paddy rice production has been on the increase. Tremendous increases in area planted, output, and productivity in paddy rice production were achieved over the last two decades and now stand at 1.09 million tons. More so, the production continued to rise higher from 1978 and since 1980.

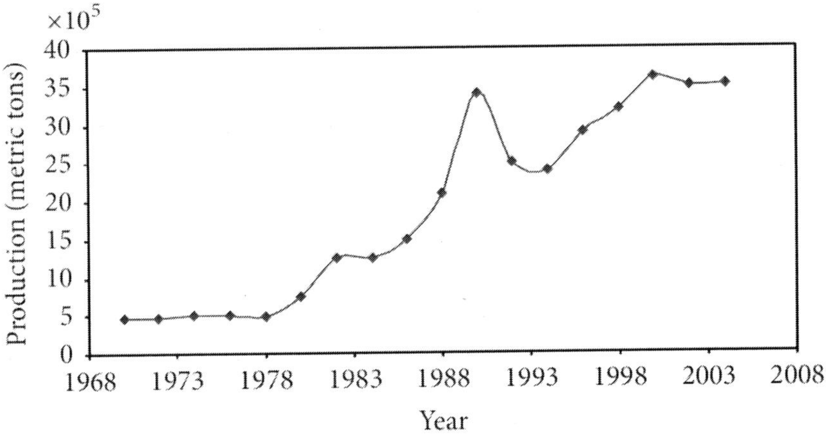

Figure 2: Rice production in Nigeria. Source: United State Department of Agriculture [16].

Nigeria has become the highest rice productions country in West Africa and the third largest in Africa, after Egypt and Madagascar [13]. The production reached the peak in 1990 in which the country was producing 3.4 million tons of rice from about 1.2 million tons [14] before it slightly fluctuated down the slope in 1993.

However, the production soared up from 1993 to 2006 where the production remains 3.8 million metric tons. The graph shows that as years increased the production also increased; this may be due to increase in population growth as well as increase in consumption trend of rice in the country. This healthy production trend would have been sustained but for the unsteady government policy on rice imports. Increased production over the last two decades could be attributed to the ban imposed on rice imports in 1985 and if this restriction had been maintained, Nigerian rice farmers would have risen to the challenge of meeting the domestic demand for the commodity. This has not been the case; however, as the government slackened rice import restriction in 1997, the resultant effect was that foreign rice flooded back onto Nigeria markets. Restrictions on rice imports were reintroduced later in 1997, and local production has increased in response to the attractive prices offered [15]. For rice production to be boosted, Nigerian government introduced institutions to monitor production and distribution of local rice. Table 1 highlights the specific objectives of each institution.

Table 1: Summary of the different Federal Government policies on rice production.

Dates of establishment	Institutions	Mandates
1970	Federal Rice Research Station (FRRS)	Development of improved varieties of grains
1972	National Accelerated Food Production Program (NAFPP)	Design, test, and transfer technology package for production of rice, maize, sorghum, millet, and wheat
1974	National Cereals Research Institute (NCRI)	Research on high-yielding rice varieties for farmers, on-farm adaptive research, seed multiplication, and training of extension staff
1976	The Operation Feed the Nation (OFN)	Self-sufficiency in domestic food supply
1978	Abakaliki Rice Project	Rice production and processing
1987	Agricultural Development Project (ADP)	Policies on rice production were implemented
1988	Nigerian Agricultural and Cooperative Bank (NACB)	Special credit schemes to boost rice production
1999	The Presidential Rice Initiative	Addressing the widening demand supply gap and attain self-sufficiency in rice production

Source: Emodi and Madukwe [15].

As the years went by, the indigenous production of rice could not meet the demand for rice. Therefore, Nigeria started importing rice into the country. Figure 3 shows clearly the rice importation figures as reported by United State Department of Agriculture [16]. This importation of rice did not follow a regular pattern but it is observed from the figure that rice importation was not up to 1 million metric ton per year since 1983–1998. However, from 2006 upwards, the importation figure soared yearly to reach the peak in 2013 with 3.5 million metric tons. The major reason is the preference of foreign to indigenous rice which has a low level of processing technology [17].

Furthermore, Figure 4 shows the importation yearly rice growth rate from 1983 to 2013. It is observed that growth rate is irregular and in some years there is no growth at all in rice importation business. From this figure, the growth rate is counted when the curves are upward ones and no growth rate occurs when curves are downward ones. This growth rate implies that the higher the growth rate is, the higher the country depends on importation of rice and the less the production of local rice is. This means that in 1997 Nigeria depended heavily on imported rice to feed its populace. This had negative consequence on the economy.

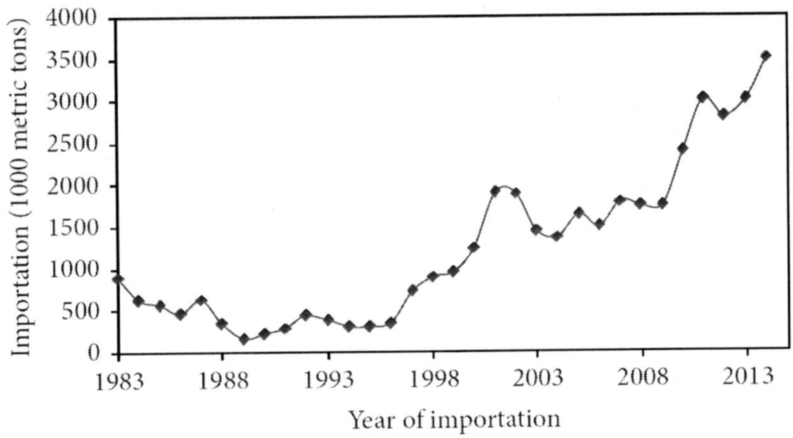

Figure 3: Rice importation in Nigeria. Source: United State Department of Agriculture [16].

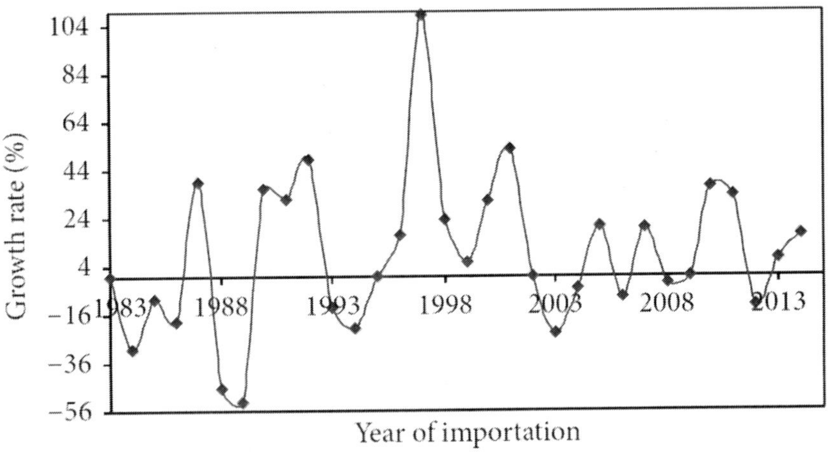

Figure 4: Rice importation growth rate in Nigeria. Source: United State Department of Agriculture [16] modified.

(d) Rice Consumption Trend in Nigeria. As reported by Imolehin and Wada [14] that in the 1960s Nigeria was almost 99 percent self-sufficient in the rice production consumed by its citizens. Over the following two decades (1970s and 1980s) self-sufficiency declined to 38 percent, leading to demand outstripping supply.

Table 2 shows the percentage of expenditure on categories of foods consumed in urban centers in Nigeria. The table shows that tubers and plantain take 23.93% followed by rice which means Nigerian spent a lot of money on rice. That is why the local rice could not be sufficient; therefore import is necessary. Philippine Rice Research Institute [18] reported that the Federal Government of Nigeria resorted to massive importation of rice. More than 540,000 tons of rice was imported in 1983 alone. Per capita rice consumption rose from 3.5 kg in 1970 to more than 14 kg in the 1990s.

Table 2: Household expenditure by commodity, urban.

Serial number	Commodities	Percentage of foods
1	Maize	3.54
2	Rice	8.65
3	Other cereals	3.98
4	Bread and similar foods	6.05
5	Tubers and plantains	23.93
6	Poultry	0.96
7	Meats	3.52
8	Sea foods	4.57
9	Dairy products	1.14
10	Oils, fats, and oil rich nuts	3.65
11	Fruits	2.42

Source: National Bureau of Statistics [26].

The same trend of household expenditure on rice is found in Table 3 for rural dwellers except that the expenditure on rice is 9.07% which is slightly higher than that of urban expenditure. The reason is that most of the other food items are produced in rural areas; hence rice may be viewed as a special diet by the rural dwellers; therefore they may commit more finance to rice than the urban dwellers. From these Tables 2 and 3, it is observed that rice constitutes a staple food among Nigerians that the demand has always exceeded the supply. Because of this reason, Nigeria has a favourable market for rice production. However, because of the poor quality of locally produced rice, most consumers prefer the imported rice from other countries such as Thailand, Bangkok, and China to the indigenous rice [19].

Table 3: Household expenditure by commodity, rural.

Serial number	Commodities	Percentage of foods
1	Maize	5.48
2	Rice	9.07
3	Other cereals	14.30
4	Bread and similar foods	2.30
5	Tubers and plantains	22.30
6	Poultry	0.36
7	Meats	2.61
8	Sea foods	4.09
9	Dairy products	1.10
10	Oils, fats, and oil rich nuts	3.80
11	Fruits	1.55

Source: National Bureau of Statistics [26].

2. TRADITIONAL METHODS OF PROCESSING RICE PADDY

The traditional methods of processing rice paddy involve soaking of the paddy in water for 2 to 3 days to soften the kernel, followed by steaming of the soaked paddy for 5–10 minutes as shown in Figure 5 and dried in the sun as shown in Figure 6, followed by pounding the dried paddy in a mortar and pestle device to remove the husk or use of simple machines for dehulling/milling as shown in Figure 7; then the grain is cleaned using a winnowing basket [20]. Though the traditional method of processing rice is simple, but tedious, it has very low outturn and results in breakages of rice kernels and incomplete removal of husks. More so, it has a short storage life as the fat in the bran develops rancidity [20].

Figure 5: Small scale rice parboiler. Source:

Figure 6: Traditional method of rice drying in Nigeria. Source: Basorun [25].

Figure 7: Small scale rice dehuller. Source: Propcom [21].

3. MODERN METHODS OF PROCESSING RICE PADDY

Propoor Opportunities in Commodity and Service Markets Project [21] carried out survey on modern rice processing and discovered that there has been improvement in rice processing in Nigeria compared to the 1990s. In modern methods, the rough rice or paddy is first cleaned to remove contaminants, and the husks are then removed by the so-called shellers; these are most commonly horizontally spaced rotating abrasive stones, but increasing use is being made of rubber roll or rubber belt made shellers. The rice and hulls are separated by aspiration and any paddy remaining with the rice is removed in a paddy separator. It is discovered that the main problem of Nigerian rice is the presence of stone in the rice grains. The reported survey shows that there has been locally produced destoner made in Kano, Nigeria, as shown in Figure 8. The machine costs N65,000 and was fabricated using angle iron and sheet metals.

Figure 8: Photograph of the destone machine. Source: Propcom [21].

Another improvement in rice processing in Nigeria reported by Propcon is the drying process. The traditional sun drying has been replaced by mechanical dryer or improved sun drying method. A typical example of mechanical dryer is as shown in Figure 9. This type of dryer is found scattered all over the country. It can process about 3000 kg and remove 50% moisture of rice in 6 hrs.

Figure 9: A mechanical dryer. Source: Propcom [21].

Figure 10: Showing the diagram of a solar dryer. Source: Propcom [21].

Apart from this mechanical dryer which uses diesel or electricity, other dryers have been developed such as solar dryer for drying rice paddy. An example of this is as shown in Figure 10 developed in National Centre Solar Energy Research, Sokoto, Nigeria. The development of this is a result of incessant power outage all over the country. This type of solar dryer is equipped with fan to enhance hot air distribution over the rice paddy.

4. CONSTRAINTS OF RICE PROCESSING IN NIGERIA

Several factors are militating against rice production/processing in Nigeria. Wilfred [22] reported that the inadequate knowledge on the use of herbicides and pesticide, postharvest handling, processing and marketing, pest and diseases, soil fertility management, irrigation and water, harvesting skills are some of the constraints facing the rice farmers. Agidi [23] iterated the labour intensity involved such as ploughing, planting, weeding, harvesting, threshing, and transportation being strenuous and laborious; the problem is worsen by lack of appropriate rice farming tools, implementation, and equipment. Besides this, the equipment available is often too expensive for the average farmer. Most farmers depend on rudimentary, labour, and time consuming hand tools such as hoes, slashers, sickles, axe, and rake for various farm operations. Other factors include lack of industrial drive due to poor government policy and high cost of production. Because local rice is of low quality, the rice merchants use this opportunity to expand on rice importation. The solution lies in the provision of incentives, machineries, and government policies to confront these challenges faced by rice processors.

5. CONCLUSION

In conclusion, Nigeria has a good climate for rice production and it has favourable market to absorb the production. There is need to improve the quality of indigenous rice to compete with foreign rice through selection and adaptation of modern rice technologies to ease labour in production and enhance nutritional qualities in rice processing and production. Government has role to play in forming strong policies that will favour production of local rice as it is being practiced in the advanced world.

REFERENCES

1. E. J. Wibberley, Cereal Husbandary, Farming Press Books, Ipswich, UK, 1st edition, 1989.
2. Agric News, Rice is Life. International Year of Rice, 2004 Declaration, vol. 1, no. 2, Bayelsa State Ministry of Agriculture & Natural Resources, 2003.
3. FAO, Food and Agriculture Organization Sustainable Rice Based Production and People's Livelihood, International Rice Commission Newsletter (Special Edition) vol. 52, International Rice Commission, FAO, Rome, Italy, 2003.
4. T. Akande, The Rice Sector in Nigeria, United Nation Crop Project (UNCP) Country Agricultural Project on Trade Liberalization in Agricultural Sector and the Environment, Geneva, Switzerland, 2003.
5. O. A. T. Ebuehi and A. C. Oyewole, "Effect of cooking and soaking on physical characteristics, nutrient composition and sensory evaluation of indigenous and foreign rice varieties in Nigeria," African Journal of Biotechnology, vol. 6, no. 8, pp. 1016–1020, 2007.
6. R. C. Schramm, Rice processing; milling and value added effects [M.S. thesis], Louisiana State University, Agricultural and Mechanical College, Baton Rouge, La, USA, 2006.
7. West Africa Rice Development Association (WARDA), Growing Lowland Rice. A Production Handbook, Africa Rice center, Abidja, Côte d'Ivoire, 2004.
8. R. M. Olanrewaju, "Climate and rice production in a part of the Niger River Basin development authority area (NRBDA). A case study of Edu and Lafiagi local government areas of Kwara State, Nigeria," Journal of Meteorology and Climate Science, vol. 8, no. 2, pp. 102–110, 2010.
9. R. T. Cruiz, Mitigating Impact of Climate Change in Rice Production, Agriculture, Philippine Rice Research institute, 2009.
10. D. V. Tran, "World rice production: main issues and technical possibilities," in Activités de Recherche sur le Riz en Climat Méditerranéen, J. Chataigner, Ed.,

Cahiers Options Méditerranéennes; n. 24(2), pp. 57–69, CIHEAM, Montpellier, France, 1997.

11. FAO, World Rice Information, Issue No.1, FAO, Rome, Italy, 1995.
12. J. F. Y. Hardcastle, "The development of rice production and research in the Federation of Nigeria," Tropical Agriculture, vol. 36, pp. 79–95, 1959.
13. WARDA, West Africa Rice Development Association Annual Report 1996, West Africa Rice Development Association, 1996.
14. E. D. Imolehin and A. C. Wada, Meeting the Rice Production and Consumption Demands of Nigeria with Improved Technologies, National Cereals Research Institute, Badeggi, Nigeria, 2000.
15. Emodi and M. Madukwe, "A review of policies, acts and initiatives in rice innovation system in Nigeria," Journal of Agricultural Extension, vol. 12, no. 2, pp. 77–80, 2008.
16. United State Department of Agriculture, Nigeria Milled Rice Imports by Year (1000 MT), 2014.
17. T. Akande, "Overview of the Nigeria rice economy," African Journal of Biotechnology, 2001.
18. Philippine Rice Research Institute, "Rice consumption," in IRRI Rice Facts, Los Baños. Philippines. Business Week, Philippine Rice Research Institute, Nueva Ecija, Philippines, 1994.
19. Idris, K. Rasaki, O. J. Hodefe, and B. Hakeem, "Consumption pattern of Ofada rice among civil servants in Abeokuta Metropolis of Ogun State, Nigeria," Journal of Biology, Agriculture and Healthcare, vol. 3, no. 6, pp. 106–112, 2013.
20. Srilakshmi, Food Science, New Age International Publisher, 3rd edition, 2003.
21. Pro-poor Opportunities in Commodity and Service Markets Project (Propcom, 2012), http://www.propcom.org/.
22. O. R. Wilfred, Final Survey Report on the Status of Rice Production, Processing and Marketing in Uganda, Japan International Cooperation Agency, Osaka, Japan, 2006.
23. G. Agidi, "The making of a practical engineer," T. K. 10, pp. 1, 11, 1993.
24. FAO, "Rice Market Monitor. Volume XVI—Issue No. 1," 2013, http://www.fao.org/economic/est/publications/rice-publications/rice-market-monitor-rmm/en/.
25. J. O. Basorun, "Expository analysis of rice processing activities in Igbemo, rural Nigeria," American Journal of Social Issues and Humanities, vol. 3, no. 2, pp. 79–81, 2013.
26. National Bureau of Statistics, "Consumption pattern in Nigeria 2009/10," Preliminary Report, National Bureau of Statistics, 2012.

CHAPTER 5

Use of Recycled Containers for Hermetic Maize Storage in East Africa

Yakubu A.[1*], Wilson L. A.[2], Bern C. J.[1] and Brumm T. J.[1]

[1]*Department of Agricultural and Biosytems Engineering, lowa State University, Ames, Iowa 50011 USA.*

[2]*Department of Food Science and Human Nutrition, Iowa State University, Ames, Iowa 50011, USA.*

ABSTRACT

Affordable, pesticide-free, and reliable maize storage containers for seed and food purposes are not available in many subsistence farming cultures. As a result, subsistence farmers lose their crop to pests and are forced to dispose of their maize for a low price right after harvest thus, robbing them of profit and food for the rest of the year. The objective of this research was to develop information to enable utilization of recycled sanitary hermetic maize storage containers. Use of these containers will allow farmers to preserve their maize for as long as they wish, using locally available resources. A market survey found edible oil containers available for sale and re-use in markets in Uganda, Tanzania, and Kenya. A laboratory study compared three cleaning methods (oil drain+45°C water(A), oil drain+ 90 to 100°C water(B), oil drain+ 90 to 100°C water + soap(C)) for cleaning soybean oil contaminated 20 L HDPE containers. Research results, indicate that using "C" will clean the containers to a cleanliness level comparable to a new container. This is an encouraging result in light of the edible oil containers available for sale and re-use in East African markets. A Ugandan field study also tested the efficacy of recycled containers for hermetic storage, and confirmed that hermetic storage using these containers is lethal to maize weevils, while preserving the quality of stored maize.

Keywords: Maize storage, hermetic storage, recycled containers.

INTRODUCTION

About 19% of maize produced in East Africa is lost during post-harvest storage, due to insects, fungi, rodents, and birds (PHL Network, 2009; Lindblad and Druben, 1980; Demissie et al., 2008a, b; Gregg and Billups, 2009). The highest losses are usually due to maize weevils (*Sitophilus zeamais*) (Holst, et al., 2000; Demissie et al., 2008a). Hermetic storage seeks to eliminate gas exchange with the outside environment and thus causes weevils to die as oxygen in the grain container is used up. It can eliminate maize weevil damage along with bird, and rodent damage (Navarro, et al., 1994; CRC, 2003; Springer, 2007; Glevitzky, et al., 2009; FAO, 2010; Yakubu, et al., 2011). If a container is to be used for hermetic maize storage, it needs to be sealable, robust, affordable, available, and free of harmful contaminants which can be transferred to stored maize. Some recycled containers may meet these criteria and be suitable for grain storage by subsistence farmers. Lindblad and Druben (1980) described the procedure to use recycled 220-L (55-gallon) oil drums for storage of grains.

Storing maize in recycled containers

Maize is grown for food. Container recycling and re-use for maize storage, therefore, requires proper selection and cleaning to prevent any toxic contents from contaminating stored maize (Schmidt and Erickson, 2008; ASM, 2009; EPA, 2009a; 2009b). Since recycled containers may have held a variety of possibly toxic materials, only food grade containers previously utilized for storage of carbonated soft drinks and triglycerides, were considered for the recycling and maize storage research (Bhatt, 2004; Shachman, 2004; Ashaye and Olusoji, 2006). But since soft drinks do not have uniform chemical contents only containers previously containing edible oil were considered in this research (Lindblad and Druben, 1980; **Cleveland et al., 2001; Murdock, et al., 2003; Adhikarinayake, 2005;** EPA, 2006; Mercer, 2006; Malik et al., 2006; Wiley-Blackwell, 2006; EPA, 2009c; Tsimihodimos, et al., 2009; CRU, 2010; EWG, 2011). "Clean" refers to being free from dirt or pollution, unadulterated, sanitary or pure (Merriam-Webster, Inc. 2010). However, according to Coats (2010) and others the end-products of some chemicals are more toxic than the starting chemicals, and containers previously used to store some chemicals cannot be cleaned adequately (Cleveland et al., 2001; EPA. 2006). To preserve maize quality, while preventing cross- contamination between stored maize and the associated rancid oil free radicals, as well as dirt, a procedure is needed for cleaning edible oil containers such that nearly 100% of the oil is removed (Andrikopoulos et al., 2002; Andrikopoulos, 2004; Choe and Min, 2006). This can give subsistence farmers the benefits of effective hermetic storage in a clean container (Carroll and Fulton, 2008; Murdock et al., 2003).

Research need

Little information was found on the availability and suitability of used edible oil containers in East Africa. A study was therefore needed on container sizes, prices, and availability in East Africa. A study was also needed to establish cleaning procedure for reused edible oil containers to ensure food safety.

Objectives

1. To determine the availability of used edible oil containers in East Africa suitable for hermetic maize storage.
2. To develop and test procedures for cleaning previously used edible oil containers.
3. To deploy and test the efficacy of the containers for hermetic storage, in the field.

MATERIALS AND METHODS

This study consisted of a market survey, laboratory research and field study.

Market survey

To explore recycled edible oil container availability for hermetic storage in East Africa, a recycled container survey form was designed and dispatched to contacts in Uganda, Kenya, and Tanzania. The purpose of the survey was to determine availability of containers in selected East African markets which can be recycled for maize storage. Containers identified were expected to be at least five L in capacity and airtight (no holes). Container properties to be identified by the survey included previous or intended use (edible oil, soft drink concentrate, etc.), volume (L), material (plastic, steel, etc.), quantity in each market, as well as price. Each surveyor was expected to carry out the survey in at least three markets. East African Markets surveyed included (1) Jua Kali Drum Dealers (Nairobi), (2) Frere Town (Mombasa) and (3) Musila Enterprises (Kikambala Village), in Kenya; (1) Mwembe, (2) Same Center, and (3) Kwasakwasa, in Tanzania; (1) Namanve Market (Mukono-Kampala), and (2) Soko Mujinga Market (Kitale), in Uganda.

Laboratory research

The laboratory part of the research conducted at Iowa State University was done in two stages (Figure 1). The first stage involved applying three treatments (A,B,C) to randomly assigned experimental units (20-L HDPE containers). Experimental units utilized were previously used (contaminated) and new (uncontaminated) 20-L (~5 gallon) HDPE containers. The complementary second stage utilized the goldfisch oil extraction method to measure leftover oil quantities in the 20-L containers for use in statistical analysis which allowed comparison of the three cleaning treatments.

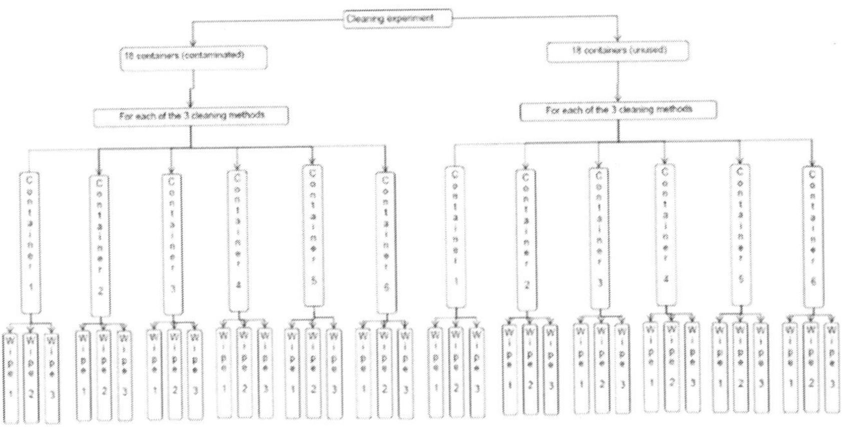

Figure 1. Cleaning experiment design flowchat.

Field study

A study simulating East African field conditions was designed and tested at Iowa State University, based on Yakubu et al. (2011), then deployed for standard field testing in Uganda (Bern et al., 2015).

Treatment definition

Cleaning of oil-contaminated containers using water, with or without soap, is a common practice in East Africa (Myers, 2006; Yakubu et al., 2011). Literature, however, suggests that cleaning, disinfecting and sanitizing containers to prevent cross-contamination of food stored within them is best done at high temperature, using soap and mechanical action, such as scrubbing and shaking (Knoxa and Walkera, 1947; Gangneux, et al., 2004; Atlas and Snyder, 2006; Sebastião, et al., 2006; CRC, 2006; Helmenstine, 2011; MTL, 2011; Patwardhan and Kelkar, 2011). Heat treatment destroys oil-splitting enzymes, arrests hydrolytic rancidity and autoxidation during vegetable oil extraction and causes oil to leach out from container surfaces (FAO, 2002). The use of soap plus hot water cleaning treatment was expected to produce similar advantages. Factors considered for this research were (i) soap and water, at two water temperatures: (control (45°C) and hot or boiling (90 to 100°C)), and (ii) container history (contaminated and new (control)). We hypothesize, therefore that cleaning with soap at high temperature (90 to 100°C), with mechanical action would be enough to clean and sanitize vegetable oil contaminated containers for maize quality preservation within them.

Experimental 20 L HDPE containers

The 36 experimental units (containers) utilized for the research each had a net weight of about 16 kg (35 lb) when full and normally contain about 15.88 kg of oil claimed to be 100% pure soybean oil (Columbus Foods, Chicago, Illinois). Eighteen were new and unused from Columbus Vegetable Oils, 30E. Oakster Street, Des Plaines, IL and eighteen were recycled used containers obtained from Ames and Des Moines, IA Chinese restaurants. They were made of high density polyethylene (HDPE), with a resin classification or recycling code of 2 (ACC, 2007; Bakers and Chefs, distributed by Sam's West, Inc. 608 S.W. 8th Street, Bentonville, AR). HDPE has a melting point of 130 to 135°C, a tensile strength of 4550 psi, and can withstand temperatures of 120°C for short periods or 110°C continuously (EOS/EDS, 2000; Antec, 2001; Dow 2009; Dynalab, 2011).

Quantification of vegetable oil remnants

Average weight of the new containers was 296.50 g. On average, contaminated containers treated with "A", "B", and "C" lost weight equaling 10.45, 14.23, and 20.06 g oil, respectively. The follow up wiping treatment using weighed cheesecloth strips to wipe container interior indicated that on average, contaminated containers treated with "A", "B", and "C" contained 0.249, 0.142, and 0.004 g, remnant oil, respectively. In cleaning contaminated containers, to which "C" was employed as treatment, enough soap was added to remove all the oil, irrespective of whether leftover oil is free flowing or bound tightly to the container internal surfaces. Therefore, the soap quantity necessary to remove nearly 100% oil contaminant was based on that. Since water at 45°C and 90 to100°C was utilized in dissolving congealed oil before draining, the oil dissolved in the water flowed out, readily reducing the leftover oil in the drained containers.

Experimental soap and water

The preliminary soap quantification involved determining the cleansing power of Ivory soap available with these containers, and quantifying the number of grams of the soap that effectively removes each gram of the vegetable oil contaminant. From an initial soap quantification experiment, it was determined that 1 g of soap removed almost 1 g of oil, from oil contaminated experimental units. However, three times that amount of soap (3 g soap/g oil) was utilized with hot water for cleaning the containers, in order to account for variability in the history of containers collected from the field. Deionized (demineralized) water was used for this research.

Oil removal treatments

Stage one of the research involved adding the assigned and calculated treatment quantities, followed by application of 3.78 L of deionized water to each EU. The mixture was then shaken at the onset and at 5 min intervals afterwards for a total of 1.5 h, and emptied. Cans were then rinsed three times and turned upside down for 48 h to allow water to flow out and dry from inside. The low temperature treatment ("A") involved just adding water at 45°C to assigned EUs, shaking at 5 min intervals, and emptying it at the end of the treatment application period. The high temperature treatment ("B") application involved addition of 3.78 L of water at 90 to 100°C to assigned EUs, while the soap treatment ("C") involved addition of the calculated quantities of soap to the assigned EUs, followed by 3.78 L of water at 90 to 100°C, and shaking. Since these treatments involve only use of soap at various temperatures, they should be repeatable in Africa.

Oil residue measurements

The second stage, follow-up, treatment involved tying about 1.7 g of absorbent cheese cloth from Prym Creative, Estopilla, Prym Consumer USA, Inc., Spartanburg, SC, to the end of a wood stick and using the cloth end to wipe each of three pre-assigned 229 mm x 229 mm areas (right, left, bottom) of the interior of each experimental container, in order to determine the level of oil remnant left. New cheesecloth was installed on the stick prior to each wiping. The oil remnant levels were determined by quantifying the oil content of the cheesecloths using the goldfisch (35001-00, LABCONCO, Kansas City, MO. 1637) hexane oil extraction method. The second stage was performed after the EUs had dried, following the initial treatment application.

RESULTS AND DISCUSSION

Survey results

The recycled container survey identified a total of 55522 containers that met the survey criteria. However, only 42208 (76%) of the containers had edible oil residue or were earmarked for edible oil storage (Table 1). These ranged in storage capacity from 5 to 20 L, with prices ranging from $0.72 (US) to $2.08 (US), based on size. On average, the number of recyclable plastic containers per market was 1142 (5 L), 967 (10 L), and 2581 (20 L), respectively.

Table 1. Used edible oil containers in selected Tanzania, Kenya and Uganda markets.

	5-L Plastic		10-L Plastic		20-L Plastic	
	Price	No. in stock	Price	No. in stock	Price	No. in stock
Kenya Market						
Market 1	-	1	-	1	-	1
Market 2	-	1	$ 0.60	200	$ 1.00	1000
Market 3	-	1	$ 0.80	200	-	1
Tanzania Market						
Market 1	$ 0.70	73	$ 1.80	92	$ 2.50	66
Market 2	$ 0.80	59	$ 1.00	64	$ 2.00	2004
Market 3	$ 0.70	48	$ 1.50	45	$ 2.00	57
Uganda Market						
Market 1	$ 1.00	100	$ 2.00	100	$ 3.00	100
Market 2	$ 0.40	10000	$ 1.00	8000	$ 2.00	20000
Total in stock		10280		8701		23227
Average Price	$ 0.72		$ 1.24		$ 2.80	
Average Containers per market		1142		967		2581

1. Not available in this market

Laboratory research results

Each treatment was assigned six containers, and three wipes (interior bottom, right, and left) were taken from each of the containers. Therefore, each of the bars in Figure 2 represent the mean of the wipes from 18 (six 20 L containers*3 wipes) container locations assigned to each of the treatments. The figure shows results obtained from cleaning recycled and new containers with three treatments (A, B, C). It indicates a strong treatment effect, and shows that while treatments "A" and "B" left behind significant ($p<.0001$) amounts of oil, treatment "C" as well as the control treatments (A, B, C, applied to unused containers) had an insignificant (p = 0.8992, 0.8633, 0.8448, and 0.8526, respectively) oil remnant, following cleaning treatment application.

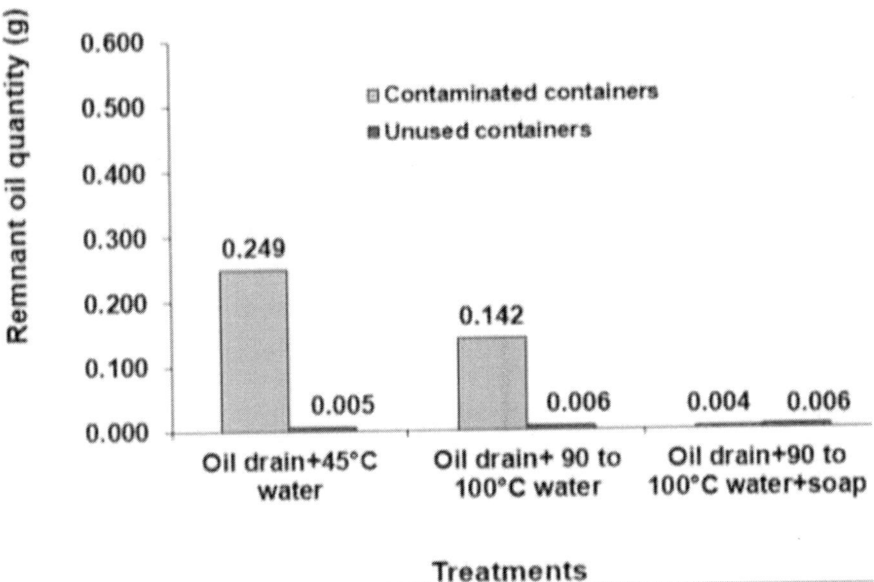

Figure 2. Treatment means bar chart (averaged over six replications).

Since analysis was done at the 0.05 significance level (Dallal, 2003), where $p < 0.05$ is sufficient evidence for accepting a hypothesis of treatment effect, the laboratory results conclude that there is a treatment difference between treatments A, B (contaminated containers) versus treatment C (contaminated containers) and all the control treatments. Any level of cross contamination between stored food and storage container surface contaminants is unacceptable (Shachman, 2004; Onsongo, et al., 2005; EPA, 2009a, b). The un-contaminated containers utilized in this research are reference standards for clean containers. Based on the results, using 3 g

(0.031 moles) of Ivory soap (99.4% pure) to clean each g of soybean oil in the contaminated containers, will remove enough oil to prevent rancidity and cross-contamination of soybean oil with maize that would be stored in them.

A total of 108 (18 wipes*6 treatments) samples were analyzed by the goldfisch oil extraction method, using 54 (108/2) wipes obtained from each of the contaminated and unused containers. The remnant oil quantity (Figure 2) shows the average of 18 (3 wipes/container for six containers) wipes obtained from six containers assigned to each treatment. Treatment "C" had the lowest oil remnant (0.004 g) following cleaning of the containers, in relation to "A" (0.249 g) and "B" (0.142 g).

Table 2. Method effect for the recycling research using three cleaning methods.

Method	Mean oil quantity (g)	Test of Significance
Oil drain + 45°C water	0.127	
Oil drain + 90 to 100°C water	0.074	p = 0.0035
Oil drain + 90 to 100°C water + soap	0.005	

Treatment C's oil remnant (0.004 g), from contaminated containers compare favorably with those for control treatments (from unused containers), which had 0.005 g, 0.006 g, and 0.006 g of oil remnants, respectively for (i) "A", (ii) "B", and (iii) "C". Method by container interactions shows a clear difference between the effect of cleaning methods on the two container types (Table 2). For contaminated containers, cleaning with "C" produced the cleanest containers, followed by "B", and "A", respectively. However, there is no distinguishable difference in the oil remnant for uncontaminated containers cleaned using all three methods. Significant container effects are also reflected in oil quantity means (Table 3), for contaminated and uncontaminated containers.

Table 3. Container effect for the recycling research using two container types.

Container	Mean oil quantity (g)	Test of Significance
Contaminated (used)	0.132	p = 0.00001
Uncontaminated (new)	0.006	

Location by method interactions

Location by method interactions show more leftover oil for location 1 (interior, container bottom), compared to location 2 and 3 (interior, left and right container sides, respectively), which are almost alike in remnant levels (Figure 3). This is because oil usually settles to the bottom of its holding container, and the bottom is expected to hold more oil for this reason. The figure also reflects differences in treatment effects, with treatment "C" being the most effective. This is obvious from the leftover oil differences between contaminated containers assigned to treatments "A", and "B" compared to those assigned to the (i) control treatments, and (ii) treatment "C" assigned to contaminated containers. "C" produced as much clean as control treatments, and since new containers and contaminated containers assigned to "C" had virtually no oil, their remnant oil levels remain close to zero. The leftover oil levels for "A" and "B" are about the same, although, slightly lower for "B". The lack of a clear trend in control treatments is expected since control EUs had no oil contaminants at the onset.

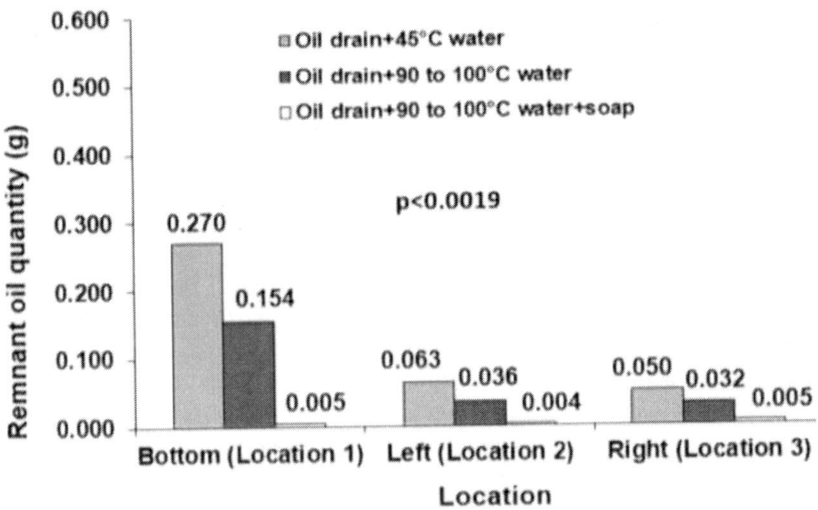

Figure 3. Location by method interactions (averaged over six replications).

Field test results

The Ugandan field study confirms the superiority of hermetic storage over open-air maize storage (Bern et al., 2015).

CONCLUSION

Based on the results of this study, the following conclusions can be drawn:

1. Used, edible oil containers are readily available in East African (Kenya, Tanzania, and Uganda) markets for use in hermetic grain storage. 5 to 20 L, recyclable plastic containers exist with prices ranging from $0.72 to $2.08, and markets had 1,142 (5 L), 967 (10 L), and 2,581 (20 L) such containers respectively, on average per market.
2. Previously used edible oil containers can be recycled following cleaning with soap plus water at 90 to100°C, and mechanical action, for safe hermetic maize storage.
3. Rancidity and associated negative health effects can be eliminated using the cleaning procedures outlined in the research, while preserving maize quality for the end-user.
4. Hermetic storage can be implemented successfully using recycled edible oil containers, to meet the needs of subsistence farmers.

REFERENCES

1. Adhikarinayake TB (2005). Thesis. Wageningen University. Chapter 7.Technical and economic feasibility of airtight ferrocement bin for on-farm storage of paddy in Sri Lanka. The Netherlands.
2. Andrikopoulos NK, Kalogeropoulos N, Falirea A, Barbagianni MN (2002). Performance of virgin olive oil and vegetable shortening during domestic deep-frying and pan-frying of potatoes. Int. J. Food Sci. Technol. 37(2):177-190.
3. Andrikopoulos NK (2004). Oxidative stressed frying fats and oils. Potential role for health. 4th International symposium on deep-frying: "healthier and tastier deep fried foods", January 11-13, Hagen, Germany, pp. 11-12.
4. ASM (American society for microbiology) (2009). Pathogens and toxins in foods: challenges and interventions. Juneja VK, Sofos JN (Eds.). Herndon, VA: ASM Press, pp. 420.
5. Atlas RM, Snyder JW (2006).Handbook of media for clinical microbiology. CRC Press, P 105.
6. ACC (American Chemistry Council) (2007).Plastic packaging resins. Plastics division.Plastic packaging resins. Arlington, VA (USA).
7. ANTEC (2001). Plastic: The lone star. Conference proceedings.Volume II. Materials. Society of Plastics Engineers Technical Conference. Boca Raton, FL: CRC press, P 2705.
8. Ashaye OA, Olusoji OC (2006). Evaluation of some quality attributes of soybean oils in Ibadan. Afr. J. Agric. Res. 1(1):017-019.
9. Bern CJ, Yakubu A, Rosentrater KA, Brumm TJ (2015). Hermetic Storage Systems for Maize (Mısır için Hermetik Depolama Sistemi). 9(68):74-81.
10. Bhatt BI (2004). Stoichiometry. India: Tata Mcgraw Hill Education, pp. 127-178.

11. Carroll NJ, Fulton JR (2008). Extension methods to teach hermetic storage in West-Central Africa. In 2008 Providence, Rhode Island, June 29–July 2, 2008. Am. Soc. Agric. Biol. Eng. P 1.
12. Choe E, Min DB (2006). Mechanisms and factors for edible oil oxidation. Comprehensive Rev. Food Sci. Food Saf. 5:169.
13. Cleveland CB, Oliver GR, Chen B, Mattsson J (2001). Risk Assessment under FQPA: Case Study with Chlorpyrifos. NeuroToxicol. 22(5):699-706.
14. Coats JR (2010). Toxicology 550. Pesticides in the environment. Iowa State University. Class notes. Spring 2010.
15. CRC (CRC Press) (2003). Handbook of postharvest technology: cereals, fruits, vegetables, tea, and spices. In: Chakraverty A, Mujumdar AS, Ramaswamy HS (Eds). Boca Raton, FL: CRC Press, pp.188.
16. CRC (2006). Handbook of lubrication and tribology: Volume I. Application and maintenance, Second Edition. George E. Totten (Eds). Boca Raton: CRC press, pp. 1-11.
17. CRU (2010). Clean water for Washington. Why the concern about agricultural contamination in groundwater?
18. Dallal GE (2003). Why P=0.05? Retrieved on December 9, 2011 from
19. Demissie G, Teshome A, Abakemal D, Tadesse A (2008a). Cooking oils and "Triplex" in the control of sitophilus zeamais motschulsky (coleoptera: curculionidae) in farm-stored maize. J. Stored Prod. Res. 44:173-178.
20. Demissie G, Tefera T, Tadesse A (2008b). Importance of husk covering on field infestation of maize by sitophilus zeamais motsch (coleoptera:curculionidea) at Bako, Western Ethiopia. Afr. J. Biotechnol. 7(20):3777-3782.
21. EPA (2006). Finalization of interim reregistration eligibility decisions (IREDs) and interim tolerance reassessment and risk management decisions (TREDs) for the organophosphate pesticides, and completion of the tolerance reassessment and reregistration eligibility process for the organophosphate pesticides pp. 44-69.
22. EPA (2009a). List of contaminants & their maximum contaminant level (MCLs). Retrieved on September 17, 2011, from
23. EPA (2009b). Food. Food safety. Food contaminants & adulteration. Data on Benzene in soft drinks and other beverages. Data through May 16, 2007.
24. EPA (2009c). Sprill prevention, control and countermeasure (SPCC) Rule Amendments. Who is subject to the SPCC rule?
25. EWG (Environmental working group) (2011). The power of information. Chemical index. Chemicals.
26. Dow (The Dow Chemical Company) (2009). Product safety assessment. High density polyethylene (HDPE) resins.
27. Dynalab (2011). Plastic properties of high density polyethylene (HDPE).
28. EOS/EDS (Electrical Overstress - Electrostatic Discharge) symposium proceedings. 22 (2000). ESD association in collaboration with IEEE, Anaheim: California. September 28-30, 1999, NY: Rome, 22: 2B.1.3.
29. FAO (2002). Small scale palm oil processing in Africa. FAO agricultural services bulletin 148:12.
30. FAO (2010). Rice: Grain storage hermetically sealed systems.Technology for agriculture. Proven technologies for smallholders.
31. Gangneux J, Noussair L, Bouakline A, Roux N, Lacroix C, Derouin F (2004). Experimental assessment of disinfection procedures for eradication of Aspergillus fumigatus in food. Blood, 104(7).
32. Glevitzky M, Dumitrel GA, Perju D, Popa M (2009). Studies regarding the use of preservatives on soft drinks stability. Chem. Bull. "POLITEHNICA" Univ. (Timisoara). 54(68)1.
33. Gregg BR, Billups GL (2009). Seed conditioning. management. Enfield, NH. Sci. publish. 1:184.
34. Helmenstine MA (2011). How Do Detergents Clean?

35. Holst N, Meikle WG, Markham RH (2000). Grain injury models for Prostephanus truncatus (Coleoptera: Bostrichidae) and Sitophilus zeamais (Coleoptera: Curculionidae) in rural maize stores in West Africa. J. Econ. Entomol. 93(4)1338-1346.
36. Knoxa R, Walkera J (1947). Bacteriological investigation of the washing and sterilization of food containers. J. Hygiene 45:151-158.
37. Lewis L, Onsongo M, Njapau H, Schurz-Rogers H, Luber G, Kieszak S, Nyamongo J, Backer L, Dahiye AM, Misore A, DeCock K, Rubin C, Kenya Aflatoxicosis Investigation Group (2005). Aflatoxin contamination of commercial maize products during an outbreak of acute aflatoxicosis in Eastern and Central Kenya. Environmental Health Perspectives: 113. Retrieved February 20, 2007 from
38. Lindblad C, Druben L (1980). Storage methods. Small farm grain storage. Volume III. Appropriate technologies for development. Manual M-2/ VITA publications manual series no. 35E. Arlington: VITA.
39. Malik VS, Schulze MB, Hu FB (2006). Intake of sugar-sweetened beverages and weight gain: a systematic review. Am J Clin Nutr. 84:274-288.
40. Merriam-webster (2010).
41. MTL (Mid-Continent Testing Laboratories, Inc.,) (2011). Limiting factors on a diesel engine's oil life. Rapid City, SD (USA): 1.
42. Mercer C (2006). Cadbury's benzene secret recall.
43. Myers D (2006). Surfactant science and technology. Hoboken: John wiley & Sons, Inc., publishing, P 122.
44. Murdock LL, Dogo SD, Ntoukam G, Kitch L, Shade RE (2003). Preservation of cowpea grain in sub-Saharan Africa—Bean/Cowpea CRSP contributions. Field Crops Res. 82:169-178.
45. Navarro S, Donahaye E, Fishman S (1994). The future of hermetic storage of dry grains in tropical and subtropical climates. In: Highley, E., Wright EJ, Banks HJ, Champ BR (Eds). Proceeds of the 6th international working conference on stored product protection held at Canberra, Australia. Cambridge: CAB International, pp. 130-138.
46. Patwardhan N, Kelkar U (2011). Disinfection, sterilization and operation theater guidelines for dermatosurgical practitioners in India. Indian J Dermatol Venereol Leprol. 77(1)83-93.
47. PHL Network (2009). Post harvest losses information system.
48. Shachman M (2004). The soft drinks companion: a technical handbook for the beverage industry. Boca Raton, FL: CRC Press, pp. xv-116.
49. Schmidt RH, Erickson DJ (2008). FSHN04-09. Sanitary Design and Construction of Food Equipment. University of Florida. IFAS extension. Florida Cooperative Extension Service, Institute of Food and Agricultural Sciences, University of Florida.
50. Springer (2007). Lentil: an ancient crop for modern times. In: Yadav SS, Mcneil D, Stevenson, PC (Eds). New York, NY: Springer, pp. 407.
51. Sebastião J, Pint S, Lanças FM (2006). Hydrolysis of corn oil using subcritical water. J. Braz. Chem. Soc.17(1)85-89.
52. Tsimihodimos V, Kakaidi M, Elisaf M (2009). Cola-induced hypokalaemia: pathophysiological mechanisms and clinical implications. Int. J. Clin. Pract. 63:833-835.
53. Wiley-Blackwell (2006). Carbonated soft drinks: formulation and manufacture. In: Ashurst PR, Steen, P (Eds). Hoboken: Wiley-Blackwell, pp. 1-79.
54. Yakubu A, Bern CJ, Coats JR, Bailey TB (2011). Hermetic on-farm storage for maize weevil control in East Africa. Afr. J. Agric. Res. 6(14):3311-3319.

CHAPTER 6

Nondestructive Maturity Assessment Tools for Commercially Viable Fruits And Vegetables in Uganda

Paul Okiror[1*], Julius Bunny Lejju[1], Joseph Bahati[2], Grace Kagoro Rugunda[1] and Collins Inno Sebuuwufu[1]

[1]*Faculty of Science, Mbarara University of Science and Technology, P. O. Box 1410, Mbarara, Uganda.*

[2] *School of Forestry, Environmental and Geographical Sciences, Makerere University, P. O. Box 7062, Kampala, Uganda.*

ABSTRACT

Fruit and vegetable maturity at harvest influences transportation and storage requirements as well as market value. However, small scale farmers in Uganda lack technologies for maturity assessment leading to high pre- and post-harvest losses and low financial returns from fruit and vegetable farms. This study, therefore, assessed the development of fruits (pineapple, passion fruit, watermelon) and vegetables (cabbage, egg plant, pumpkin and tomato), determined optimal maturity indices and fabricated and tested nondestructive tools for maturity assessment. Propagation trials and testing of tools were undertaken at Nangabo and Kangulumira sub counties in central Uganda. The findings show that eggplant, passion fruit, pineapple and pumpkin underwent 3 distinct development stages during which their diameter, length and outer colour changed. Cabbage remained green, but its bulb diameter and length varied with maturity. Watermelon had 4 development stages with significant ($P \leq 0.05$) changes in morphology and outer colour. Tomato fruits had 6 distinct outer colour changes. Age and colour were maturity indices for passion fruit. Fruit age and diameter were the maturity indices for watermelon. Age, diameter, length and colour were the maturity indices for pineapple, tomato and eggplant. Cabbage and pumpkin share indices including: age, diameter and length. A farm record book (FRB) was designed for documenting phenology and maturity stages of studied fruits and vegetables. A calibrated calliper (CC) and Calibrated tape (CT) were fabricated for assessing the morphological development in pineapple, tomato, eggplant, cabbage, pumpkin and watermelon. Customized colour charts (CCC) were designed for monitoring colour changes as passion fruits, pineapples, tomatoes and eggplants mature. On-farm trials show that CCCs were the most effective tools for monitoring passion fruit (80%), pineapple (64%), tomato (60%), eggplant (68%). Similar studies involving several cultivars and maturity determination tools are, therefore, recommended.

Keywords: Fruit, Kangulumira, maturity index, maturity determination tools, Nangabo, vegetable.

INTRODUCTION

Fruits and vegetables should be harvested with precision to ensure that their maturity meets or exceeds the minimum level acceptable to the consumer at the time they are consumed (Reid, 2002). According to Okiror et al. (2017b), development and use of non-destructive tools for monitoring the maturity of fruit and vegetable is a critical intervention towards determining optimal harvest time and minimizing pre-and postharvest losses. This is so because the maturity level at harvest is vital to the development of good flavor and taste quality in the fruit when fully ripe (Dadzie and Orchard, 1997). Xudong et al. (2009) assert that fruit and vegetable maturity influences market value, transportation and storage requirements. Thus, it is important for individuals harvesting fruit to have effective methods of determining maturity (Kader and Mitcham, 2008).

Several methods have been developed for the nondestructive determination of fruit and vegetable maturity (Slaughter, 2009). The nondestructive technologies are based on aroma, colour, defects, shape, size, firmness, composition and density (Chen, 1996; Abbott, 1999; Butz et al., 2005). Other destructive tools including laboratory determination of total soluble solids, total titrable acidity, starch content, moisture content and protein content were designed and evaluated for commercial manufacturers (Slaughter, 2009; AOAC, 2000).

The existing methods are expensive and technically robust and thus unapplicable to small-scale Ugandan farmers in the determination of the optimal time of fruit and vegetable harvest (Kato, 2011; Muzaale, 2014). This could be one of the key factors for increasing postharvest losses and food insecurity in Uganda (IPC, 2017). It is, therefore, critical to work with farmers to develop farmer low-cost and technologically adoptable maturity assessment tools that can minimize postharvest losses while at the same time ensuring financially sustainable returns for growers (Hanrahan and Röder, 2017).

This study was an attempt to fill a gap from previous scholars including Kader and Mitcham (2008), Slaughter (2009) and Zhang and McCarthy (2012) who focused on high cost technologies such as near infrared and magnetic resournance imaging, colorimeters and starch testers that may not be affordable to small scale farmers in Uganda. Hanrahan and Röder (2017), assert that for a technology to be useful for any operation, associated instruments should be accurate and reliable, especially when operated under field conditions by personnel with minimum training such as peasants in Uganda. In addition to the initial purchase price, the cost of labor (training, performing of task and maintenance) should be modest (Hanrahan and Röder, 2017).The specific objectives of this study were therefore to (i) investigate the development of fruits (pineapple, passion fruit and watermelon) and vegetables (cabbage, eggplant, pumpkin and tomato), (ii) determine the optimal maturity indices, and (iii) fabricate and test nondestructive tools for fruit and vegetable physical maturity assessment using low-skill and low-cost techniques in Uganda.

MATERIALS AND METHODS

Description of study area

Study fruits (pineapple, passion fruit and watermelon) and vegetables (cabbage, eggplant, pumpkin and tomato) were propagated to determine maturity indices and design nondestructive maturity assessment tools, in Makerere University Agricultural Research Institute Kabanyolo (MUARIK), Nangabo Sub County, Wakiso district, central Uganda (Figure 1). MUARIK is located on coordinates 0°27'60"N, 32°36'24"E and at an altitudinal range of 1,250 to 1,320 m above mean sea level (Yost and Eswaran, 1990).

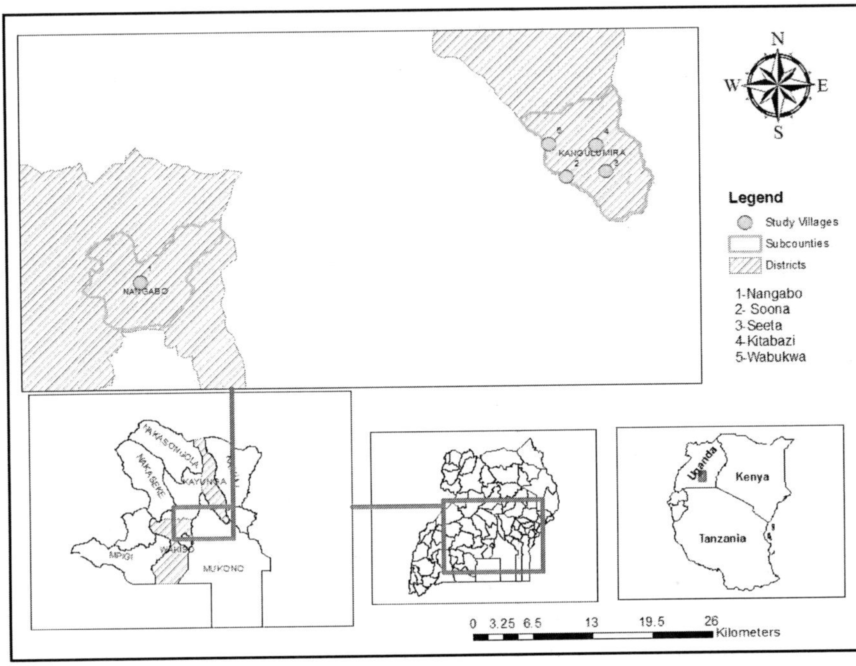

Figure 1. Map of Uganda showing the location of the study sites.

Komutunga and Musiitwa (2001) reported Kabanyolo to be part of the Lake Victoria basin that receives an average annual precipitation of 1,218 mm and slightly drier periods in June to July and December to February of the year. The average annual temperature is 21.5°C. Kabanyolo soils are formed on residuum and colluvium from quartzites, gneiss and basement complex rocks. A recent study by Okiror et al. (2017a) reported Kabanyolo soils to be acidic, with the pH ranging from 6.08-6.2, and deficient in organic matter and most essential minerals. The texture of MUARIK soil is predominantly clayey. On the side slopes, colluvium enriched with lateritic gravel is common (Yost and Eswaran, 1990). Training and research in agricultural sciences are the main activities conducted at MUARIK by the Makerere University College of Agricultural and Environmental Sciences and collaborating institutions (Okiror et al., 2017a). However, as part of the Buganda surface, the predominant farming system around MUARIK is the banana-coffee system (Kisamba-Mugerwa, 2001).

On-farm testing of nondestructive maturity assessment tools was carried out with fruit and vegetable famers in Kangulumira Sub County, Kayunga district (Figure 1). Kagulumira is located on 0°34'54"N, 33°1'46"E within 1,070 m above mean sea level. Kangulumira sub county is part of Kayunga district that experiences a bimodal rainfall pattern with peaks between

March and May and September to December of the year (NEMA, 2016). The rainfall is evenly distributed within the district. Subsistence agriculture employs almost 96% of the population (UBOS, 2016). The major food crops include bananas, sweet potatoes, cassava, maize, beans and ground nuts. Coffee is the main cash crop, but due to the coffee wilt disease, its production is declining. The farmers have opted for fruits (papaya, pineapples, watermelon, mangoes and passion fruit) and vegetables (amaranth, cabbages, eggplants, pumpkins, tomato) to diversify their sources of income and enhance household food security (UBOS, 2016).

Determination of maturity indices

The propagation trials were run between May and August 2015 at MUARIK. Save for pineapple that was propagated using suckers, other fruits (passion fruit and watermelon) and vegetables (cabbage, eggplant, pumpkin and tomato) seedlings were produced following procedures described by Pinho et al. (2011). The soils were ploughed and loosened using hand hoes. Manual watering was done on days in which the site did receive rains as done by most small-scale farmers in Uganda (Kato, 2011). The trials were run between May and August 2015. Seedlings were transplanted manually using hand hoes into pre-made holes in 3 randomized blocks on 26 May 2015. As is the case with most local small scale farmers (Muzaale, 2014), study fruits and vegetables were established in the field without application of any fertilization at transplanting. The cultivars that were studied are presented in Table 1.

Table 1. Study fruit and vegetable cultivars.

Fruit/vegetable	Experimental code	Scientific name	Family	Cultivar	Seed source
Fruit					
Passion fruit	PSF	*Passiflora edulis* Sims	Passifloraceae	Possum Purple	His Grace Agrochemicals (U)
Watermelon	WTM	*Citrullus lanatus* (Thub.)	Cucurbitaceae	Sugar Baby	Balton (U)
Pineapple	PNP	*Ananus comosus* (L.) Merr.	Bromeliaceae	*NK	Kangulumira, Kayunga district
Vegetable					
Tomato	TMT	*Solanum lycopersicum* L.	Solanaceae	Ghalia 281	Nsanja (U)
Eggplant	EGP	*Solanum melongena* L.	Solanaceae	Back Beauty	East African Seeds Limited
Cabbage	CAB	*Brassica oleracea* L.	Brassicaceae	Copenhagen Market Cabbage	East African Seeds Limited
Pumpkin	PMK	*Cucurbita pepo* L.	Cucurbitaceae	NK	Mecron Ent. (U)

*NK=Not known.

Plants were tagged and data on leafing, root collar diameter, plant height, pest and disease incidence and weeding and pruning were recorded until flower emergence at 8 weeks (watermelon), 10 weeks (passionfruits), 13 weeks (pineapples), 14 weeks (eggplant and pumpkin) and 15 weeks after transplanting tomato. After flower emergence, fruit development was monitored as described by Pinho et al. (2011). On each study plant, randomly selectedfruits were tagged and fruit diameter, height, color and pest and disease incidence were recorded daily. Similar data was collected from cabbage samples.

Two (2) randomly selected fruits and vegetables were harvested from each of the respective maturity stages (Table 2) but in consideration of the three (3) slope positions in the three (3) propagation blocks, labelled and packed in a cool box (Marina 24S) to avoid manual contact. The harvest sampling of passion fruit (N=54), pineapple (N=54), watermelon (N=72), eggplant (N=54), pumpkin (N=54) and tomato (N=108) was based on outer colour, while cabbage (N=54) selection depended on the size of the bulb. Samples were harvested in the morning based on among other characteristics; (i) uniformity of color, (ii) size and (iii) absence of disease and injury (Pinho et al., 2011). The samples were labeled, packed in a cool box (Marina 24S) and delivered to Makerere University Food Science and Technology Laboratory for physico-chemical and nutritional analysis following Okia et al. (2013). The laboratory tests were undertaken to determine the morphological indicators that correlate with chemical parameters such as total soluble solids, protein content, total titrable acidity, pH and carbohydrate content.

Maturity assessment tools

Nondestructive and noninvasive tools were fabricated based on maturity indices for study fruits and vegetables. Local Artisans and Fine Artists in Nangabo Sub County were engaged to develop Farm Record Books (FRB), Calibrated Callipers (CC), Calibrated Tapes (CT) and Customized Colour Charts (CCC). The calibrations were based on fruit and vegetable development and maturation data (diameter, height, age and colour) and physico-chemical and nutritional analyses results on total soluble solids, carbohydrate content, total titrable acidity, pH and protein content. FRBs were designed to enhance farmers' capacity to document and archive the planting or sowing dates, leafing, flowering, fruiting, maturity, quantity and quality of fruits and vegetables (Utegi and Utegi, 2014). While the CCs and CTs were fabricated using locally cut steel for assessing the morphological (diameter and length) development of pineapple, tomato, eggplant, cabbage and pumpkin (Muchui et al., 2010). Customized colour charts (CCC) were designed for monitoring colour changes in passion fruit, pineapple, tomato and eggplant (Dadzie and Orchard, 1997). Tools were calibrated using the fruit and vegetable planting or sowing dates, leafing, flowering, fruiting, diameters, lengths and colours. Thus, FRBs, CCCs, CTs and CTs were developed for each study fruit and vegetable.

Table 2. Fruit and vegetable development[1].

Fruit/vegetable	Parameter	Unit	Maturity stage[2]					
			1	2	3	4	5	6
Fruit								
PSF[3]	Diameter	(cm)	4.77±0.24	4.35±0.24	4.20±0.24			
	Length	(cm)	5.10±0.24	4.63±0.24	4.55±0.24			
	Fruit age	(wks)	1.00±1.25	3.00±1.25	4.00±1.25			
	Observed colour	(USDA, 1991)	Green	Turning	Purple			
WTM	Diameter	(cm)	3.50±4.91	9.00±4.91	13.30±4.91	16.60±4.91		
	Length	(cm)	5.60±7.80	14.30±7.80	21.20±7.80	26.40±7.80		
	Fruit age	(wks)	1.00±1.48	3.00±1.48	4.00±1.48	5.00±1.48		
	Observed colour	(USDA, 1991)	Green	Light green	Yellowish green	Light yellow		
PNP	Diameter	(cm)	6.50±0.25	6.80±0.25	7.10±0.25			
	Length	(cm)	6.80±0.42	7.10±0.42	7.80±0.42			
	Fruit age	(wks)	1.00±2.06	4.00±2.06	6.00±2.06			
	Observed colour	(USDA, 1991)	purple	Light purple	Yellow			
Vegetable								
TMT	Diameter	(cm)	4.35±0.16	4.30±0.16	4.70±0.16	4.65±0.16	4.40±0.16	4.35±0.16
	Length	(cm)	5.30±0.19	5.55±0.19	5.15±0.19	5.15±0.19	4.95±0.19	5.10±0.19
	Veg. age	(wks)	1.00±1.34	3.00±1.34	4.00±1.34	4.43±1.34	4.72±1.34	4.86±1.34
	Observed colour	(USDA, 1991)	Green	Breaker	Turning	Pink	Light red	Red
EGP	Diameter	(cm)	7.85±1.63	11.10±1.63	7.45±1.63			
	Length	(cm)	10.15±2.48	12.30±2.48	6.30±2.48			
	Veg. age	(wks)	1.00±1.25	3.00±1.25	4.00±1.25			
	Observed colour	(USDA, 1991)	Purple	Bronze	Dark purple			
CAB	Diameter	(cm)	6.30±1.73	8.80±1.73	10.50±1.73			
	Length	(cm)	6.15±2.05	9.30±2.05	11.10±2.05			
	Veg. Age	(wks)	1.00±1.82	2.00±1.82	3.00±1.82			
	Observed colour	(USDA, 1991)	Green	Green	Green			
PMK	Diameter	(cm)	14.50±0.56	15.85±0.56	15.00±0.56			
	Length	(cm)	8.05±3.78	17.25±3.78	13.00±3.78			
	Veg. Age	(wks)	1.00±1.25	3.00±1.25	4.00±1.25			
	Observed colour	(USDA, 1991)	Green	Turning	Yellow			

[1]Data means±standard deviation; [2]The distinct maturity stages observed in this study were: 6(TMT), 4(WTM) and 3 for PSF, PNP, EGP, CAB and PMK. Beyond these stages, the study fruits and vegetables began to ripen and/or go bad. [3]PSF=Passion fruit, WTM=Watermelon, PNP=Pineapple, TMT=Tomato, EGP=Eggplant, CAB=Cabbage, PMK=Pumpkin.

On-farm validation of maturity assessment tools

To validate the non-destructive maturity assessment tools, on-farm tests were undertaken with 20 fruit and vegetable farmers in 4 villages in Kangulumira Sub County, Kayunga district in December 2016. Kangulumira sub county was selected because it is among the reknown areas for fruit and vegetable production in Uganda (Kato, 2011). The farmers were randomly selected from Kitabazi, Seeta, Soona and Wabukwa villages based on their Local Leaders' lists as done by Agea (2010). Each farmer was tasked to use the tool and report whether it was effective or non-effective in determining fruit and/or vegetable maturity. Priority ranking was based on the number of farmers that cosidered the tool effective. A similar approach was used for farmer-led priority ranking in eastern and northern Uganda (Okia, 2010). The tools that were subjected to on-farm validation included FRB, CC, CT and CCC.

Data analysis

Data obtained from the monitoring of fruit and vegetable growth, physico-chemical and nutritional laboratory tests and on-farm validation of maturity assessment tools were entered in MS Excel (Vers. 2010) computer software. The Pearson's correlations (r) were computed to determine the associations between morphological (age, diameter, length and colour) and nutritional maturity indices (total soluble solids, carbohydrate content, total titrable acidity, pH and protein content). Using MS Excel (Vers. 2010), analysis of variance (ANOVA) was run at 5% significance level to ascertain the most significant maturity indicators that were subsequently used to design the maturity assessment tools. Microsoft excel was used to generate frequencies and percentages that guided the priorititzation and ranking of maturity assessment tools.

RESULTS AND DISCUSSION

Fruit and vegetable development

From the propagation experiment, three distinct development stages were observed for passion fruit during which its morphology (diameter and length) and outer colour changed (Table 2). The highest diameter (4.77 cm) was observed at green stage, followed by turning (4.35 cm) and the least was at purple passion fruits (4.20 cm). Fruit length also decreased with maturity of passion fruits, from 5.10 cm at green stage, 4.63 (turning) to 4.55 cm in purple samples. Watermelon had four (4) development stages with significant changes in the diameter, length and outer colour of the fruit. Fruit diameter and length increased as the fruit matured. The highest diameter of 16.60 cm was observed in the 4th stage (light yellow), followed by 13.30 cm in the 3rd stage (yellowish green) and the least (3.50 cm) was recorded in the 1st stage of fruit development. Fruit length increased from 5.60 cm at green stage, 21.20 (yellowish green) to 26.40 cm in light yellow watermelons (Table 2). The morphological changes in a watermelon are similar to what was observed in pineapple samples. Fruit diameter increased from 6.5 (purple), 6.8 (light purple) to 7.1 cm to yellow stage, while fruit length changed from 6.8, 7.1 to 7.8 in the purple, light purple and yellow pineapples, respectively.

Tomato fruit had six (6) highly ordered development stages during which its morphology (diameter and length) and outer colour changed (Table 2). The peak diameter (4.70 cm) was at turning, followed by 4.65 cm at pink stage and the lowest diameter (4.30 cm) was in the breaker. Fruit length was highest (5.55 cm) in the breaker, followed by green (5.30 cm) and lowest (4.95 cm) in the light red stage. Eggplant changed through three stages including purple (diameter =7.85 cm, length = 10.15 cm), bronze (diameter = 11.10 cm, length = 12.30 cm) to dark purple (diameter = 7.45 cm, length = 6.30 cm). Cabbage bulbs remained green through three distinct morphological changes. At stage 1, diameter was 6.3 cm and length was 6.15 cm. In stage 2, cabbage bulbs were 8.8 cm wide and 9.3 cm long. The 3rd stage of cabbage had an average diameter of 10.5 cm and 11.1 cm long. Pumpkin took three stages to mature including green, turning to yellow in which fruit diameter changed from 14.5 cm, 15.85 cm to 15.0 cm and length increased from 8.05 to 17.25 cm then dropped to 13.00 cm, respectively (Table 2).

Similar observations have been reported in other studies. For example Wu and Kubota (2008), observed tomato fruit enlarge with time after anthesis during the green stage, reach maximum size at around the end of the green stage and hardly change in size after the breaker stage through the red stage. Additionally, the changes in fruit diameter and length usually arise from variations in the cellular structure and internal structure of the fruit (Zhang and McCarthy, 2012). Similarly, this study reveals the unique changes in the diameter and length in the fruits (pineapples, passion fruit and watermelon) and vegetables (cabbage, eggplant, pumpkin and tomato) that were investigated.

In addition, fruit and vegetable maturity correlates with color which usually reflects the biochemical changes during ripening (Zhang and McCarthy, 2012). For example, tomatoes progressed from green to breaker, light pink, light red to red when fully mature. To enhance acceptance of fruit by consumers, tomatoes should be harvested at breaker stage for distant markets and fully ripe for local markets (Dadzie and Orchard, 1997). Other fruit and vegetable colour changes that were monitored in this study are relevant in the planning of optimal harvest regimes (Muchui et al., 2010), especially by small scale farmers in Uganda.

Maturity indices

The maturity indices for study fruits and vegetables are presented in Table 3. Fruit age ($P \leq 0.05$) and colour ($P \leq 0.05$) were significant for passion fruits. It also emerged that fruit age ($P \leq 0.05$) and fruit diameter ($P \leq 0.05$) were significant for watermelon. In addition, fruit age ($P \leq 0.05$), diameter ($P \leq 0.05$), length ($P \leq 0.05$) and colour ($P \leq 0.05$) were significant for pineapples (Table 3). Tomato maturity can be gauged from the changes in its age ($P \leq 0.05$), diameter ($P \leq 0.05$), length ($P \leq 0.05$) and colour ($P \leq 0.05$). Similarly, eggplant development can be monitored based on age ($P \leq 0.05$), diameter ($P \leq 0.05$), length ($P \leq 0.05$) and colour ($P \leq 0.05$). Cabbage and pumpkin shared maturity indices including age ($P \leq 0.05$), diameter ($P \leq 0.05$) and length ($P \leq 0.05$) (Table 3).

Table 3. Morphological maturity indices.

Fruit/vegetable	Maturity index*			
	Age	Diameter	Length	Colour
Fruit				
PSF[1]	0.04*			0.01**
WTM	0.35*	0.05*		
PNP	0.04*	1E-10**	0.021*	1E-11**
Vegetable				
TMT	1E-10**	0.05**	1E-13**	0.02*
EGP	1E-9**	0.03*	0.43*	1E-07**
CAB	0.02*	0.02*	0.04*	
PMK	0.03*	0.02*	0.04*	

[1]PSF = Passion fruit, WTM = Watermelon, PNP = Pineapple, TMT = Tomato, EGP = Eggplant, CAB = Cabbage, PMK = Pumpkin; Morphological parameters with * and ** were significant maturity indices at 0.05 and 0.01 alpha level, respectively when correlated with nutritional composition (total soluble solids, carbohydrate content, total titrable acidity, pH and protein content) of the samples.

The revelation of age as a maturity indicator for the sample fruits and vegetables observed in this study (Table 3) is supported by findings from other previous studies. According to Hanrahan and Röder (2017), the time of day and prevailing weather conditions influence product maturity and shelf life of fruits and vegetables. Further more, Mattheis and Fellman (1999) opin that fruits and vegetables harvested at an early stage of maturity are susceptible to shriveling and mechanical damage and develop poor flavor and taste, despite having long storage life. In contrast, harvesting at an advanced stage of maturity produces fruits and vegetables that have good taste and flavour, but have a short storage life and are not suitable for transporting for long distances (Dadzie and Orchard, 1997). This implies that fruit and vegetable age can be monitored as an indicator of maturity.

Colour is a useful indicator of fruit and vegetable maturity despite Zhang and McCarthy (2012) argument that during tomato processing, the fruit fed to the processing line are usually a mixture of tomatoes of multiple cultivars with varying maturity conditions. Thus, as much as colour is a significant index of maturity, it may not be reliable for a mixture of cultivars. Molyneux et al. (2004) stated that fruit or vegetable skin color may vary tween cultivars despite the cultivars falling within the same maturity stage. However, Dadzie and Orchard (1997) support external color because its assessment is noninvasive and nondestructive and does not require high skilled staff.

Morphological parameters such as diameter and length are equally vital maturity indicators. Muchui et al. (2010) demonstrated the relevance of the changes in fruit length and diameter in maturity determination. It was therefore recommended that the use of noninvasive tools be promoted among rural farmers to improve harvesting regimes, minimize postharvest losses and improve income and food security (Muchui et al., 2010; Robinson, 1996).

Table 4. Maturity assessment tools.

Fruit/vegetable	Maturity index and measurement tool			
	Age	Diameter	Length	Colour
Fruit				
PSF[1]	*FRB[1]	CC		CCC
WTM	FRB	CC		
PNP	FRB	CC	CT	CCC
Vegetable				
TMT	FRB	CC	CT	CCC
EGP	FRB	CC	CT	CCC
CAB	FRB	CC	CT	
PMK	FRB	CC	CT	

[1]PSF=Passion fruit, WTM=watermelon, PNP=pineapple, TMT=tomato, EGP=eggplant, CAB=cabbage, PMK=pumpkin; FRB=farm record book; CC=calibrated calliper; calibrated tape; CCC=customized colour chart.

Maturity assessment tools

Based on the significant maturity indices in Table 3, the following tools were developed to enhance small-holder farmers' capacity to assess fruit and vegetable maturity. A farm record book (FRB) was designed for documenting the planting or sowing dates, leafing, flowering, fruiting, maturity, quantity and quality of fruits and vegetables (Table 4). A calibrated caliper (CC) and calibrated tape (CT) were developed with local artisans for assessing the morphological (diameter and length) development of pineapple, tomato, eggplant, cabbage and pumpkin. Customized colour charts (CCCs) were designed for monitoring colour changes in passion fruit, pineapple, tomato and eggplant (Table 4).

As shown in Table 2, fruit and vegetable maturity can be determined by the number of weeks after fruit onset. Therefore, farmers with proper records can predict the optimal time of harvest and plan for transportation and marketing of fruit and vegetable products. The FRB should be used concurrently colour charts, callipers and diameter tapes to determine the optimal time of harvest. A farm record book refers to a tool for systematic documentation of all activities and transactions regarding all aspects of farm operations (Batte and Foster, 2008). Okojie and Ayinde (2012), categorized farm records into inventory, production (e.g. phenology, yield and maturity), expenditure and income and special supplementary records. To make FRBs useful, the records must be accurate, neat and complete and should be filled-out as soon as the operation or transaction occurs (Okojie and Ayinde, 2012). Utegi and Utegi (2014) analyzed the activities of a farmer and showed the relevance of farm records and accounting in agricultural production.

As much as FRBs are recommended, their adoption may face challenges including illiteracy and limited campaign by government agencies towards record keeping in Uganda. These challenges could be addressed through (i) issuance of low-interest bank loans to farmers as an incentive for farm record keeping, (ii) illiterate farmers employing relatives who are educated as data clerks, (iii) low taxation charges from government for farmers who keep farm records and (iv) distribution of fertilizers and subsidized farming tools to only farmers who keep records (Utegi and Utegi, 2014). Just like in most African countries, most small scale farmers in Uganda are less-educated, unwilling and financially unable to employ Clerks. Famers undertake mental record-keeping and accounting. There is, therefore, need to encourage documentation through farmer extension services (Winkler, 1938) to enhance farmers' capacity to plan produce harvesting, transportation and marketing.

Calibrated callipers (CC) and calibrated tapes (CT) were developed with local artisans for assessing the morphological (diameter and length) development of pineapple, tomato, eggplant, cabbage, pumpkin and watermelon. Slaughter (2009), recognizes shape and size as characteristics for which nondestructive methods for assessing fruit and vegetable maturity. It is preferable to use noninvasive and nondestructive tools such as CC and CTs that can show structural variations as fruit or vegetable mature (Zhang and McCarthy, 2012). Besides, the CCs and CTs can be fabricated and maintained using locally available steel or woody material by artisans within the farmer community.

Customized colour charts (CCC) were designed for monitoring colour changes in passion fruits, pineapples, tomatoes and eggplants. The CCCs are noninvasive and nondestructive and can be used to assess fruit maturity and vegetable in the field, market or inspection points (Dadzie and Orchard, 1997). In addition, Slaughter et al. (2006) used nondestructive optical measurements in the visible region to measure maturity of peaches. Similarly, the USDA colour classification chart for fresh fruits is one of the most handy tools for assessing maturity and classifying fruits (USDA, 1991). Besides colour charts, other tools have been used to determine maturity from colour. For example, Malevski et al. (1977) used a colorimeter to assess skin colour of fruits. Jha et al. (2006, 2007) used a portable spectrophotometer to measure the average skin color and reflectance spectra of fruits and Ornelas-Paz et al. (2008) nondestructively measured the skin color of mangoes during ripening using a colorimeter. However, Malevski et al. (1977) assert that outer skin color measured with a colorimeter at an arbitrarily selected site on the fruit was an unreliable index of maturity. This study, therefore, recommends the use of low cost and low-tech CCCs as tools for fruit and vegetable maturity assessment by small-scale farmers in Uganda.

On-farm validation of maturity assessment tools

On-farm trials of the maturity tools was undertaken with 20 fruit and vegetable farmers in 4 villages in Kangulumira Sub County, Kayunga district (Table 5). It emerged that calibrated colour charts (CCCs) were the most effective tools for assessing the maturity of passion fruit (80%), followed by farm record books (20%). For watermelon, the farmers appraised calibrated callipers (68%) and farm registers (32%). Pineapple maturity assessment tools that were highly ranked by farmers included calibrated colour charts (64%), farm registers (24%) and calibrated callipers (08%).

Relatedly, calibrated colour charts (60%), farm registers (28%) and calibrated tapes (04%) were preferred for tomato monitoring by the study group. For eggplant assessment, the farmers prioritized calibrated colour charts (68%) and calibrated callipers (20%) and detested farm registers (4%). The tools preferred for monitoring the maturity of cabbage were calibrated callipers (44%), farm record books (32%) and the least was calibrated tapes (24%). Similarly, calibrated callipers (40%), farm record books (36%) and calibrated tapes (24%) were validated for pumpkin maturity measurement (Table 5).

The effectiveness of CCCs reported by farmers is not surprising because external fruit and vegetable appearance can be used to predict the internal characteristics (Slaughter, 2009). In addition, colour charts have been used by the United States Department of Agriculture for rapid maturity assessment for decades (USDA, 1991). The noninvasive and nondestructive nature of colour charts could also be another factor influencing farmers' choice. In addition, the use of CCCs does not require highly skilled personnel and is, therefore, appropriate for illiterate farmers (Dadzie and Orchard, 1997).

Calibrated callipers (CCs) and calibrated diameter tapes (CTs) were validated for cabbage, eggplant, pineapple, pumpkin and tomato maturity measurement (Table 5). The fabricated CCs and CTs measure the outer diameter and length of the sample and are therefore nondestructive to the fruits or vegetable products. According to Zhang and McCarthy (2012) and Dadzie and Orchard (1997), it is advisable for small holder farmers to use easy-to-use and low cost maturity assessment tools. Perhaps the farmers noticed the simplicity and ease of fabrication and maintenance of the CCs and CTs that they prioritized to use them over other tools in their fruit and vegetable farms.

Farm record books were preferred for monitoring all the study fruits and vegetables by the sampled farmers (Table 5). The preference for FRBs is not uncommon among farmers especially literate ones. World over, farm record keeping is a reknown best practice. Other scholars including Batte and Foster (2008), Okojie and Ayinde (2012) and Utegi and Utegi (2014) agree that FRBs should be used to document inputs, practices, outputs and for growth and maturity assessment in fruit and vegetable gardens. Perhaps this could be one of the reasons Winker (2008) makes emphasis on the integration of farm record management in agricultural extension programmes.

CONCLUSIONS

This study demonstrates that passion fruit undergoes three distinct development stages during which its morphology (diameter and length) and outer colour change. Watermelon was observed to undergo four (4) development stages with significant changes in the morphology (diameter and length) and outer colour of the fruit. Pineapple fruit diameter and

Table 5. On-farm validation of maturity assessment tools.

Fruit/vegetable	Priority ranking of maturity tools by farmers		
	Name	% effectiveness (N=20)	Rank
Fruit			
PSF[1]	FRB	20	2
	CCC	80	1
WTM	FRB	32	2
	CC	68	1
PNP	FRB	24	2
	CC	08	3
	CT	04	4
	CCC	64	1
Vegetable			
TMT	FRB	28	2
	CC	08	3
	CT	04	4
	CCC	60	1
EGP	FRB	04	4
	CC	20	2
	CT	08	3
	CCC	68	1
CAB	FRB	32	2
	CC	44	1
	CT	24	3
PMK	FRB	36	2
	CC	40	1
	CT	24	3

[1]PSF = Passion fruit, WTM = watermelon, PNP = pineapple, TMT = tomato, EGP = eggplant, CAB = cabbage, PMK = pumpkin; FRB = farm record book; CC = calibrated calliper; calibrated tape; CCC = customized colour chart.

length increased as its outer colour progressed from purple, light purple to yellow stage. Tomato fruits undergo six (6) distinct development stages during which the morphology (diameter and length) and outer colour change. Eggplant diameter and length also changed through three stages including purple, bronze to dark purple. Cabbage remains green throughout three distinct morphological. Pumpkin takes three colour changes to mature including green, turning to yellow with variations in fruit diameter and length (Table 2).

Fruit age and colour were significant (*P≤0.05*) indices for passion fruits. Whereas fruit age and diameter were significant (*P≤0.05*) for watermelon. Age, diameter, length and colour are the maturity indices for pineapple, tomato and eggplant. Cabbage and pumpkins share the same maturity indices including: age, diameter and length (Table 3).

Furthermore, study findings indicate that a FRB is handiest tool for recording planting dates, leafing, flowering, fruiting and maturity of fruits and vegetables. A calibrated calliper (CC) and calibrated tape (CT) will be useful to for assessing the morphological (diameter and length) development of pineapple, tomato, eggplant, cabbage and pumpkin. Customized colour charts (CCC) are important for montitoring colour changes in passion fruit, pineapple, tomato and eggplant (Table 4). On-farm trial tests of the maturity tools with farmers showed CCCs as the most effective tools for assessing the maturity of passion fruit (80%), pineapple (64%), tomato (60%), eggplant (68%). 44, 40 and 68% of the farmers preferred CCs for monitoring the maturity of cabbage, pumpkins and watermelons, respectively (Table 5).

REFERENCES

1. Abbott JA (1999). Quality measurement of fruits and vegetables. Postharvest Biol. Technol. 15:207-225
2. Agea JG (2010). Use and potential of wild and semi-wild food plants in alleviating household poverty and food insecurity: A case study of Bunyoro-Kitara Kingdom, Uganda. Ph.D Thesis. School of Environment, Natural Resources & Geography, Bangor University, Bangor, U K. 360+xiv pp
3. AOAC (2000). Association of Official Analytical Chemists. Official methods of analysis of the AOAC international. 17. ed. Washington, DC: AOAC
4. Batte M, Foster L (2008). Ohio Farm Computer Usage. Farm Management Update, Winter 2003-04.20
5. Butz P, Hofmann C, Tauscher B (2005). Recent developments in noninvasive techniques for fresh fruit and vegetable internal quality analysis. J. Food Sci. 70(9):R131-R141
6. Chen P (1996). Quality evaluation technology for agricultural products. Seoul, Korea: Agricultural Machinery Engineering.In. Proc. Intl. Conf. November 12-15, pp. 171-204

7. Dadzie BK, Orchard JE (1997). Routine Post Harvest Screening of Banana/Plantain Hybrids. Criteria and Methods. INIBAP Technical Guidelines 2. IPGRI, Rome
8. Hanrahan I, Röder S (2017). New Tools to Help Determine Maturity of Tree Fruit. Washington State University Tree Fruit. TIANNA.DUPONT(http: //treefruit.wsu.edu/news/new-tools-to-help-determine-maturity-of-tree-fruit/)
9. IPC (2017). Report of the integrated food security phase classification Analysis for Uganda. Prepared by Uganda IPC Technical Working Group, January 2017. 80p
10. Jha SN, Chopra S, Kingsly ARP (2007). Modeling of color values for nondestructive evaluation of maturity of mango. Food Eng. 78:22-26
11. Jha SN, Kingsly ARP, Chopra S (2006). Non-destructive determination of firmness and yellowness of mango during growth and storage using visual spectroscopy. Biosyst. Eng. 94(3):397-402
12. Kader A, Mitcham B (2008). Optimum Procedures for Ripening Mangoes. In. Fruit Ripening and Ethylene Management: 47-48. Univ. Calif. Postharvest Technology Research and Information Center Publication Series #9
13. Kato J (2011). Green house tomato growers root for market. Agribusiness Magazine, New Vision, January 11th, 2011
14. Kisamba-Mugerwa W (2001). Social Background. In: Mukiibi (Ed). Agriculture in Uganda – General information. NARO- Uganda, pp.186-199
15. Komutunga ET, Musiitwa F (2001). Climate. In. Mukiibi, J.K, Ed., Agriculture in Uganda , Volume 1: General Information, Fountain Publishers, Kampala, pp. 21-32
16. Malevski Y, Gomez-Brito L, Peleg M, Silberg M (1977). External color as maturity index of mango. J. Food Sci. 42:1316-1318
17. Molyneux SL, Lister C, Savage GP (2004). An investigation of the antioxidant properties and colour of glasshouse grown tomatoes. Int. J. Food Sci. Nutr. 55:537-545
18. Muchui MN, Njoroge CK, Kahangi EM, Onyango CA (2010). Determination of Maturity Indices of Tissue Cultured Bananas (*Musa* spp.) 'Williams' and 'Grande Naine'. Proc IC on Banana & Plantain in Africa. Eds.: T. Dubois et al. Acta Hort. 879, ISHS 2010
19. Muzaale F (2014). Growing good quality tomatoes. Farming Magazines, Daily Monitor. Wednesday January 8th, 2014
20. NEMA (2016). State of the Environment Report for Uganda 2014. National Environment Management Authority (NEMA), Kampala
21. Okia CA (2010). *Balanites aegyptiaca*: A resource for improving nutrition and income of dryland communities in Uganda. Phd thesis, University of Wales, Bangor, United Kingdom. 310 p
22. Okia CA, Kwetegyeka J, Okiror P, Kimondo JM, Teklehaimanot Z, Obua J (2013). Physico-Chemical Characteristics and Fatty Acid Profile of Desert Date Kernel Oil. Afr. Crop Sci. J. 21(3):723-734
23. Okiror P, Lejju JB, Bahati J, Rugunda GK, Sebuuwufu CI, Mulindwa P, Ocan JJ (2017a). Suitability of Kabanyolo Soils for Fruit and Vegetable Production. Open J. Soil Sci. 7:19-33
24. Okiror P, Lejju JB, Bahati J, Rugunda GK, Sebuuwufu CI (2017b). Maturity Indices of Tomato (*Solanum lycopersicum* L.), cv. Ghalia 281 in Central Uganda. Afr. J. Agric. Res. 12(14):1196-1203
25. Okojie LO, Ayinde IA (2012). Course materials for AEM 302, Principles of Farm Management. Open Courseware, Unoversity of Agriculture, Abeokuta, Nigeria
26. Ornelas-Paz JJ, Yahia EM, Gardea AA (2008). Changes in external and internal color during postharvest ripening of 'Manila' and 'Ataulfo' mango fruit and relationship with carotenoid content determined by liquid chromatography-APcI+-time-of flight mass spectrometry. Postharvest Biol. Technol. 50:145-152
27. Pinho L, Almeida AC, Costa CA, Paes MCD, Glória MBA, Souza RM (2011). Nutritional properties of cherry tomatoes harvested at different times and grown in an organic cropping. Hortic. Bras. 29:205-211

28. Reid MS (2002). Maturation and maturity indices. In. Postharvest technology of horticultural crops, A. A. Kader, ed. Univ. of California, Oakland, CA, USA: ANR Publication 3311. pp. 55-62
29. Robinson JC (1996). Bananas and Plantains. 2nd edition. University Press, Cambridge
30. Slaughter DC (2009). Nondestructive Maturity Assessment Methods for Mango: A Review of Literature and Identification of Future Research Needs. Biological and Agricultural Engineering, University of California, Davis, January 2009
31. Slaughter DC, Crisosto C (2006). Nondestructive determination of internal quality in clingstone peaches. ASABE Paper 066180. St. Joseph, Mich. USA: ASABE
32. UBOS (2016). The 2016 Statistical Abstract. Uganda Bureu of Statistics, Kampala, Uganda.
33. USDA (1991). United States standards for grades of fresh tomatoes. USDA, Agricultural Market Service, Washington, DC. (http://www.ams.usda. gov/AMSv1.0/getfile?dDocName=STELPRDC5050331)
34. Utegi M, Utegi EN (2014). The Importance of Farm Records And Accounting In Agricultural Production. Katsina-Ala Multidiscip. J
35. Winkler MM (2008). Farm Accounting from the Viewpoint of the Farm Manager. J. ASFMRA 2(1):27-56
36. Wu M, Kubota C (2008). Effects of high electrical conductivity of nutrient solution and its application timing on lycopene, chlorophyll and sugar concentrations of hydroponic tomatoes during ripening. Sci. Hortic. 116:122-129
37. Xudong S, Hailiang Z, Yande L (2009). Nondestructive assessment of quality of Nanfeng mandarin fruit by a portable near infrared spectroscopy. Int. J. Agric. Biol. Eng. 2(1):65-71
38. Yost D, Eswaran H (1990). Major Land Resource Areas of Uganda.World Soil Resources. Soil Conservation Service-USDA. Washington D.C, USA. 227p
39. Zhang L, McCarthy MJ (2012). Measurement and evaluation of tomato maturity using magnetic resonance imaging. Postharvest Biol. Technol. 67:37-43

CHAPTER 7

Non-Conventional Tools to Preserve and Prolong the Quality of Minimally-Processed Fruits and Vegetables

Maria Rosaria Corbo*, Daniela Campaniello, Barbara Speranza, Antonio Bevilacqua and Milena Sinigaglia

Department of the Science of Agriculture, Food and Environment, University of Foggia, Via Napoli, 25, 71122 Foggia FG, Italy

ABSTRACT

The main topic of this paper is a focus on some non-conventional tools to preserve the microbiological and physico-chemical quality of fresh-cut fruits and vegetables. The quality of fresh-cut foods is the result of a complex equilibrium involving surface microbiota, storage temperature, gas in the headspace and the use of antimicrobials. This paper proposes a short overview of some non-conventional approaches able to preserve the quality of this kind of product, with a special focus on some new ways, as follows: (1) use of edible or antimicrobial-containing coatings (e.g., chitosan-based coatings) on fruits or vegetables; (2) alternative modified atmospheres (e.g., high O_2-modified atmosphere packaging (MAP)) or the use of essential oils in the headspace; (3) conditioning solutions with antimicrobials or natural compounds for fruit salad; and (4) biopreservation and use of a probiotic coating.

Keywords: fresh-cut; conditioning liquid; coatings; spoiling microorganisms; probiotics

1. INTRODUCTION

In recent years the attention towards a healthy diet is considerably growing; the consumer is paying more attention to the importance of the freshness and healthiness of foods. Consumption of fruits and vegetables is considered a successful way to contrast the current tendency toward obesity (especially in children). It is known that fruits and vegetables contain essential nutrients and are the major source of various antioxidants (vitamin C, vitamin E and β-carotene) that are necessary for the health, growth and development of the children. The World Health Organization (WHO), Food and Agriculture Organization (FAO), United States Department of Agriculture (USDA) and European Food Safety Authority (EFSA) recommended an increase of fruit and vegetable consumption to decrease the risk of cardiovascular diseases, cancer and ageing [1].

In addition, the consumer profile is changing as an effect of a stressed lifestyle: the number of working women and singles is gradually increasing; they are short on time and demand minimally-processed foods (also called ready-to-use) to save time on food preparation. To meet the consumer needs, the products should have an adequate shelf-life; it should be at least 4–7 days or even 21 days, depending on the products.

The International Fresh-cut Produce Association (IFPA) defines fresh-cut products as fruits or vegetables that have been trimmed and/or peeled and/or cut into a 100% usable product that is bagged or pre-packaged to offer consumers high nutrition, convenience and flavor, while still maintaining its freshness [2]. However, it is known that the processing of fruits and vegetables promotes physiological deterioration, biochemical changes and microbial degradation, beginning with raw materials, through processing methods and ending with packaging factors that affect the quality and shelf-life. These changes are due to tissue wounding, which includes browning, weight loss, accelerated respiration rate, off-flavor development, texture breakdown and increased susceptibility to microbial spoilage [3].

Thus, the search for new inexpensive and effective methods (such as chemical-based washing treatments, physical treatments, hurdle technology and packaging requirements) able to minimize these negative effects is of great interest to all of the stakeholders involved in the production and distribution of fresh fruits and vegetables.

The aim of this paper is to give insight into some non-conventional approaches able to ensure the microbiological stability and quality of minimally-processed fruits and vegetables with a special focus on some new ways currently available: (1) the use of edible or antimicrobial containing coatings (e.g., chitosan-based coatings) on fruits or vegetables; (2) alternative modified atmospheres (e.g., high O_2-modified atmosphere packaging (MAP)) or the use of essential oils in the headspace; (3) conditioning solutions with antimicrobials or natural compounds for fruit-salad; and (4) biopreservation and the use of probiotic coatings.

2. USE OF EDIBLE OR ANTIMICROBIAL-CONTAINING COATINGS (E.G., CHITOSAN-BASED COATINGS) ON FRUITS OR VEGETABLES

Edible coatings (ECs) and edible films belong to the modern food protection system; over the past few years, interest in the use of edible coatings for perishable foods has considerably increased due to their advantages and potential applications [4]. Edible coatings and edible films are terms that are often used interchangeably; however, a distinction is necessary.

An edible coating is defined as a thin layer of edible material applied to the surface of foods in addition to or as a substitution for natural protective coatings, able to form a barrier to moisture, oxygen and solute movement for the food [5,6,7,8,9,10].

On the other hand, edible film is defined as a thin layer of edible material formed on a product surface as a coating or placed (pre-formed) on or between food components [11].

Thus, an edible film is a thin skin, which has been pre-formed (for example, by casting a biopolymer solution separately from the food to form a film later applied to the food), whilst an edible coating is a suspension or an emulsion, which is applied directly to the food surface and later forms a film [12].

Edible coatings and films do not replace traditional packaging materials, but provide an additional factor to be applied for food preservation; ECs are consumed along with the food, thus the composition must conform to the regulations applied to the food product.

One of the advantages in using edible coatings and films is the reduction of water loss, considered one of the main factors in the deterioration of perishable foods. In fact, this thin layer protects fruits and/or vegetables against moisture loss, maintaining the texture and extending the shelf-life of the product, forming a protective barrier. On the other hand, when edible coatings are poor in water vapor barrier properties, a weight or moisture loss of the product could be recovered.

Numerous benefits result when edible coatings are applied, and these are summarized in Figure 1.

In addition, edible coatings may enhance sensory characteristics, can be consumed along with the food, provide additional nutrients and include quality-enhancing antimicrobials. Furthermore, they may reduce the cost and also the amount of traditional packaging used [13].

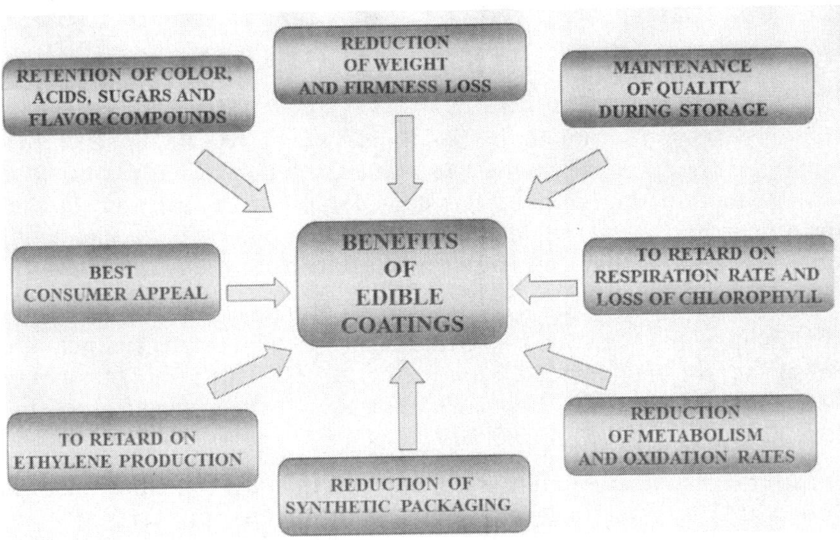

Figure 1. Main benefits of edible coatings.

2.1. Edible Coating Materials

Numerous varieties of fruits and vegetables are characterized by cuticle, a natural waxy layer on the surface, which, generally, has a low permeability to water vapor. To enhance the barrier properties of cuticle and/or substitute it when processing operations remove it, edible coatings could be applied.

Water and ethanol (or a blend of these compounds) are the solvents generally used for edible coating production.

Various coating materials could be added, but it is necessary that they possess some specific requirements (water solubility, hydrophobic of hydrophilic nature, easy formation of coatings, good sensorial properties) to obtain the resulting coating. In addition, an essential requirement is the plasticizing capacity to provide films with good flexibility properties; for this purpose, glycerol is generally used, followed by sorbitol, polyethylene glycol (PEG) and sugars [13].

Hydrocolloids have a good aptitude for forming continuous and cohesive matrices thanks to the hydrogen bonding of their polymeric chains.

Polysaccharides, proteins and lipids, alone or in combination, can also be used to produce edible films and coating. Lipids (together with waxes and fatty acids) do not have a suitable stand-alone filmmaking nature; they are usually opaque and relatively inflexible; the resulting films could also be quite fragile and unstable (rancidity) [9]. For this reason, lipids are generally incorporated into hydrocolloids-based films formulations [13].

The main polysaccharide, protein and lipid compounds used in edible coatings are reported in Table 1.

Polysaccharides include starch, dextrins, pullulan, cellulose and derivatives, alginate, carrageenan, gums, pectins and chitosan. Polysaccharides render transparent and homogeneous edible films; these films are oxygen, aroma and oil barriers (due to their tightly-packed, ordered hydrogen-bonded network structure and low solubility), but are not effective moisture barriers due to their hydrophilic nature. However, when applied in the form of high-moisture gelatinous coatings, they can retard the moisture loss of food [15].

Proteins can be obtained from animal sources, such as casein and whey protein (the main milk protein fractions: 80% and 20%, respectively), and from plant sources, such as zein, gluten and soy proteins [34].

Different proteins are able to form films and coatings; this ability depends on their molecular weight, conformations, electrical properties (charge *vs.* pH), flexibilities and thermal stabilities [18]. Nevertheless, proteins have been studied less extensively than polysaccharides.

Generally, protein-based films have good oxygen, carbon dioxide and lipid barrier properties and mechanical properties; on the other hand, the poor water vapor resistance limits their application; this can be attributed to the inherent hydrophilicity of proteins.

Table 1. Main polysaccharides, proteins and lipid compounds used for edible coatings.

Main Component	Description
Polysaccharides	**Starch** contains amylose ((1→4)-α-D-glucopyranosyl) and amylopectin (amylose branched with side units of D-glucopyranosyl linked by α-1,6-glycosidic bonds). Amylose has a film-forming ability, rendering strong, isotropic, odorless, tasteless and colorless films. It is inexpensive, widely available and easy to handle. Due to its good oxygen barrier, starch is used for coating fruits and vegetables characterized by high respiration rates, thus suppressing respiration and retarding the oxidation of the coated products.
	Dextrins derive from starch and are characterized by a smaller molecular size. Coatings resulting from dextrins provide a better water vapor resistance than starch coatings [14].
	Pullulan derives from starch and is edible and biodegradable; relative films are transparent, elastic, odorless and tasteless [15]. Pullulan-based coatings have shown potential for preserving fresh strawberries and kiwifruits because of their barriers to moisture, O_2 and CO_2 [16]. Pullulan films themselves do not demonstrate antimicrobial activity [17].
	Cellulose is the structural material of plant cell walls. It is composed of linear chains of (1→4)-β-D-glucopyranosyl units. Cellulose ethers (carboxymethylcellulose (CMC); methylcellulose (MC); hydroxypropyl cellulose (HPC); hydroxypropyl methylcellulose (HPMC)) are obtained by partial substitution of hydroxyl groups in cellulose by ether functions. Cellulose-derived films are tough, flexible, totally transparent, water soluble presence and are resistant to fats and oils [15,18], but are too expensive. Crosslinking treatments decrease the water solubility of cellulose ethers [19]. Cellulose derivative-based coatings were applied to some fruits and vegetables for providing barriers to oxygen, oil or moisture transfer [15]. CMC coatings help: (i) to retain the original firmness and crispness of apples, berries, peaches, celery, lettuce and carrots when used in a dry coating process; (ii) to preserve flavor components of some fresh fruits and vegetables; and (iii) to reduce oxygen uptake without increasing the carbon dioxide level in the internal environment of coated apples and pears by simulating a controlled atmosphere environment [15]. Li and Barth [20] observed that, after 3 weeks of storage, cellulose-based edible coating-treated carrots maintained a fresh appearance and higher carotene retention compared to controls. **Alginates** are extracted from brown seaweeds. They are sodium salts of alginic acid, which is a linear (1→4) linked polyuronic acid containing poly-β-D-manopyranosyluronic acid (M) blocks, poly-α-L-gulo pyranosyluronic acid (G) blocks and M–G blocks containing both polyuronic acids. Alginates possess good film-forming properties, producing uniform, transparent films with poor water resistance. Alginates react irreversibly with polyvalent metal cations (for example, calcium ions) to produce water-insoluble polymers. Alginate coatings are good oxygen barriers and retard lipid oxidation in fruits and vegetables. They show good ability to reduce weight loss and natural microflora counts in minimally-processed carrots [21]. Calcium alginate coatings reduce shrinkage, oxidative rancidity, moisture migration, oil absorption and sealing-in volatile flavors, improving appearance and color and reducing the weight loss of fresh mushrooms in comparison with uncoated ones [22].
	Carrageenan is extracted from red seaweeds. It is a complex mixture of several water-soluble galactose polymers. There are 3 different carrageenans (lambda-, iota- and kappa-carrageenan), differing in sulfate ester content and the distribution of 3,6-anhydro-α-D-galactopyranosyl residues [13]. Carrageenan-based coatings were applied to fresh apples for reducing moisture loss, oxidation or disintegration of the apples [23,24]. In combination with ascorbic acid, carrageenan-based coatings resulted in positive sensory results and reduction of microbial levels on minimally-processed apple slices [24].

Polysaccharides	**Gums** include exudate gums (arabic, tragacanth and karaya), seed gums (locust bean and guar) and microbial fermentation gums (xanthan and gellan). Xanthan gum provides uniform coatings. Gum arabic was used for coating pecan nut halves to eliminate a moist and oily appearance [25].
	Pectins are polymers mainly composed of (1→4) α-D-galactopyranosyluronic acid units naturally esterified with methanol. According to their content of methyl esters or the degree of esterification (DE), pectins are divided into high methoxyl (HM, DE > 50%) or low methoxyl (LM, DE < 50%). The DE has a decisive effect on pectin solubility and gelation properties. The literature on pectin-based coatings is poor. Maftoonazad *et al.* [26] evaluated the protective effect of a pectin-based edible emulsion coating on the activity of *Lasiodiplodia theobromae*, a plant pathogen, in avocados, finding that the pectin-based coating was effective at controlling the spread and severity of stem end rot in avocados.
	Chitosan is a natural polysaccharide prepared by the alkaline deacetylation of chitin β-1,4-*N*-acetylglucosamine (found in fungi, arthropods and marine invertebrate); commercially, it is produced from exoskeletons of crustacean, such as crab, shrimp and crawfish. Chitosan is used for its film-forming ability; it is environmentally friendly, due to its biodegradability, biocompatibility, antimicrobial activity, non-toxicity and versatile chemical and physical properties. When applied on fruit and vegetables, chitosan-based coatings have shown effectiveness in delaying ripening and decreasing respiration rates [27,28].
	Casein forms transparent, flavorless and flexible films. Casein-based emulsion films (emulsified with lipid-based materials) were more effective than pure caseinate films in controlling moisture loss of fruits and vegetables [29].
Proteins	**Whey proteins** produce transparent, flavorless and flexible films, similar to caseinate films. Whey protein-based films possess excellent oxygen barrier properties comparable to the synthetic polymer films [30]; they also are good grease barriers [31,32].
	Zein is the principal protein of corn. It is characterized by water vapor permeabilities. Applied on vegetable and fruits, zein-based coatings are able to retard the ripening of tomatoes, to maintain the color and firmness of broccoli florets, to provide an adhesive and stable coating, to reduce the growth of *Listeria monocytogenes* on cooked sweet corn and to maintain the gloss and other qualities of apples [15].
Proteins	**Soy proteins (SP)** are extracted from defatted protein meal. Soy protein coatings generally have poor moisture resistance and water vapor barrier properties due to the inherent hydrophilicity of the protein and the addition of hydrophilic plasticizers; whilst they are potent oxygen barriers. SP coatings are used to preserve the freshness of apple slices and/or to retard the senescence process of kiwifruit [15].
	Most **fatty acids** derived from vegetable oils are considered GRAS (generally recognized as safe). Lipid-based coatings are compatible with other coating-forming agents; in addition, they have high water vapor and gas barrier properties. However, lipid-based coatings present a greasy surface and undesirable organoleptic properties, such as waxy taste and lipid rancidity [33]. Finally, some lipid materials are unstable when subjected to temperature changes.
Lipids	**Waxes** (carnauba wax, beeswax, paraffin wax and others) have been used as protective coatings for blocking moisture transport, reducing surface abrasion during fruit handling, controlling the browning of the skin in fruits by improving mechanical integrity and controlling the internal gas composition of the fruits. Wax coatings are applied on fruits (citrus, apples, cucumbers) and vegetables (tomatoes, asparagus, beans, beets, carrots, celery, eggplant, peppers, potatoes, radishes, squash and turnips.
	Resin coatings are effective at reducing water loss, but are the least permeable to gases; thus, fruit can suffer undesirable changes, such as anaerobic respiration and flavor changes.

As protein films are generally brittle and susceptible to cracking (due to the strong cohesive energy density of the polymers), an improvement of their properties could be attained by adding other components; for example, the addition of compatible plasticizers could improve their extensibility [35].

Edible lipid coatings include neutral lipids, fatty acids, waxes and resins. These compounds are effective in providing a moisture barrier and improving the surface appearance. Triglycerides or neutral lipids form a continuous stable layer on the food surface, thanks to their high polarity.

The growing interest addressed toward edible coatings leads to the formulation of new compositions consisting of blends of polysaccharides, proteins and lipids. The combination could be proteins and carbohydrates, proteins and lipids and carbohydrates and lipids [13], and it is based on the fact that each polymer has characteristic functions, that, when combined with each other, enhance the final functionality of the coating [15], improving the mechanical properties and, with emulsifiers, stabilizing composite coatings and improving coating adhesion.

Edible coatings can incorporate several compounds, such as: plasticizers (glycerol, acetylated monoglyceride, polyethylene glycol and sucrose), antimicrobials (bacteriocins (nisin)), enzymes (lysozyme, peroxidase and lactoperoxidase), essential oils (cinnamon, oregano, lemongrass, clove, rosemary, tea tree, thyme and bergamot), nitrites and sulfites [36], as well as synthetic antioxidants (butylated hydroxyanisole, butylated hydroxytoluene, propyl gallate, octyl gallate, dodecyl gallate, ethoxyquin, ascorbyl palmitate and tertiary butyl hydroquinone) and natural antioxidants (tocopherols, tocotrienols, ascorbic acid, citric acid, carotenoids). In these cases, edible films and coatings act as carriers of active compounds that, applied to the surface of fruits and vegetables, lead to the extension of shelf-life, the reduction of the risk of foodborne pathogenic microorganisms' growth on cut surfaces [37] and the improvement of the quality, stability and safety of coated foods [15].

2.2. Edible Coating: A Focus on Chitosan

Physical and chemical damage accrued during minimal processing of fruits and vegetables causes disruption of the plant tissues, and the exudates become ideal substrates for the growth of several microorganisms (pathogens, molds and bacteria). Natural biodegradable compounds with antimicrobial activity are recognized as safe (generally recognized as safe (GRAS)) and environmentally friendly, and chitosan is one such compound. Chitosan is able to create a semi-permeable film on the fruit surface, which results in limiting respiration and/or transpiration and in reducing weight loss [38,39]. Furthermore, its compatibility with other substances and its capability to induce host resistance to pathogens [40] have prompted its application as a coating on fruits and vegetables [41].

Chitosan has been widely used in controlling weight-loss in fresh strawberries (*Fragaria x ananassa*) and raspberries (*Rubus idaeus*), mango (*Mangifera indica*), litchi, blueberries and other fruit and vegetables [42]. Meng *et al.* [43] and Romanazzi *et al.* [44] reported that postharvest application of chitosan coating has a good control effect on decay of grapes. Chitosan owes its antimicrobial activity to its polycationic nature, which allows the interaction and formation of polyelectrolyte complexes with acid polymers produced at the surface of microbial cells, increasing their permeability and causing cell death [45].

Several factors affect the antimicrobial activity: type of chitosan, degree of acetylation, molecular weight, concentration, target microorganism, pH of the medium and presence of other additives or food components [46].

Benhabiles *et al.* [47] demonstrated that by reducing the number of steps for the synthesis and chemical reagents, chitosan coating was effective at improving the quality of strawberries by delaying changes in weight loss and the appearance of molds. These authors used three different chitosan coatings (chitosan C1 obtained by the classical method, chitosan C2 without decoloration and chitosan C3 without the decoloration and deproteinization steps) on strawberries, and they observed the best quality for the strawberries coated with C3 (1%), which was obtained through a lesser number of steps.

In a recent work, chitosan (in acid and water solution) exhibited its antibacterial activity against *Burkholderia seminalis*, an apricot fruit rot pathogen [48]. Lou and his coworkers demonstrated that acid-solution chitosan at a concentration of 2 mg/mL inhibited *B. seminalis*, while water-solution chitosan showed limited inhibition activity.

Chitosan has been proven to be a natural compound with antifungal activity for a wide varieties of postharvest fruits [38,49]. As reported by Bautista-Banos *et al.* [38], the level of inhibition of fungi is highly correlated to chitosan concentration. It is known that the polycationic nature of this compound is the key to its antifungal properties and that the length of the polymer chain enhances its antifungal activity. An additional explanation includes the possible effect that chitosan might have on the synthesis of certain fungal enzymes. El Ghaouth *et al.* [50] studied the antifungal effect of chitosan against *Botrytis cinerea* and *Rhizopus stolonifer*. These authors hypothesized that the mechanisms by which chitosan coating reduced the decay of strawberries appear to be related to its fungistatic properties, rather than to its ability to induce defense enzymes. A further confirmation of chitosan's ability to control fungal growth was reported by Park *et al.*[51]. They obtained a reduction of 2.5 and 2 log CFU/g in the counts of *Cladosporium* sp. and *Rhizopus* sp., respectively, on strawberries coated with a chitosan-based edible film, just after the coating application. A reduction in the counts of aerobic and coliform microorganisms during storage has been also reported. Chien *et al.* [52] investigated the effects of coating with low and high molecular weight chitosan on the decay of citrus and the maintenance of its quality. A concentration of 0.2% low molecular weight chitosan (LMWC) exhibited its antifungal activity in controlling the growth of *Penicillium digitatum*and *Penicillium italicum*. LMWC coating resulted also in being able to improve firmness, titratable acidity, ascorbic acidity and the water content for citrus stored at 15 °C for 56 days.

González-Aguilar *et al.* [53] have also reported a reduction in mesophilic aerobic microorganism count when fresh-cut papayas were coated with an edible coating based on chitosan of low and medium molecular weight. These researchers also observed a complete inhibition of yeast and molds throughout the storage (14 days at 5 °C). As reported by Ali *et al.*[54], chitosan preserved papaya fruit, delaying the ripening process by reducing the respiration rate. These results could be the reason for the delayed senescence and reduced tendency to decay [40]. Chitosan had also made improvements in the taste, peel and pulp color, texture and flavor of treated papaya fruit, but the sensory features of the papaya fruits coated with a 1.5% chitosan concentration demonstrated overall superiority after five weeks of storage.

Pilon *et al.* [55] made an alternative use of chitosan. They obtained chitosan nanoparticles and demonstrated that chitosan, used as a coating based on nanoparticles, reduced the microbial growth in fresh-cut apples. The samples treated with chitosan-tripolyphosphate (CS-TPP) nanoparticles (10 nm) showed higher antimicrobial activity against mesophilic and psychrotrophic bacteria, as well as molds and yeasts than conventional chitosan coating and control [55].

Numerous papers have demonstrated that chitosan-based coatings inhibit microbial growth on fresh produce, increasing shelf-life.

In a recent paper, Benhabiles *et al.* [56] reported that chitosan coatings and a chitosan derivative (N,O-carboxymethyl chitosan (NOCC)) coating improved the quality of tomato fruits (through delaying ripening, reducing weight loss and retaining fruit firmness) and extended the shelf-life. No microbial growth was observed during storage.

Assis and Pessoa [57] and Han *et al.* [58] proposed chitosan for extending the shelf-life of sliced apples and fresh strawberries, respectively. Chien *et al.* [52] reported the effectiveness of chitosan coating (at a concentration of 0.5%, 1% and 2% (w/v)) for prolonging the quality and extending the shelf-life of sliced mango fruit through a delay in the growth of mesophilic aerobic bacteria.

Durango *et al.* [59] and Devlieghere *et al.* [60] used a chitosan-based coating to cover carrots and lettuce, respectively, observing a reduction in the respiration rate and ethylene production, as well as a decrease in firmness loss. In particular, Durango *et al.* [59] controlled the growth of mesophilic aerobes, yeasts, molds and psychrotrophs of minimally-processed carrots during the first five days of storage at 15 °C using an edible yam starch coating containing chitosan. Campaniello *et al.* [28] observed that chitosan coating in combination with low temperature and suitable packaging was able to control browning and decay in strawberry fruits. Chitosan affected the microbial growth (psychrotrophic, lactic acid bacteria and yeasts) and did not affect the visual appearance. pH and thickness values remained unchanged by chitosan coating, whereas color was positively influenced.

Pushkala *et al.* [61] investigated chitosan-based powder coating on radish shreds, demonstrating the favorable effects of two different forms of chitosan (purified chitosan (CH) and chitosan lactate (CL)) on shelf-life extension of radish shreds by a minimum of 3 d over the control. Both CH- and CL-coated samples enhanced the microbial quality and sensory acceptability of the radish shreds, exhibiting a lower degree of weight loss, respiration rate, titrable acidity, % of soluble solids, a higher content of phytochemicals, moisture and pH, compared to control samples. The treated samples also exhibited lower exudate volume, lesser browning and lower microbial load compared to the control.

Sometimes, a chitosan-only coating demonstrated certain defects (including limited inhibition to some microorganisms that led fruit to decay and a poor coating structure); thus, chitosan was combined with other substances to improve its performance [62]. Chitosan coatings combined with organic acids are easy to handle, biodegradable and cause no harm to the coated fruit and/or vegetable [62]. For example, Yu *et al.* [63] combined 1% phytic acid (known as inositol hexakisphosphate (IP6), inositol polyphosphate) with 1% chitosan to preserve fresh cut lotus root. This composite coating decreased the weight loss rate and malondialdehyde (MDA) content of fresh-cut lotus root, postponed browning, restrained the activities of peroxidase (POD), polyphenol oxidase (PPO) and phenylalanine ammonia-lyase (PAL) and maintained the content of vitamin C and polyphenol at a relatively high level.

Chitosan combined with natamycin significantly decreased fresh melon decay and weight loss caused by *Alternaria alternata* and *Fusarium semitectum*, two strains of spoilage fungi [64].

Zhang *et al.* [65] reported that chitosan was able to inhibit the growth of *Botrytis cinerea* and *Rhizopus* sp. by increasing the activities of various defense enzymes, such as β-1,3-glucanase (in orange, strawberries and raspberries) and phenylalanine ammonia-lyase (PAL) (in strawberries andtable grapes).

As expected, the antimicrobial activity of chitosan is also dependent on the food matrix; generally, the antimicrobial activity of chitosan is higher at a low pH because more of its amino groups are protonated; thus, it is able to interact with the negatively-charged surfaces inhibiting bacterial growth. Since the antimicrobial activity of chitosan is dependent on the charges on chitosan and the electrostatic forces, each food component could influence these interactions, affecting its activity.

Regarding this issue, Devlieghere *et al.* [60] published an interesting work on the interaction of chitosan and food components. The authors evaluated the effect of different food components (starch, proteins, NaCl and fat) on the antimicrobial activity of chitosan, following the growth of *Candida lambica*, a spoiling yeast strain, in a laboratory medium.

The authors reported that the higher concentrations of starch (30% (w/v)) inhibited the antimicrobial activity of chitosan, hypothesizing that it was due to a protective effect of starch or to electrostatic interactions when the starch was charged by modification. Proteins influenced the antimicrobial activity of chitosan depending on the pH of the medium, as their charges depend on the combination of the iso-electric point (IEP) of the proteins and the pH of the medium. If the pH is lower than the IEP, proteins and chitosan are positively charged; thus chitosan can exert its antimicrobial activity, as the interactions between both will be restricted. If the pH is higher than the IEP of the protein, the antimicrobial activity of chitosan is inhibited: proteins are negatively charged and neutralize most of the positive charges on the chitosan; thus, it cannot interact with the negatively-charged microbial surfaces [60].

NaCl reduces the antimicrobial activity of chitosan, because it interferes with the electrostatic forces between chitosan: Cl^- ions neutralize the positive charges on the chitosan, and on the other hand, the Na^+ ions compete with chitosan for the negative charges on the cell surface. Devlieghere *et al.* [60] observed an improved solubility of chitosan by adding NaCl at different concentrations to the medium. The authors explained that this behavior was probably due to a shielding effect of NaCl against the positive charges, which led to a coiled structure of chitosan and less interactions with components in the media. Finally, the influence of fat on the antimicrobial activity of chitosan was found negligible.

The addition of the essential oils to enhance chitosan antimicrobial action is common.

The active agents embedded into composite films may be released into or absorb substances from the packaged food or its surrounding environment. The interactions that can occur between the food product, the film or coating used as packaging and the surrounding environment are governed by different mass transfer processes: migration, adsorption, absorption and permeation [66]. For example, through these processes, the essential oils included in the coating could be transferred to the product and modify its organoleptic and nutritional characteristics by interacting with peptides, vitamins, *etc.* It is recognized that some important food components, such as vitamins, minerals and other nutrients, can also be sequestered with the consequent modification of food properties and functionality. In addition, when transferred from the coating to the food to pursue their protective action (antioxidants, antimicrobials, *etc.*), these compounds could also lose their effectiveness. It is, in fact, well recognized that generally, the efficacy of various antimicrobial compounds may be reduced by food components; for example, essential oils (EOs) and/or their components have a significant antimicrobial activity *in vitro*, but higher amounts are required (1%–3%) to achieve the same effect in foods. The presence of fat, protein, carbohydrate, water, salt, antioxidants, preservatives, other additives and pH reaction strongly influence the effectiveness of the natural compounds in foods.

High levels of fat and/or protein in foodstuffs protect bacteria from the action of EOs in some way. Some authors suggested that the fat provides a protective layer around the bacteria, or the lipid fraction could absorb the antimicrobial agent, thus decreasing its concentration and effectiveness in the aqueous phase [67]. Patrignani *et al.* [68] reported that the low fat content of vegetables may contribute to the successful use of EOs. Due to their lipophilic nature, EOs could share fat, missing the microbial target.

The pH is an important factor affecting the activity of oils. At a low pH, the hydrophobicity of some EOs increases due to their ability to dissolve more easily in the lipid phase of the bacterial membrane, thus enhancing the antimicrobial action.

Carbohydrates in foods do not protect bacteria from this action, whilst high water and/or salt level seems to facilitate the action of EOs.

The antibacterial activity of chitosan-based films combined with lemon (LO), tea tree (TTO) or bergamot essential oils (BO) was tested against two Gram-positive bacteria (*Staphylococcus aureus* and *L. monocytogenes*) and one Gram-negative bacteria (*Escherichia coli*) [69]. CH-EO composite films exhibited a significant antimicrobial activity against the three pathogens tested. The nature and concentration of the

EOs, the film matrices and the interactions between CH and EOs affected the antimicrobial activity of the films. When TTO was added, CH exhibited the highest antimicrobial activity.

Chitosan-cinnamon oil coating extended the shelf-life of sweet pepper: after storage at 8 °C for 35 days, samples treated with chitosan-oil coatings showed a lower percentage of infected peppers and good sensory acceptability [70]. Sessa *et al.*[71] studied a novel approach to preserve vegetable products through modified chitosan edible coatings containing nanoemulsified natural antimicrobial compounds (lemon, mandarin, oregano or clove essential oils). A modified chitosan edible coating combined with lemon essential oil resulted in a remarkable increase in antimicrobial activity, with respect to other essential oils. This combination prolonged the shelf-life of rucola leaves from 3 to7 days, in comparison to the untreated samples.

Moreover, the modified chitosan containing the nanoemulsified antimicrobial caused a significantly longer shelf-life also in comparison to a coating made of modified chitosan or essential oil alone. Thanks to this novel treatment, it was possible to prolong the shelf life of rucola leaf vegetables to about 10–14 days, without alteration of the organoleptic properties of the product, preventing the loss of firmness and color changes and preserving palatability during storage.

More recently, Randazzo *et al.* [72] used chitosan-based and methylcellulose-based films added to several EOs derived from citrus fruits (orange, mandarin and lemon) to perform the antilisterial assay and concluded that chitosan films containing essential oil from lemon were the most effective at reducing *L. monocytogenes* counts.

A chitosan-methyl cellulose-based film was used as a coating on cantaloupe fruit. The application of a chitosan (1.5% w/v)/methyl cellulose (0.5% w/v) film on fresh-cut cantaloupe reduced populations of *E. coli* inoculated on fresh-cut cantaloupe by more than 5 log CFU/piece in 8 days at 10 °C. Furthermore, a reduction of 3 log CFU/piece of *Saccharomyces cerevisiae* in 4 days of storage at 10 °C was reported [73].

Krasaekoopt and Mabumrung [74] observed that the incorporation of 1.5% and 2% chitosan in the methylcellulose coating, applied on fresh-cut cantaloupe, produced a better microbiological quality in the final product. This coating reduced the growth of mesophilic aerobes, psychrotrophs, lactic acid bacteria, yeast and molds and prevented the multiplication of *E. coli* and *Salmonella* strains.

3. ALTERNATIVE MODIFIED ATMOSPHERES (E.G., HIGH O_2-MAP) OR THE USE OF ESSENTIAL OILS IN THE HEADSPACE

The final operation in producing minimally-processed fruits and vegetables is packaging.

Packaging performs several functions: it protects fresh-cut fruits and vegetables against deteriorative effects, contains the products, communicates to the consumers as a marketing tool and provides consumers products with ease of use and convenience. Fresh fruits and vegetables continue to respire, consuming oxygen and producing carbon dioxide and water vapor; thus, the atmosphere surrounding the package changes into another composition before sealing in vapor-barrier materials [75].

Modified atmosphere packaging (MAP) is a technique used for prolonging the shelf-life of fresh or minimally-processed foods [76]. MAP reduces the unwanted metabolic reactions and helps to protect processed products against contamination by microorganisms, thus slowing down the process of ageing [77]. MAP can be vacuum packaging (VP) and controlled atmosphere packaging (CAP). VP removes most of the air, or air is removed by vacuum or flushing and replaced with another gas mixture, before the product is packaged; the headspace atmosphere and product could change during storage, without additional manipulation of the internal environment. Differently, CAP continuously controls the environment to maintain a stable gas atmosphere within the package, as well as the temperature and humidity are monitored; it is generally used to control the ripening and spoilage of fruits and vegetables [76]. The composition, microflora, pH and the organoleptic characteristics of fruits and vegetables are extremely variable; therefore, it is impossible to define a univocal mixture of gases. It is necessary to maintain an optimum balance of oxygen and carbon dioxide; thus, an appropriate MAP must be studied for each product. Many factors affect the MAP of fresh produce and are summarized in Figure 2.

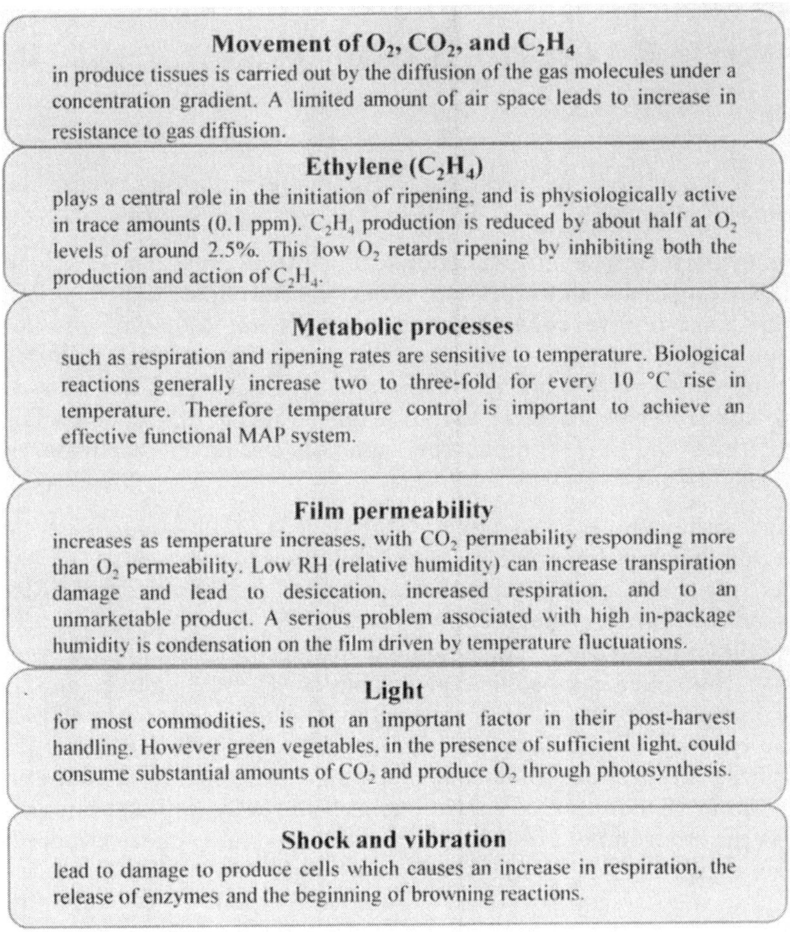

Figure 2. Factors affecting MAP of fresh products [76].

3.1. Gases Used in MAP

The main gases used in MAP are CO_2, O_2 and N_2; carbon monoxide (CO) and sulfur dioxide (SO_2) are also used. Noble gases, generally, are used for some products, such as coffee and snacks, but recently have been used also for minimally-processed apples and kiwi fruits [76]. The choice of gas depends on the food product being packed. Singly or in combination, these gases are commonly used to balance safe and shelf-life extension with optimal sensorial properties of the food. Table 2 reports the main gases used in MAP.

Low levels of oxygen and high levels of carbon dioxide reduce the produce respiration rate, delay senescence and, consequently, extend the shelf-life of fruits and vegetables. Once the package is closed, the composition of the gases inevitably changes due to produce respiration and the gas permeability of the film. If the oxygen levels are too low, fermentative processes are favored.

Table 2. Main gases used in MAP.

Gases	Characteristics and Functions
Carbon Dioxide (CO_2)	It is a colorless gas with a slight pungent odor at very high concentrations. It is an asphyxiant and slightly corrosive in the presence of moisture. CO_2 is lipid soluble and also dissolves readily in water to produce carbonic acid (H_2CO_3), which causes an increase of the acidity of the solution, reducing the pH; this significantly affects the microbiology of packed foods. CO_2 has bacteriostatic and fungistatic properties. This bacteriostatic effect depends on the concentration and partial pressure of CO_2, the volume of headspace gas, the type of microorganism, the age and load of the initial bacterial population, the microbial growth phase, the growth medium used, the storage temperature, the acidity, the water activity and the type of product being packaged [78–81]. As MAP contains CO_2, it is not an advisable option for these products where spoilage yeasts are stimulated by high levels of CO_2. In addition, CO_2 is not effective towards *Clostridium perfringens* and *Clostridium botulinum*. Carbon dioxide is generally effective in fresh fruits and vegetables where the normal spoilage organisms consist of aerobic, Gram-negative psychrotrophic bacteria. To guarantee the maximum antimicrobial effect, low storage temperatures for MAP are recommended, because CO_2 is most soluble at low temperatures. The fat and moisture of the product affect the absorption of CO_2. Fruits and vegetables can suffer physiological damage due to high CO_2 levels.
Oxygen (O_2)	Oxygen is a colorless and odorless gas and is highly reactive. It has a low solubility in water and promotes several deteriorative reactions in foods (fat oxidation, browning reactions and pigment oxidation). Most of the common spoilage bacteria and fungi require oxygen for growth; therefore, if the pack atmosphere contains a low concentration of oxygen, the shelf-life of foods is prolonged. As an alternative to low oxygen concentrations, superatmospheric O_2 concentrations ($\geq$70 kPa) have been proposed.
Nitrogen (N_2)	Nitrogen is an inert, tasteless, colorless and odorless gas and is relatively un-reactive. It has a lower density than air, is non-flammable and has a low solubility in water and other food constituents. N_2 is used to balance CO_2 and O_2 gases during food packaging, thus preventing packaging collapse that could occur when high concentrations of CO_2 are used. In addition, N_2 is able to delay oxidative rancidity and inhibit the growth of aerobic microorganisms without affecting the growth of anaerobic bacteria. Nitrogen can also indirectly influence the microorganisms in perishable foods by retarding the growth of aerobic spoilage microorganisms.
Carbon Monoxide (CO)	CO is a colorless, tasteless and odorless gas; it has a low solubility in water, but it is relatively soluble in some organic solvents. CO has been licensed for use in the USA to prevent browning in packed lettuce. It is highly reactive and very inflammable, thus its commercial application has been limited. Finally, CO has little inhibitory effect on microorganisms.
Noble Gases	These include helium (He), argon (Ar), xenon (Xe) and neon (Ne). These gases are not very reactive and are used in numerous food applications, e.g., potato-based snack products.

It is known that fresh-cut processing involves several forms of damage:

- Increases respiration rates;
- Causes major tissue disruption: enzymes and substrates, normally contained within the vacuole, are mixed with other cytoplasmic and nucleic substrates and enzymes;
- Increases wound-induced C_2H_4, water activity and surface area per unit volume, which lead to an accelerated water loss and favored microbial growth;
- Causes flavor loss;
- Causes cut surface discoloration or color loss;
- Causes decay;
- Causes increased rate of vitamin loss;
- Causes rapid softening;
- Causes shrinkage and a shorter storage life.

Thus, MAP facilitates the maintenance of fresh-cut products.

Actually, the safety of minimally-processed fruits and vegetables is mainly based on the correct chilling chain and hygienic practices, which seem to not be able to guarantee a sufficient degree of safety; in fact, the number of documented outbreaks of human infections associated with the consumption of these products has considerably increased. On these bases, traditional MAP is not enough to ensure the improvement of the quality and safety characteristics. In this context, many researchers have studied alternative tools, for improvement of MAP for minimally-processed fruits and vegetables; for example, the use of non-conventional atmospheres, including active packaging containing some natural antimicrobial compounds or MAP with high O_2, has been proposed.

3.2. MAP and Natural Antimicrobial Compounds

Plants and plant products are generally considered natural alternatives to improve the shelf-life and the safety of foods, since they are characterized by a wide range of GRAS volatile compounds, which are used as food flavoring agents [76]. They are able to inhibit numerous microorganisms; thus they are used as components of biological means for prolonging the shelf-life of post-harvest or minimally-processed fruits and vegetables [76].

Literature data indicate that volatile compounds can represent a useful tool to increase the shelf-life of plant products. Corbo *et al.* [82] evaluated the effects of hexanal and *trans*-2-hexenal on the shelf-life of fresh sliced apples. The authors added *trans*-2-hexenal to the gas mixture containing 70% N_2 and 30% CO_2 and found a significant extension of the shelf-life also when *Pichia subpelliculosa* (a spoilage yeast) was inoculated and abuse storage temperatures were used, although it had a weak negative effect on color retention.

Lanciotti *et al.* [83], postulated that future trends in the use of natural compounds, such as hexanal, 2-(E)-hexenal and hexyl acetate, would be focused on the use of specific active packaging able to release the active molecules in the head space slowly over time.

The authors reported that 150 ppm of hexanal, 20 ppm of 2-(E)-hexenal and 150 ppm of hexyl acetate displayed a bactericidal effect on *L. monocytogenes* and caused a significant extension of the lag phase of *E. coli* and *Salmonella* Enteritidis inoculated at levels of 10^4–10^5 CFU/g in fresh sliced apples packaged in ordinary or modified atmosphere.

Campaniello *et al.* [84] investigated the possibility of combining hexanal and MAP (65% N_2, 30% CO_2 and 5% O_2) on minimally-processed cactus pear fruits. The hexanal showed an antimicrobial effect against *Enterobacteriaceae*, normally contaminating minimally-processed fruits, both in the control and the modified atmosphere. The inclusion of the antimicrobial compound in the atmosphere determined an improvement of the original color retention; as well as a mesophilic selection favoring *Pantoea* spp., which have antagonistic activity against molds responsible for the decay of fruits during post-harvest phase.

Siroli *et al.* [85] proposed the use of several antimicrobial compounds (citron EO, hexanal, 2-(E)-hexenal, citral and carvacrol) alone or in combination in order to increase the shelf-life and quality parameters (texture and color) of sliced apples packaged in active modified atmosphere (7% O_2 and 0% CO_2), into medium permeability bags and stored at 6 °C. In all of the samples, the spoilage yeast threshold was not attained within the 35 days of storage. When treated with citral/2-(E)-hexenal and hexanal/2-(E)-hexenal, the sample showed a good color retention. This latter combination also improves on the retention of firmness, which was the best throughout 35 days of storage.

Furthermore, plant essential oils, constituted mainly by terpenoids, are used for their antimicrobial activity against many microorganisms. Most of the essential oils are GRAS; nevertheless, their use is often limited because of a high impact on the organoleptic characteristics of food products.

The activity of oils from *Labiatae* and citrus fruits, as well as the action of single constituents have been studied in order to better understand the cell targets of the molecules. Due to their antimicrobial effect, citrus essential oils could represent good candidates to improve the shelf-life and the safety of minimally-processed fruits. Essential oils, such as citrus, mandarin, cider, lemon and lime, were able to increase the shelf-life and safety of minimally-processed fruit salads without any impact on the sensory characteristics, even when the product was inoculated with spoilage or pathogenic bacterial species [76].

Citrus essential oils (EOs) exhibited their antimicrobial effect against a range of food poisoning-causing bacteria. A blend of citrus EO vapor against vancomycin-resistant (VRE) and vancomycin-susceptible (VSE) *Enterococcus faecium* and *Enterococcus faecalis* on lettuce and cucumber was assessed. Food samples were subjected to the vapor for 45 s in a 600-L vapor chamber at 25 °C. Results showed that microbial cell load was reduced and that no significant changes in taste were observed [86].

Tian *et al.* [87] tested the antifungal activity of essential oil extracted from the fruits of *Cicuta virosa L.* var. *latisecta Celak*(CVEO) against *Aspergillus niger*, *Aspergillus flavus*, *Aspergillus oryzae* and *Al. alternata* strains inoculated on cherry tomatoes. The samples were pre-treated with ethanol and wounded with a sterilized cork borer; then, each fruit was separately inoculated with 10 μL of a spore suspension containing 1×10^6 spores/mL of each fungal strain. CVEO (dissolved separately in 0.5 mL of 5% of Tween-20) was pipetted aseptically onto filter paper discs respectively placed into individual weighing bottles (without lids) to produce the concentrations of 200, 100 and 50 μL/mL. The essential oil was vaporized inside the containers spontaneously at 18 °C. The authors reported that at 200 μL/mL, CVEO showed the lowest percentages of decayed cherry tomatoes for all fungi compared to the control, as well as the highest inhibition of fungal infection.

In a recent paper, Vitoratos *et al.* [88] studied the antifungal activity of several essential oils obtained from oregano (*Origanum vulgare* L. ssp. *hirtum*), thyme (*Thymus vulgaris* L.) and lemon (*Citrus limon* L.) plants against *Bo. cinerea*inoculated in tomatoes, strawberries and cucumbers. All of the fruits were pre-treated and packaged. Different concentrations of essential oils were placed in small glass containers placed in the bottom of the package. Oregano and lemon oils were very effective at controlling disease of infected fruit by *Bo. cinerea* in tomatoes, strawberries and cucumbers. In particular, in tomatoes, *Bo. cinerea* was inhibited by oregano essential oils at 0.30 μL/mL; moreover, lemon essential oils also induced a significant reduction of grey mold. An amount of 0.05 μL/mL of lemon essential oils leads to a complete inhibition of *Bo. cinerea* in strawberries, whilst in cucumber, it leads to a reduction (39%) of the infected fruits.

3.3. Potential Applications of High O_2

It is known that low O_2 modified atmosphere packaging is universally accepted to prolong the shelf-life of minimally-processed fruits and vegetables; however, the growth of some anaerobic pathogen strains might be allowed or even stimulated. In addition, when O_2 levels are too low, this induces anaerobic respiration, leading to fermentation processes and to the production of undesirable metabolites [89].

The application of high O_2 concentrations (>70% O_2) could overcome the disadvantages of low O_2 modified atmosphere packaging for some ready-to-eat fruits and vegetables. High O_2 was found to be particularly effective at inhibiting enzymatic discoloration, preventing anaerobic fermentation reactions and inhibiting microbial growth [90]. It is hypothesized that high O_2 levels may cause substrate inhibition of the enzyme polyphenol oxidase (PPO), which is the enzyme responsible for initiating discoloration on the cut surfaces of processed products [91].

High O_2-modified atmosphere packaging has been used in some vegetables and fruits, such as minimally-processed carrots, strawberry, minimally-processed baby spinach, fresh-cut mango, apple slices, and so on [89].

Amanatidou *et al.* [92] screened numerous microorganisms (Pseudomonas fluorescens, Enterobacter agglomerans, Aureobacterium strains 27, Candida guilliermondii, Candida sake, *Salmonella* Typhimurium, *Salmonella* Enteritidis, *E. coli*, *L. monocytogenes*, *Leuconostoc mesenteroides* var. *mesenteroides*, *Lactobacillus plantarum* and *Lactococcus lactis*) associated with the spoilage and safety of minimally-processed vegetables. High oxygen levels (80%–90%) caused a reduction in the growth rate of microbial targets at 8 °C, whilst the lag phase was prolonged (it was more evident at 90% O_2 concentrations). Amanatidou *et al.* [92] and Kader and Ben Yehoshua [93] hypothesized that high O_2 concentrations lead to intracellular generation of reactive oxygen radical species (ROS, O_2^-, H_2O_2, OH^-), causing damage to the vital cellular macromolecules, thus reducing cell viability when oxidative stresses overwhelm cellular protection systems.

Jacxsens *et al.* [90] studied the effect of high oxygen-modified atmosphere packaging (*i.e.*, >70% O_2) on microbial growth and the sensorial qualities of mushroom slices, grated celeriac and shredded chicory endive and observed that high O_2atmospheres were effective at inhibiting enzymatic browning of the tested vegetables. In addition, an improvement of the microbial quality (as a reduction in yeast growth) was obtained.

In an interesting paper, Allende *et al.* [94] studied the effect of superatmospheric oxygen packaging on sensorial quality, spoilage and *L. monocytogenes* and *Aeromonas caviae* growth in fresh processed mixed salads. An initial O_2 concentration of 95 kPa was combined with two plastic films (low and high barrier impermeability for O_2). Packaged salads were stored up to eight days at 4 °C and at temperatures simulating the chilled distribution chain.

The authors reported that superatmospheric O_2 does not affect all microorganisms in the same way: high oxygen levels affected lactic acid bacteria and *Enterobacteriaceae*, which were inhibited in both plastic films (with low and high permeability). On the contrary, the growth of yeast and *A. caviae* seemed to be stimulated, whereas the growth of psychrotrophic bacteria and *L. monocytogenes* was not affected. The general appearance was maintained for longer, and the shelf-life of the mixed salads was prolonged by using O_2 concentrations higher than 60 kPa throughout the storage period.

Chunyang *et al.* [89] studied the effect of high oxygen-modified atmosphere packaging on fresh-cut onion quality at room temperature. Onion slices were packaged in a high-barrier film package of 70 μm in thickness and stored in five different modified atmospheres (100% O_2; 95% O_2/5% CO_2; 80% O_2/20% CO_2, 75% O_2/25% CO_2; and air). Results showed that the fresh-cut onions packaged in air had a short shelf-life, because the respiration rate was quickened due to mechanical damage, and microorganisms were not inhibited. High O_2 reduced the weight loss, the respiration rate, the total reducing sugar loss and the total titrable acidity increase of the fresh-cut onions. The best modified atmosphere seemed to be 80% O_2/20% CO_2, where total bacteria counts increased slowly. High oxygen affected also the sensory characteristics: sensory quality was acceptable up to five days, while air-packaged fresh-cut onions were not acceptable after two days of storage at room temperature.

Day [91] also confirmed that the most effective high O_2 gas mixtures were found to be 80%–85% O_2/15%–20% CO_2. This had the most noticeable sensory quality and antimicrobial benefits on a range of freshly-prepared produce items.

In fact, recommended optimal headspace gas levels immediately after freshly-prepared produce package sealing are: 80%–95% O_2 and 5%–20% N_2. After package sealing, headspace O_2 levels will decline, whereas

CO_2 levels will increase during storage, due to the intrinsic respiratory nature of fresh product.

The levels of O_2 and CO_2 within packages are influenced by numerous variables:

- Intrinsic produce respiration rate (which itself is affected by temperature, atmospheric composition, produce type, variety, cultivar, maturity and the severity of preparation);
- Packaging film permeability;
- Package volume;
- Surface area and fill weight;
- Produce volume/gas volume ratio;
- Degree of illumination.

It would be preferable to maintain headspace levels of $O_2 > 40\%$ and CO_2 between 10% and 25% to maximize the benefits of high O_2 MAP. This can be achieved by:

- Lowering the temperature of storage;
- Selecting produce having a lower intrinsic respiration rate;
- Minimizing cut surface tissue damage;
- Reducing the ratio of produce volume/gas;
- Using a packaging film able to maintain high levels of O_2 whilst allowing excess CO_2 to go out
- By incorporating an innovative active packaging sachet that can absorb excess CO_2 and emit an equal volume of O_2.

A further measure to maintain a gas mixture $O_2 > 40\%/CO_2$ 10%–25% could be obtained by introducing the highest level of O_2 (balanced with N_2) possible just prior to prepared produce package sealing. Generally, it is not necessary to have any levels of CO_2 in the initial gas mixture, since it will increase rapidly during chilled storage.

Numerous products (iceberg lettuce, sliced mushrooms, broccoli florets, lettuce, baby-leaf spinach, Lollo Rossa lettuce, flat-leaf parsley, cubed swede, coriander, raspberries, strawberries, grapes and oranges) processed with high O_2 MAP reported beneficial effects on sensory quality if compared to industry-standard air packing and low O_2 MAP [91].

Finally, the recommended packaging material for high O_2-modified atmosphere is 30-μm orientated polypropylene (OPP), which has sufficient O_2 barrier properties (to maintain high in-pack O_2 levels >40%), and it is sufficiently permeable to ensure that CO_2 did not rise above 25% after 7–10 days of storage at 5–8 °C.

Other packaging materials suitable for high O_2 MAP of fresh prepared produce are:

- Laminations or extrusions of OPP with low density polyethylene (LDPE);
- Ethylene-vinyl acetate (EVA);
- Polyvinyl chloride (PVC).

or other medium to very high O_2 permeability films.

4. CONDITIONING SOLUTIONS WITH ANTIMICROBIALS OR NATURAL COMPOUNDS FOR FRUIT SALAD

Coatings, dipping, spraying, *etc.*, are the main applications of antimicrobial compounds, as natural alternatives to chemical additives, whilst little is known about the use of antimicrobials as filling liquids.

An initial work by Senesi *et al.* [95] on the use of rectified apple juice as a filling liquid to increase the quality and shelf-life of fresh-cut apples was followed, more recently, by D'Amato *et al.* [96]. The authors evaluated the possibility of using a chitosan, honey and pineapple juice solution as filling liquids to prolong the microbiological shelf-life of a fruit-based salad.

"Granny Smith" apples, "Gialla" first crop cactus pear fruits and "Regina" table grapes were washed; then, apples and cactus pears were peeled and sliced, while grapes were cut without seeds. Mixed fruits were placed in conical frustum-shaped cups, filled with four different solutions (sterile distilled water as the control; 30% of acacia honey and 70% of distilled water; 50% of pineapple juice and 50% of distilled water; 50% of low molecular weight chitosan solution and 50% of sterile distilled) and stored at 4, 8 and 12 °C.

The authors observed that the use of the natural antimicrobial compounds, as a filling liquid, affected the microbiological shelf-life of salad.

At 4 °C, mesophilic and psychrotrophic bacteria did not reach the established limit (1×10^6 CFU/g) during the whole storage period (14 days), in samples with honey and chitosan solutions, whereas it ranged from about 6.34–7.81 days and from 9.38 to 11.49 for the control and samples with pineapple juice, respectively.

At 8 °C, all of the samples were stored for 10 days. For the samples treated with honey and chitosan solutions, psychrotrophic bacteria did not reach the established limit during the storage, while in control and pineapple juice, it was of 5–6 days. Concerning mesophilic bacteria, microbiological shelf-life was strongly reduced in the control samples and in pineapple juice (about 3.7–4.3 days), whilst in chitosan and honey, it was over six days. In addition, in the samples with chitosan solution, the microbiological shelf-life, calculated by using the growth of yeasts, was over seven days and, by using lactic acid bacteria, did not reach the established limit.

At 12 °C, the samples were stored for six days. The microbiological shelf-life of fruit salad without antimicrobial compounds and with pineapple juice was always very low (1.5 and 4 days, respectively).

In the samples honey added, mesophilic and psychrotrophic bacteria did not reach the limit of 1×10^6 CFU/g. For lactic acid bacteria and yeasts, microbiological shelf-life reached 3.82 and 1.81 days, respectively.

Chitosan exhibited its high antimicrobial activity towards psychrotrophic bacteria, which did not reach the established limit, while no advantage resulted for mesophilic bacteria and lactic acid bacteria and yeasts.

The authors concluded that:

- Honey inhibited mesophilic and psychrotrophic bacterial growth; the antimicrobial activity of this natural antimicrobial compound was less effective on lactic acid bacteria and yeasts at all storage temperatures;
- Chitosan exerted a high antimicrobial activity for all microbial groups considered, particularly at a low temperature of storage;
- Pineapple juice was not effective towards all microbial groups, probably due to the high amount of nutrients.

Further investigations are needed to improve the antimicrobial efficacy of honey and chitosan for potential commercial applications. The literature is poor in articles addressing this topic; thus, it would be advisable to explore the use of antimicrobial compounds as filling liquids.

5. BIOPRESERVATION AND THE USE OF PROBIOTIC COATINGS

In recent years, biological preservation has emerged as a promising strategy to extend the shelf-life and to improve the microbiological safety of foods [97], since it fits well with the diffuse desire to preserve foods by natural means. Several bacteria and yeasts have been already identified as bioprotective agents [98], and different studies have been carried out on their application to fresh-cut fruits and vegetables [99]. Table 3 summarizes the most recent studies performed on this topic.

In general, lactic acid bacteria (LAB) have shown the greatest potential as biocontrol agents of several minimally-processed foods, because they are widely used in fermented foods and have a long history of safe use [100]. However, several other bacteria and yeasts, often selected among the naturally-occurring microflora, including strains of *Pseudomonas syringae*, *Pseudomonas graminis*, *Gluconobacter asaii*, *Candida* spp., *Dicosphaerina fagi*, *Metschnikowia pulcherrima* and *C. sake*, have been proposed as biocontrol agents in fresh-cut fruits and vegetables [101,102,103,104].

Table 3. Overview of the most recent studies performed on the use of biocontrol microorganisms to biopreserve minimally-processed products.

Minimally-Processed Product	Proposed Microorganisms	Reference
Biopreservation		
Apples	*Leuconostoc mesenteroides*, *Leuconostoc citreum*	[103]
Apples	*Pseudomonas graminis*	[102,105]
Apples	*Candida sake*	[101]
Apples	*Candida* sp., *Gluconobacter asaii*, *Dicosphaerina fagi*, *Metschnikowia pulcherrima*	[106]
Apples	*Enterobacteriaceae*	[107]
Apples	*Lactobacillus plantarum*	[99]
Apples	*Lactococcus lactis*	[108]
Iceberg lettuce	*Leuconostoc mesenteroides*, *Leuconostoc citreum*	[103]
Iceberg lettuce	*Leuconostoc* spp.	[104]
Lamb's lettuce	*Lactobacillus plantarum*, *Lactobacillus casei*	[99]
Melon	*Pseudomonas graminis*	[109]
Melon	*Pseudomonas graminis*	[110]
Peaches	*Enterobacteriaceae*	[107]
Scarola salad	*Lactobacillus casei*	[111]
Probiotic Biopreservation		
Apple	*Lactobacillus rhamnosus* GG	[112]
Apple	*Lactobacillus rhamnosus* GG	[113]
Apple	*Bifidobacterium lactis* Bb-12	[114]
Papaya	*Bifidobacterium lactis* Bb-12	[114]

The success of LAB in preventing the growth and activity of undesirable microorganisms is due to a large diversity of mechanisms of action related to the production of antimicrobial compounds, organic acids, hydrogen peroxide, bacteriocins and diacetyl [103,104,115]. The combination of low pH values and antibacterial activities of organic molecules produced by LAB remains the main mechanism for biopreservation [97]. Several bacteriocin-producing LAB have been shown to be effective against spoilage and pathogenic microorganisms in minimally-processed fruits and vegetables [116,117]. However, the direct application of bacteriocins to fresh-cut products did not provide completely satisfactory results probably due to the adsorption or inactivation of the added compound into the food product [116]; on the other hand, the direct use of living bacteriocinogenic bioprotective strains could lead to a more effective protection of the food product by circumventing the mentioned problems because of the localized and constant delivery of the antibacterial compound, which will add to other advantages, like space colonization by the strain [116,117]. The inhibitory properties against contaminating foodborne pathogens and spoilage microorganisms could also consist of a mere competition for nutrients (vitamins, minerals, trace elements and peptides), and therefore, via competition or antibiosis, LAB are able to function as a hurdle to pathogen growth and survival [118].

In the last decade, different LAB species were proposed as biocontrol agents in minimally-processed fruits and vegetables (see Table 3). In 2004, Scolari and Vescovo [111] performed several challenge tests on salad leaves by simultaneously inoculating *Lactobacillus casei* and various pathogens (*S. aureus*, *Aeromonas hydrophila*, *E. coli* and *L. monocytogenes*). A significant inhibitory effect by the LAB towards all of the pathogenic strains was observed and confirmed by the same authors during a subsequent study about the influence of *Lb. plantarum* on the growth of *S. aureus* [119]. Trias *et al.* [103] found five strains of LAB (some *Leuconostoc* spp. strains) that were able to inhibit *L. monocytogenes* and *Salmonella* Typhimurium in cut iceberg lettuce leaf, but were not effective in reducing the amount of *E. coli*. More recently, Siroli *et al.* [108] showed the good performance of a nisin-producing strain of *Lc. lactis*, which was able to inhibit *L. monocytogenes*, *E. coli* and the total mesophilic species when added at a level of 7 log CFU/mL in the washing solution of minimally-processed lamb's lettuce and combined or not with thyme essential oil. In 2015, the same authors proposed two other LAB strains (*Lb. plantarum* V7B3 and *Lb. casei* V4B4) to be used as biocontrol agents alone or in combination with thyme essential oil (EO) in lamb's lettuce [99]. In this work, the use of the *Lb. plantarum* V7B3 strain (6 log CFU/mL) during the washing phase of fresh-cut lettuce increased product shelf-life and safety; *L. monocytogenes* and *E.*

coli viabilities were significantly reduced over the nine days of refrigerated storage. Promising results were also obtained for biopreservation of minimally-processed golden delicious apples packaged in a modified atmosphere alone or in combination with natural antimicrobials (2-(E)-hexenal/hexanal and 2-(E)-hexenal/citral) [99]: a strain of *Lb. plantarum* (CIT3) was able to increase the safety of sliced apples, when inoculated at levels of 6–7 log CFU/g in the washing dipping solution, both alone or in combination with natural antimicrobials.

Other non-lactic acid bacteria and yeasts were also proposed as biocontrol agents (Table 3) For example, the growth of *L. monocytogenes* and *Salmonella* Enterica in fresh-cut apples has been prevented using fungal antagonists [106]. When inoculated at a low level, *L. monocytogenes* cell loads were greatly reduced (from 5.7 to 6.0 log units after seven days) by strains of *G. asaii*, *Candida* spp., *D. fagi* and *M. pulcherrima*. At high pathogen inoculum levels, only *G. asaii* and *Candida* spp. reduced the cell load of *L. monocytogenes* population to non-detectable levels. Abadias *et al.* [101] found that the application of the fungal postharvest antagonist *C. sake* CPA-1 reduced the growth of a mixture of *E. coli* strains in fresh-cut apples at 25 °C, whereas Alegre *et al.* [107] isolated a new strain of *Enterobacteriaceae* that reduced the growth of *Salmonella*, *L. monocytogenes* and *E. coli* O157:H7 on fresh-cut apples and peaches. *P. graminis* was found be able to reduce or slow down the development of foodborne pathogens on minimally-processed fresh-cut apples and peaches [105]; in this case, the inhibitory effect of the antagonists on the foodborne pathogens was not instantaneous and became apparent after 6 days at 5 °C. This strain was also proposed for fresh-cut melon, reducing *Salmonella* and *L. monocytogenes* growth during storage at 5, 10 and 20 °C [109]. The strain effectiveness depended on the pathogens' concentration and on storage temperature. At a low pathogen concentration and 20 °C, *L. monocytogenes* growth was reduced between 2.1 and 5.3 log CFU/g after two days of storage and *Salmonella* growth between 2.0 and 7.3 log CFU/g.

Although research on the use of biocontrol agents in minimally-processed fruits and vegetables has increased in recent decades, the standardization of a biopreservative approach is still difficult to realize. Independent of the species and strains proposed, in fact, the various studies available in the literature clearly highlight that the efficacy of biocontrol agents is affected by different factors, such as the inoculation level, the presence of other bacteria, the physico-chemical and compositional features of the products and the storage conditions. Further investigations are required, especially considering that process conditions have to be taken into account, during the scaling up at the industrial level.

An interesting modern challenge is to incorporate probiotic bacteria into coated processed fruits and vegetables in order to improve their shelf-life (biopreservation), while providing new non-dairy functional foods. As is well accepted, edible films/coatings may serve as carriers of food additives, such as anti-browning agents, antimicrobials, colorants, flavors, nutrients and spices [24]. Some authors [114,120] proved that the coatings were also good carriers for antioxidant agents, such as cysteine, glutathione and ascorbic and citric acids; thus, the addition of LAB to obtain functional edible films and coatings could have successful implications. Several studies were already conducted on different fruits, as is shown in Table 3. This new trend has arisen from more considerations. As just mentioned, the inclusion into coatings of cultures with inhibitory properties could improve the shelf-life and safety of minimally-processed products, while reducing the need to use increasing levels of chemical additives [118]. In addition, the steps of minimally-processing vegetables, such as peeling and cutting, promote the release of cellular content rich in minerals, sugars, vitamins and other nutrients, creating ideal conditions for microbial growth: this characteristic allows the use of fruit and vegetable food products as probiotic carriers. According to Soccol *et al.* [121], in fact, minimally-processed fruits and vegetables are very good matrices providing ideal substrates for probiotics, since they contain minerals, vitamins, antioxidants and fibers. Moreover, due to their cellulose content, fruits, such as apples and pears, may also exert a protective effect on the probiotic microorganisms during passage through the intestinal tract [122], allowing these microorganisms to reach the colon and benefit the host. It has been reported that the optimum probiotic growth temperature is between 35 and 40 °C, and the best pH is between 6.4 and 4.5, ceasing when a pH of 4.0–3.6 is reached [123]. This situation could be solved by using some supports, such as agar, polyacrylamide, calcium pectate gel, chemically-modified chitosan beads and alginates, to provide a physical barrier against unfavorable conditions [124,125,126].

One of the first reports about this issue is the study of Tapia *et al.* [114]; fresh-cut apple and papaya cylinders were coated with 2% (w/v) alginate or gellan film-forming solutions containing viable Bifidobacteria, namely *Bifidobacterium lactis* Bb-12. The bifidus-containing coatings were more permeable to water vapor than the corresponding films without probiotics, and gellan coatings were more resistant to water transfer than the alginate ones. The most important result was that the edible coatings were efficient in supporting *Bb. lactis* Bb-12 on fresh-cut apple and papaya. In 2010, Rößle *et al.* [112] applied a probiotic microorganism (*Lactobacillus rhamnosus* GG; LGG) to fresh-cut apple wedges (cultivar Braeburn). All samples were able to maintain a probiotic load of about *ca.* 10^8 CFU/g over 10 storage days, which is sufficient for a probiotic effect, and this is comparable to the counts of probiotic bacteria in commercially-available dairy products. It is also important to underline that the physico-chemical properties of the apple wedges containing LGG compared to the control (without probiotics) remained stable over the observation period. In a subsequent work [113], the effectiveness as a biocontrol agent of the same probiotic strain against *Salmonella* and *L. monocytogenes* on minimally-processed apples throughout storage, as well as its effect on apple quality and natural microflora was evaluated. The obtained results showed that *Salmonella* was not affected by co-inoculation with LGG, whereas the *L. monocytogenes* population was 1-log unit lower in the presence of probiotic population maintained over recommended levels for probiotic action (10^6 CFU/g) along 14 days.

Although still unexplored, this new challenge appears to be highly advantageous, since minimally-processed fruits and vegetables are a food category rich in nutrients, intended for consumption by all individuals and widely accepted among consumers [127]. Thus, edible coatings, including probiotic bacteria, when applied to minimally-processed products, open new possibilities to improve their shelf-life and safety while providing innovative functional foods.

6. CONCLUDING REMARKS

The use of non-conventional tools to preserve and prolong the quality of minimally-processed fruits and vegetables seems to be the future trend to obtain improved final products. In particular, the application of the antimicrobial compounds into coatings and/or filling liquids should be further investigated due to their environmental friendliness and versatility. A combined approach of these technologies, through additive and synergistic interactions, could be useful to obtain innovative and safe minimally-processed products.

REFERENCES

1. Allende, A.; Tomas-Barberan, F.A.; Gil, M.I. Minimal processing for healthy traditional foods. *Trends Food Sci. Tech.* **2006**, *7*, 179–186.
2. Lamikanra, O. Preface. In *Fresh-Cut Fruits and Vegetables*; Lamikanra, O., Ed.; CRC Press: Boca Raton, FL, USA, 2002.

3. Gonzales-Aguilar, G.A.; Ayala-Zavala, J.F.; de La Rosa, L.A.; Ivarez-Parrilla, E.A. Preserving quality of fresh-cut product using safe technology. *J. Verbrauch. Lebensm.* **2010**, *5*, 65–72.
4. Boesso Oriani, V.; Molina, G.; Chiumarelli, M.; Pastore, G.M.; Dupas Hubinger, M. Properties of cassava starch-based edible coating containing Essential Oils. *J. Food Sci.* **2014**, *79*, E189–E194.
5. McHugh, T.H.; Senesi, E. Apple wraps: A novel method to improve the quality and extend the shelf life of fresh-cut apples. *J. Food Sci.* **2000**, *65*, 480–485.
6. Nisperos-Carriedo, M.O.; Baldwin, E.A.; Shaw, P.E. Development of an edible coating for extending postharvest life of selected fruits and vegetables. *Proc. Fla. State Hort. Soc.* **1991**, *104*, 122–125.
7. Lerdthanangkul, S.; Krochta, J.M. Edible coating effects on post harvest quality of green bell peppers. *J. Food Sci.***1996**, *61*, 176–179.
8. Avena-Bustillos, R.J.; Krochta, J.M.; Saltveit, M.E. Water vapor resistance of red delicious apples and celery sticks coated with edible caseinate-acetylated monoglyceride films. *J. Food Sci.* **1997**, *62*, 351–354.
9. Guilbert, S.; Gontard, N.; Gorris, L.G.M. Prolongation of the shelf-life of perishable food products using biodegradable films and coatings. *LWT—Food Sci. Technol.* **1996**, *29*, 10–17.
10. Smith, S.; Geeson, J.; Stow, J. Production of modified atmospheres in deciduous fruits by the use of films ad coatings. *Hort. Sci.* **1987**, *22*, 772–776.
11. Krochta, J.M.; Mulder-Johnston, C.D. Edible and biodegradable polymer films: Challenges and opportunities. *Food Technol.* **1997**, *51*, 61–74.
12. Souza, B.W.S.; Cerqueira, M.A.; Teixeira, J.A.; Vicente, A.A. The use of electric fields for edible coatings and films development and production: A review. *Food Eng. Rev.* **2010**, *2*, 244–255.
13. Campos, C.A.; Gerschenson, L.N.; Flores, S.K. Development of edible films and coatings with antimicrobial activity. *Rev. Food Bioprocess. Technol.* **2010**, *4*, 849–875.
14. Allen, L.; Nelson, A.I.; Steinberg, M.P.; McGill, J.N. Edible corn-carbohydrate food coatings. Development and physical testing of starch-algin coating. *Food Technol.* **1963**, *17*, 1437–1442.
15. Lin, D.; Zhao, Y. Innovations in the development and application of edible coatings for fresh and minimally processed fruits and vegetables. *Compr. Rev. Food Sci. Food Saf.* **2007**, *6*, 60–75.
16. Diab, T.; Biliaderis, C.G.; Gerasopoulos, D.; Sfakiotakis, E. Physicochemical properties and application of pullulan edible films and coatings in fruit preservation. *J. Sci. Food Agric.* **2001**, *81*, 988–1000.
17. Synowiec, A.; Gniewosz, M.; Kraśniewska, K.; Przybył, J.L.; Bączek, K.; Węglarz, Z. Antimicrobial and antioxidant properties of pullulan film containing sweet basil extract and an evaluation of coating effectiveness in the prolongation of the shelf life of apples stored in refrigeration conditions. *Inn. Food Sci. Emer. Technol.* **2014**, *23*, 171–181.
18. Vargas, M.; Pastor, C.; Chiralt, A.; McClements, D.J.; González-Martínez, C. Recent advances in edible coatings for fresh and minimally processed fruits. *Crit. Rev. Food Sci.* **2008**, *48*, 496–511.
19. Coma, V.; Deschamps, A.; Martial-Gros, A. Bioactive packaging materials from edible chitosan polymer-antimicrobial assessment on dairy related contaminants. *J. Food Sci.* **2003**, *68*, 2788–2792.
20. Li, P.; Barth, M.M. Impact of edible coatings on nutritional and physiological changes in lightly processed carrots. *Postharvest Biol. Tec.* **1998**, *14*, 51–60.
21. Amanatidou, A.; Slump, R.A.; Gorris, L.G.M.; Smid, E.J. High oxygen and high carbon dioxide modified atmospheres for shelf life extension of minimally processed carrots. *J. Food Sci.* **2000**, *65*, 61–66.
22. Hershko, V.; Nussinovitch, A. Relationships between hydrocolloid coating and mushroom structure. *J. Agric. Food Chem.* **1998**, *46*, 2988–2997.

23. Bryan, D.S. Prepared citrus fruit halves and method of making the same. U.S. Patent 3,707,383, 26 December 1972.
24. Lee, J.Y.; Park, H.J.; Lee, C.Y.; Choi, W.Y. Extending shelf-life of minimally processed apples with edible coatings and antibrowning agents. *Lebens. Wissen. Technol.* **2003**, *36*, 323–329.
25. Arnold, F.W. Infrared Roasting of Coated Nutmeats. U.S. Patent 3,383,220, 14 May 1968.
26. Maftoonazad, N.; Ramaswamy, H.S.; Moalemiyan, M.; Kushalappa, A.C. Effect of pectin-based edible emulsion coating on changes in quality of avocado exposed to *Lasiodiplodia theobromae* infection. *Carbohyd. Polym.* **2007**, *68*, 341–349.
27. Vargas, M.; Albors, A.; Chiralt, A.; Gonzalez-Martinez, C. Quality of cold-stored strawberries as affected by chitosan-oleic acid edible coatings. *Postharvest Biol. Tec.* **2006**, *41*, 164–171.
28. Campaniello, D.; Bevilacqua, A.; Sinigaglia, M.; Corbo, M.R. Chitosan: Antimicrobial activity and potential applications for preserving minimally processed strawberries. *Food Microbiol.* **2008**, *25*, 992–1000.
29. Krochta, J.M.; Pavlath, A.E.; Goodman, N. Edible films from casein-lipid emulsions for lightly-processed fruits and vegetables. In *Engineering and food, Preservation Processes and Related Techniques*; Spiess, W.E., Schubert, H., Eds.; Elsevier Science Publishers: New York, NY, USA, 1990; pp. 329–340.
30. Trezza, T.A.; Krochta, J.M. Application of edible protein coatings to nut and nut-containing food products. In *Protein-Based Films and Coatings*; Gennadios, A., Ed.; CRC Press: Boca Raton, FL., USA, 2002; pp. 527–549.
31. Chan, M.A.; Krochta, J.M. Grease and oxygen barrier properties of whey-protein isolate coated paperboard. *Solutions***2001**, *84*, 57.
32. Lin, S.Y.; Krochta, J.M. Plasticizer effect on grease barrier properties of whey protein concentrate coatings on paperboard. *J. Food Sci.* **2003**, *68*, 229–233.
33. Guilbert, S. Technology and application of edible protective films. In *Food Packaging and Preservation*; Mathlouthi, M., Ed.; Elsevier Applied Science Publishers: London, UK, 1986; pp. 371–394.
34. Gennadios, A.; McHugh, T.H.; Weller, G.L.; Krochta, J.M. Edible coatings and films based on proteins. In *Edible Coatings and Films to Improve Food Quality*; Krochta, J.M., Baldwin, E.A., Nisperos-Carriedo, M.O., Eds.; Technomic Publishing Co., Inc.: Lancaster, PA, USA, 1994; pp. 201–277.
35. Lim, L.T.; Mine, Y.; Britt, I.J.; Tung, M.A. Formation and properties of egg white protein films and coatings. In *Protein-Based Films and Coatings*; Gennadios, A., Ed.; CRC Press: Boca Raton, FL, USA, 2002; pp. 233–252.
36. Franssen, L.R.; Krochta, J.M. Edible coatings containing natural antimicrobials for processed foods. In *Natural Antimicrobials for Minimal Processing of Foods*; Roller, S., Ed.; CRC Press: Boca Raton, FL, USA, 2003; pp. 250–262.
37. Martín-Belloso, O.; Rojas-Graü, M.A.; Soliva-Fortuny, R. Delivery of flavor and active ingredients using edible films and coatings. In *Edible Films and Coatings for Food Applications*; Embuscado, M.E., Huber, K.C., Eds.; Springer: New York, NY, USA, 2009; pp. 295–313.
38. Bautista-Banos, S.; Hernandez-Lauzardo, A.N.; Velazquez-del Valle, M.G.; Hernandez-Lo, M.; Ait Barka, E.; Bosquez-Molina, E.; Wilson, C.L. Chitosan as a potential natural compound to control pre and postharvest diseases of horticultural commodities. *Crop. Prot.* **2006**, *25*, 108–118.
39. Jiang, Y.M.; Li, J.R.; Jiang, W.B. Effects of chitosan coating on shelf life of cold stored litchi fruit at ambient temperature. *LWT-Food Sci. Technol.* **2005**, *38*, 757–761.
40. Shiekh, R.A.; Malik, M.A.; Al-Thabaiti, S.A.; Shiekh, M.A. Chitosan as a novel edible coating for fresh fruits. *Food Sci. Technol. Res.* **2013**, *19*, 139–155.
41. Shahidi, F.; Arachchi, J.K.V.; Jeon, Y. Food application of chitin and chitosan. *Trends Food Sci. Tech.* **1999**, *10*, 37–51.

42. Moncayo, D.; Buitrago, G.; Algecira, N. The surface properties of biopolymer-coated fruit: A review. *Ing. Invest.* **2013**, *33*, 11–16.
43. Meng, X.H.; Li, B.Q.; Liu, J.; Tian, S.P. Physiological responses and quality attributes of table grape fruit to chitosan preharvest spray and postharvest coating during storage. *Food Chem.* **2008**, *106*, 501–508.
44. Romanazzi, G.; Nigro, F.; Ippolito, A.; di Venere, D.; Salerno, M. Effects of pre- and postharvest chitosan treatments to control storage grey mold of table grapes. *J. Food Sci.* **2002**, *67*, 1862–1867.
45. Ziani, K.; Fernandez Pan, I.; Royo, M.; Maté, J. Antifungal activity of films and solutions based on chitosan. *Food Hydrocoll.* **2009**, *23*, 2309–2314.
46. Aider, M. Chitosan application for active bio-based films production and potential in the food industry: Review. *LWT Food Sci. Technol.* **2010**, *43*, 837–842.
47. Benhabiles, M.S.; Drouiche, N.; Lounici, H.; Pauss, A.; Mameri, N. Effect of shrimp chitosan coatings as affected by chitosan extraction processes on postharvest quality of strawberry. *Food Meas.* **2013**, *7*, 215–221.
48. Lou, M.M.; Zhu, B.; Muhammad, I.; Li, B.; Xie, G.L.; Wang, Y.L.; Li, H.Y.; Sun, G.C. Antibacterial activity and mechanism of action of chitosan solutions against apricot fruit rot pathogen *Burkholderia seminalis*. *Carbohydr. Res.***2011**, *346*, 1294–1301.
49. Liu, L.; Liu, C.K.; Hicks, K.B. Composite films from pectin and fish skin gelatin or soybean flour protein. *J. Agric. Food Chem.* **2007**, *55*, 2349–2355.
50. El Ghaouth, A.; Arul, J.; Grenier, J.; Asselin, A. Antifungal activity of chitosan on two post-harvest pathogens of strawberry fruits. *Phytopathology* **1992**, *82*, 398–402.
51. Park, S.I.; Stan, S.D.; Daeschel, M.A.; Zhao, Y. Antifungal coatings on fresh strawberries (*Fragaria × ananassa*) to control mold growth during cold storage. *J. Food Sci.* **2005**, *70*, 202–207.
52. Chien, P.J.; Sheu, F.; Lin, H.R. Coating citrus (*Murcott tangor*) fruit with low molecular weight chitosan increases postharvest quality and shelf life. *Food Chem.* **2007**, *100*, 1160–1164.
53. Gonzalez-Aguilar, G.A.; Valenzuela-Soto, E.; Lizardi-Mendoza, J.; Goycoolea, F.; Martinez-Tellez, M.A.; Villegas-Ochoa, M.A.; Monroy-Garcia, I.N.; Ayala-Zavala, J.F. Effect of chitosan coating in preventing deterioration and preserving the quality of fresh-cut papaya "Maradol". *J. Sci. Food Agric.* **2009**, *89*, 15–23.
54. Ali, A.; Muhammad, M.T.M.; Sijam, K.; Siddiqui, Y. Effect of chitosan coatings on the physicochemical characteristics of *Eksotika*. II papaya (*Carica papaya* L.) fruit during cold storage. *Food Chem.* **2011**, *124*, 620–626.
55. Pilon, L.; Poliana, C.; Spricigo, M.M.; de Moura, M.R.; Assis, O.B.G.; Mattoso, L.H.C.; Ferreira, M.D. Chitosan nanoparticle coatings reduce microbial growth on fresh-cut apples while not affecting quality attributes. *Int. J. Food Sci. Technol.* **2015**, *50*, 440–444.
56. Benhabiles, M.S.; Tazdait, D.; Abdi, N.; Lounici, H.; Drouiche, N.; Goosen, M.F.A.; Mameri, N. Assessment of coating tomato fruit with shrimp shell chitosan and N,O-carboxymethyl chitosan on postharvest preservation. *Food Meas.***2013**, *7*, 66–74.
57. Assis, O.B.; Pessoa, J.D. Preparation of thin films of chitosan for use as edible coatings to inhibit fungal growth on sliced fruits. *Braz. J. Food Technol.* **2004**, *7*, 7–22.
58. Han, C.; Lederer, C.; McDaniel, M.; Zhao, Y. Sensory evaluation of fresh strawberries (*Fragaria ananassa*) coated with chitosan-based edible coatings. *J. Food Sci.* **2005**, *70*, S172–S178.
59. Durango, A.M.; Soares, N.F.F.; Andrade, N.J. Microbiological evaluation of an edible antimicrobial coating on minimally processed carrots. *Food Control* **2006**, *17*, 336–341.

60. Devlieghere, F.; Vermeulen, A.; Debevere, J. Chitosan: Antimicrobial activity, interactions with food components and applicability as a coating on fruit and vegetables. *Food Microbiol.* **2004**, *21*, 703–714.
61. Pushkala, R.; Raghuram, P.K.; Srividya, N. Chitosan based powder coating technique to enhance phytochemicals and shelf life quality of radish shreds. *Postharvest Biol. Tec.* **2013**, *86*, 402–408.
62. Jianglian, D.; Shaoying, Z. Application of chitosan based coating in fruit and vegetable preservation: A review. *J. Food Process. Technol.* **2013**, *4*, 227:1–227:4.
63. Yu, Y.W.; Li, H.; Jinhua, D.; Ren, Y.Z. Study of natural film with chitosan combining phytic acids on preservation of fresh-cutting lotus root. *J. Chin. Inst. Food Sci. Technol.* **2012**, *12*, 131–136.
64. Cong, F.; Zhang, Y.; Dong, W. Use of surface coatings with natamycin to improve the storability of Hami melon at ambient temperature. *Postharvest Biol. Technol.* **2007**, *46*, 71–75.
65. Zhang, H.; Li, R.; Liu, W. Effects of chitin and its derivative chitosan on postharvest decay of fruits: A Review. *Int. J. Mol. Sci.* **2011**, *12*, 917–934.
66. Salgado, P.R.; Ortiz, C.M.; Musso, Y.S.; di Giorgio, L.; Mauri, A.N. Edible films and coatings containing bioactives. *Curr. Opin. Food Sci.* **2015**, *5*, 86–92.
67. Speranza, B.; Corbo, M.R. Essential Oils for Preserving Perishable Foods: Possibilities and Limitations. In *Application of Alternative Food Preservation Technologies to Enhance Food Safety and Stability*; Bevilacqua, A., Corbo, M.R., Sinigaglia, M., Eds.; Bentham Publisher: Saif Zone, Sharjah, 2010; pp. 35–57.
68. Patrignani, F.; Siroli, L.; Serrazanetti, D.I.; Gardini, F.; Lanciotti, R. Innovative strategies based on the use of essential oils and their components to improve safety, shelf-life and quality of minimally processed fruits and vegetables. *Trends Food Sci. Technol.* **2015**.
69. Sánchez-González, L.; Cháfer, M.; Hernández, M.; Chiralt, A.; González-Martínez, C. Antimicrobial activity of polysaccharide films containing essential oils. *Food Control.* **2011**, *22*, 1302–1310.
70. Xing, Y.; Li, X.; Xu, Q.; Yun, J.; Lu, Y.; Tang, Y. Effects of chitosan coating enriched with cinnamon oil on qualitative properties of sweet pepper (*Capsicum annuum* L.). *Food Chem.* **2011**, *124*, 1443–1450.
71. Sessa, M.; Ferraria, G.; Donsìb, F. Novel edible coating containing Essential Oil nanoemulsions to prolong the shelf life of vegetable products. *Chem. Eng. Trans.* **2015**.
72. Randazzo, W.; Jimenez-Belenguer, A.; Settanni, L.; Perdones, A.; Moschetti, M.; Palazzolo, E.; Guarrasi, V.; Vargas, M.; Germanà, M.A.; Moschetti, G. Antilisterial effect of citrus essential oils and their performance in edible film formulations. *Food Control.* **2016**, *59*, 750–758.
73. Sangsuwan, J.; Rattanapanone, N.; Rachtanapun, P. Effect of chitosan/methyl cellulose films on microbial and quality characteristics of fresh-cut cantaloupe and pineapple. *Postharvest Biol. Technol.* **2008**, *49*, 403–410.
74. Krasaekoopt, W.; Mabumrung, J. Microbiological evaluation of edible coated fresh-cut cantaloupe. *Kasetsart. J. Nat. Sci.* **2008**, *42*, 552–557.
75. Corbo, M.R.; Speranza, B.; Campaniello, D.; D'Amato, D.; Sinigaglia, M. Fresh-cut fruits preservation: Current status and emerging technologies. In *Current Research, Technology and Education topics in Applied Microbiology*; Mendez-Villas, A., Ed.; Formatex Research Center: Badajoz, Spain, 2011; pp. 1143–1154.
76. Corbo, M.R.; Bevilacqua, A. Alternative Modified Atmosphere for Fresh Food Packaging. In *Application of Alternative Food Preservation Technologies to Enhance Food Safety and Stability*; Bevilacqua, A., Corbo, M.R., Sinigaglia, M., Eds.; Bentham Publisher: Saif Zone, Sharjah, 2010; pp. 196–204.
77. Siddiqui, M.W.; Chakraborty, I.; Ayala-Zavala, J.F.; Dhua, R.S. Advances in minimal processing of fruits and vegetables: A review. *J. Sci. Ind. Res.* **2011**, *70*, 823–834.
78. Church, N. Developments in modified-atmosphere packaging and related technologies. *Trends Food Sci. Technol.* **1994**, *5*, 345–352.

79. Farber, J.M. Microbiological Aspects of Modified-Atmosphere Packaging Technology—A Review. *J. Food Protect.* **1991**, *54*, 58–70.
80. Phillips, C.A. Review: Modified Atmosphere Packaging and Its Effects on the Microbiological Quality and Safety of Produce. *Int. J. Food Sci. Technol.* **1996**, *31*, 463–479.
81. Church, I.J.; Parsons, A.L. Modified atmosphere packaging technology: A Review. *J. Sci. Food Agric.* **1995**, *67*, 143–152.
82. Corbo, M.R.; Lanciotti, R.; Gardini, F.; Sinigaglia, M.; Guerzoni, M.E. Effects of hexanal, (E)-2-hexenal, and storage temperature on shelf life of fresh sliced apples. *J. Agric. Food Chem.* **2000**, *48*, 2401–2408.
83. Lanciotti, R.; Gianotti, A.; Patrignani, F.; Belletti, N.; Guerzoni, M.E.; Gardini, F. Use of natural aroma compounds to improve shelf-life and safety of minimally processed fruits. *Trends Food Sci. Technol.* **2004**, *15*, 201–208.
84. Campaniello, D.; D'Amato, D.; Corbo, M.R.; Sinigaglia, M. Combined action of hexanal and MAP on minimally processed cactus pear fruit. *Int. J. Food Sci.* **2005**, 1–10.
85. Siroli, L.; Patrignani, F.; Serrazanetti, D.I.; Tabanelli, G.; Montanari, C.; Tappi, S.; Rocculi, P.; Gardini, F.; Lanciotti, R. Efficacy of natural antimicrobials to prolong the shelf-life of minimally processed apples packaged in modified atmosphere. *Food Control.* **2014**, *46*, 403–411.
86. Fisher, K.; Phillips, C.; McWatt, L. The use of an antimicrobial citrus vapour to reduce *Enterococcus* sp. on salad products. *Int. J. Food Sci. Technol.* **2009**, *44*, 1748–1754.
87. Tian, J.; Ban, X.; Zeng, H.; He, J.; Huang, B.; Wang, Y. Chemical composition and antifungal activity of essential oil from *Cicuta virosa* L. var. *latisecta* Celak. *Int. J. Food Microbiol.* **2011**, *145*, 464–470.
88. Vitoratos, A.; Bilalis, D.; Karkanis, A.; Efthimiadou, A. Antifungal Activity of Plant Essential Oils against *Botrytis cinerea*, *Penicillium italicum* and *Penicillium digitatum*. *Not. Bot. Horti Agrobot.* **2013**, *41*, 86–92.
89. Chunyang, H.; Xiqing, Y.; Fei, L.; Binxin, S. Effect of High Oxygen Modified Atmosphere Packaging on Fresh-Cut Onion Quality. In Proceedings of the 17th IAPRI World Conference on Packaging, Tianjin, China, 12–15 October 2010.
90. Jacxsens, L.; Devlieghere, F.; Van der Steen, C.; Debevere, J. Effect of high oxygen modified atmosphere packaging on microbial growth and sensorial qualities of fresh-cut produce. *Int. J. Food Microbiol.* **2001**, *71*, 197–210.
91. Day, B.P.F. New modified atmosphere techniques fresh prepared fruit and vegetables. In *Fruit and Vegetables Processing*; Jongen, W., Ed.; Woodhead Publishing Limited: Cambridge, UK, 2002; pp. 310–330.
92. Amanatidou, A.; Smid, E.J.; Gorris, L.G.M. Effect of elevated oxygen and carbon dioxide on the surface growth of vegetable-associated micro-organisms. *J. Appl. Microbiol.* **1999**, *86*, 429–438.
93. Kader, A.; Ben-Yehoshua, S. Effects of superatmospheric oxygen levels on postharvest physiology and quality of fresh fruits and vegetables. *Postharvest Biol. Technol.* **2000**, *20*, 1–13.
94. Allende, A.; Jacxsens, L.; Devlieghere, F.; Debevere, J.; Artés, F. Effect of Superatmospheric Oxygen Packaging on Sensorial Quality, Spoilage, and *Listeria monocytogenes* and *Aeromonas caviae* Growth in Fresh Processed Mixed Salads. *J. Food Protect.* **2002**, *65*, 1565–1573.
95. Senesi, E.; De Regibus, P. Use of rectified apple juice as filling liquid to increase quality and shelf-life of fresh-cut apples (Succo di mela rettificato come liquido di governo per prolungare la curabilita' di spicchi di mela di IV gamma). *Ind. Aliment.* **2002**, *41*, 139–146.
96. D'Amato, D.; Sinigaglia, M.; Corbo, M.R. Use of chitosan, honey and pineapple juice as filling liquids for increasing the microbiological shelf life of a fruit-based salad. *Int. J. Food Sci. Technol.* **2010**, *45*, 1033–1041.

97. Galvez, A.; Abriouel, H.; Benomar, N.; Lucas, R. Microbial antagonists to food-borne pathogens and biocontrol. *Curr. Opin. Biotechnol.* **2010**, *21*, 142–148.
98. Vermeiren, L.; Devlieghere, F.; Debevere, J. Evaluation of meat born lactic acid bacteria as protective cultures for the biopreservation of cooked meat products. *Int. J. Food Microbiol.* **2004**, *96*, 149–164.
99. Siroli, L.; Patrignani, F.; Serrazanetti, D.I.; Tabanelli, G.; Montanari, C.; Gardini, F.; Lanciotti, R. Lactic acid bacteria and natural antimicrobials to improve the safety and shelf-life of minimally processed sliced apples and lamb's lettuce. *Food Microbiol.* **2015**, *47*, 74–84.
100. Carr, F.J.; Chill, D.; Maida, N. The lactic acid bacteria: A literature survey. *Crit. Rev. Microbiol.* **2002**, *28*, 281–370.
101. Abadias, M.; Usall, J.; Alegre, I.; Torres, R.; Vinas, I. Fate of *Escherichia coli* in apple and reduction of its growth using the postharvest biocontrol agent *Candida Sake* CPA-1. *J. Sci. Food Agric.* **2009**, *89*, 1526–1533.
102. Alegre, I.; Vinas, I.; Usall, J.; Anguera, M.; Altisent, R.; Abadias, M. Antagonistic effect of *Pseudomonas graminis* CPA-7 against foodborne pathogens in fresh-cut apples under simulated commercial conditions. *Food Microbiol.* **2013**, *33*, 139–148.
103. Trias, R.; Badosa, E.; Montesinos, E.; Baneras, L. Bioprotective *Leuconostoc* strains against *Listeria monocytogenes* in fresh fruits and vegetables. *Int. Microbiol.* **2008**, *127*, 91–98.
104. Trias, R.; Baneras, L.; Badosa, E.; Montesinos, E. Bioprotection of Golden Delicious apples and Iceberg lettuce against foodborne bacterial pathogens by lactic acid bacteria. *Int. J. Food Microbiol.* **2008**, *123*, 50–60.
105. Alegre, I.; Vinas, I.; Usall, J.; Anguera, M.; Figge, M.J.; Abadias, M. Control of foodborne pathogens on fresh-cut fruit by a novel strain of *Pseudomonas graminis*. *Food Microbiol.* **2013**, *34*, 390–399.
106. Leverentz, B.; Conway, W.S.; Janisiewicz, W.; Abadias, M.; Kurtzman, C.P.; Camp, M.J. Biocontrol of the food-borne pathogens *Listeria monocytogenes* and *Salmonella* Enterica serovar poona on fresh-cut apples with naturally occurring bacterial and yeast antagonists. *Appl. Environ. Microb.* **2006**, *72*, 1135–1140.
107. Alegre, I.; Vinas, I.; Usall, J.; Anguera, M.; Figge, M.J.; Abadias, M. An *Enterobacteriaceae* species isolated from apples controls foodborne pathogens on fresh-cut apples and peaches. *Postharvest Biol. Technol.* **2012**, *74*, 118–124.
108. Siroli, L.; Patrignani, F.; Salvetti, E.; Torriani, S.; Gardini, F.; Lanciotti, R. Use of a nisin-producing *Lactococcus lactis*strain, combined with thyme essential oil, to improve the safety and shelf-life of minimally processed lamb's lettuce. In Proceeding of the 11th International Symposium on lactic acid bacteria, Egmond aan Zee, The Netherlands, 31 August–4 September 2014.
109. Abadias, M.; Altisent, R.; Usall, J.; Torres, R.; Oliveira, M.; Vinas, I. Biopreservation of fresh-cut melon using the strain *Pseudomonas graminis* CPA-7. *Postharvest Biol. Technol.* **2014**, *96*, 69–77.
110. Plaza, L.; Altisent, R.; Alegre, I.; Vinas, I.; Abadias, M. Changes in the quality and antioxidant properties of fresh-cut melon treated with the biopreservative culture *Pseudomonas graminis* CPA-7 during refrigerated storage. *Postharvest Biol. Technol.* **2016**, *111*, 25–30.
111. Scolari, G.; Vescovo, M. Microbial antagonism of *Lactobacillus casei* added to fresh vegetables. *Int. J. Food Sci.* **2004**, *16*, 465–475.
112. Rößle, C.; Auty, M.A.E.; Brunton, N.; Gormley, R.T.; Butler, F. Evaluation of fresh-cut apple slices enriched with probiotic bacteria. *Innov. Food Sci. Emerg.* **2010**, *11*, 203–209.
113. Alegre, I.; Viñas, I.; Usall, J.; Anguera, M.; Abadias, M. Microbiological and physicochemical quality of fresh-cut apple enriched with the probiotic strain *Lactobacillus rhamnosus* GG. *Food Microbiol.* **2011**, *28*, 59–66.

114. Tapia, M.S.; Rojas-Graü, M.A.; Rodríguez, F.J.; Ramirez, J.; Carmona, A.; Martín-Belloso, O. Alginate and Gellan-based edible films for probiotic coatings on fresh-cut fruits. *J. Food Sci.* **2007**, *72*, 190–196.
115. Cleveland, J.; Montville, T.J.; Nes, I.F.; Chikindas, M.L. Bacteriocins: Safe, natural antimicrobials for food preservation. *Int. J. Food Microbiol.* **2001**, *71*, 1–20.
116. Allende, A.; Martınez, B.; Selma, V.; Gil, M.I.; Suarez, J.E.; Rodrıguez, A. Growth and bacteriocin production by lactic acid bacteria in vegetable broth and their effectiveness at reducing *Listeria monocytogenes in vitro* and in fresh-cut lettuce. *Food Microbiol.* **2007**, *24*, 759–766.
117. Randazzo, C.L.; Pitino, I.; Scifo, G.O.; Caggia, C. Biopreservation of minimally processed iceberg lettuces using a bacteriocin produced by *Lactococcus lactis* wild strain. *Food Control.* **2009**, *20*, 756–763.
118. Schuenzel, K.M.; Harrison, M.A. Microbial antagonists of foodborne pathogens on fresh, minimally processed vegetables. *J. Food Protect.* **2002**, *65*, 1909–1915.
119. Scolari, G.; Vescovo, M.; Zacconi, C.; Bonade, A. Influence of *Lactobacillus plantarum* on *Staphylococcus aureus* growth in a fresh vegetable model system. *Eur. Food Res. Technol.* **2004**, *218*, 274–277.
120. Rojas-Graü, M.A.; Sobrino-López, A.; Tapia, M.S.; Martín-Belloso, O. Browning inhibition in fresh-cut Fuji apple slices by natural antibrowning agents. *J. Food Sci.* **2006**, *71*, 59–65.
121. Soccol, C.R.; Vandenberghe, L.P.S.; Spier, M.R.; Medeiros, A.B.P.; Yamaguishi, C.T.; Lindner, J.D.; Pandey, A.; Soccol, V.T. The potential of probiotics. *Food Tech. Biotechnol.* **2010**, *48*, 413–434. []
122. Kourkoutas, Y.; Kanellaki, M.; Koutinas, A.A. Apple pieces as immobilization support of various microorganisms. *LWT-Food Sci. Technol.* **2006**, *39*, 980–986.
123. Shah, N.P. Functional cultures and health benefits. *Int. Dairy J.* **2007**, *17*, 1262–1277.
124. Kailasapathy, K. Microencapsulation of probiotic bacteria: Technology and potential application. *Curr. Issues Intest. Microbiol.* **2002**, *3*, 39–49.
125. Guerin, D.; Vuillemard, J.C.; Subirade, M. Protection of *Bifidobacteria* encapsulated in polysaccharide-protein gel beads against gastric juice and bile. *J. Food Protect.* **2003**, *66*, 2076–2084.
126. Kourkoutas, Y.; Xolias, V.; Kallis, M.; Bezirtzoglou, E.; Kanellaki, M. *Lactobacillus casei* cell immobilization on fruit pieces for probiotic additive, fermented milk and lactic acid production. *Process. Biochem.* **2005**, *40*, 411–416.
127. Saad, S.M.I.; Bedani, R.; Mamizuka, E.M. Benefícios à Saude dos Probióticos e Prebióticos. In *Probióticos e Prebióticos em Alimentos: Fundamentos e Aplicações Tecnológicas*; Saad, S.M.I., Cruz, A.G., Faria, J.A.F., Eds.; Editora Varela: São Paulo, Brazil, 2011; pp. 51–84, In Portuguese.

CHAPTER 8

Meatloaf with Semi-dry Vegetables: A Study of Processing and Preservation

Incoronato AL, Conte A, Gammariello D And Del Nobile Ma*

Service Center for Applied Research, University of Foggia, via Napoli, 25-71122 Foggia, Italy

ABSTRACT

This study evaluated the synergic effects of MAP, essential oils (nutmeg and laurel) and Na-lactate on shelf-life extension of a ready-to-cook fresh meal based on meatloaf and semi-dried vegetables. Microbial and sensory qualities, pH and gas composition were assessed during the storage at 4°C. Surface treatments of meat with Na-lactate solution alone or combined with the essential oils, under MAP, increased the shelf life to more than 5 and 7 days, respectively, compared to the control that remained acceptable for about 2 days. Both treatments were effective in suppressing spoilage bacteria, maintaining good surface color and preventing development of off-odor.

PRACTICAL APPLICATION: The current work represents an interesting study to develop new convenience food based on meat and vegetables with proper shelf life. The technology to realize the product was presented and some simple preservation strategies able to prolong the shelf life are also proposed. Both microbial and sensory quality were taken into account during storage, thus giving more detailed information on main quality factors that are responsible for product acceptability.

Keywords: Antimicrobial compounds; MAP; **Meat**; Ready-to-cook meal; Shelf life

INTRODUCTION

The meat loaf is a typical dish with an European origin. Meatloaf is made up of ground meat mixed with other ingredients, formed into a loaf shape, then baked or smoked. Meatloaf is usually made from ground beef, although pork, veal and poultry can be also used. The meatloaf is mainly distributed by the industry as whole piece but it can be possible to display meat slices packed in family-sized portions as ready-to-cook product. Slices preparation involves manipulation, and generally handling of food product can provoke contamination from the environment, handlers, equipment, and so forth. An additional problem is the temperature abuse which could occur during distribution, thus interrupting the refrigerated chain. Meat can be compromised by a large number of pathogenic and spoilage microorganisms during primary and further processing [1]. The contamination of sterile animal muscle is a direct consequence of slaughtering and dressing of animal carcasses. Wide ranges of microorganisms from different sources are transferred on meat surfaces that are rich in nutrients and becomes good substrate for microbial proliferation.

The number of microorganisms on fresh meat surfaces changes during chill storage following a typical microbial growth pattern. Counts of bacteria in meat are in the range of 10^2-10^5 cfu/cm^2, but only around 10% are able to grow. The initial lag phase is attributed to microbial adaptation to changing conditions (chill temperatures and surface desiccation). Ensuing logarithmic growth takes place after cells have accommodated to the new environmental setting and adapted their metabolism. When numbers exceed 10^7 cells per cm^2, the first spoilage signs are detected as off-odors. With respect to health and economic problems caused by these bacteria, it is very important to reduce the initial microbial population on meat. Sanitization of meat surfaces prior to grinding or prior to the production of batters may improve the control of food-borne pathogens or spoilage microorganisms. Various strategies have been developed to reduce the level of bacteria on surface of animals' carcass, such as washing and sanitizing with hot water, chlorinated water, food grade acids and salts [2].

Other studies have been carried out to optimize the technologies in different kind of meats, as beef, pork and poultry. Thus, applications of electrolyzed oxidizing water [3], high pressure in combination with antimicrobials [4], and surface sanitizers such as chlorine dioxide, cetyl-pyridinium chloride, and lactic acid [5,6] could extend shelf life of meats. Another surface treatment that is known to improve shelf life is the dipping in solutions of natural compounds. Compounds of plant origin and organic acid solutions or their salts actually appear to be the most promising means to control meat decay, above all when these techniques are used in combination with modified atmosphere packaging (MAP). The synergic effects of water-soluble rosemary extract and MAP on sensory and microbiological extension of turkey meatballs was also investigated by Karpinska-Tymoszyk [7]. The addition of rosemary extract gave rise to an improvement of sensory quality, in particular when stored under modified atmosphere. An increase of the preservative effect of MAP and essential oils (Laurus nobilis) has been shown in round chicken breast meat stored at 4°C [8]. Zeitoun [9] found the synergistic effect of MAP, lactic acid and sodium lactate antimicrobial activity on fresh chicken legs. Gammariello et al. [10,11] demonstrated that the combined effect of sodium lactate and MAP improved the shelf-life of ready-to-cook fresh skewer, made up of turkey meat. The salt of lactate is frequently used as an antimicrobial in meat products, due to its beneficial properties when applied at appropriate concentrations. The addition of lactate to food products with neutral pH offers good prospects for shelf-life prolongation. In general, Gram-positive bacteria were more sensitive toward lactate than Gram-negative bacteria under optimum growth conditions (pH 6.5, 20°C) [12]. In this context, research on the effects of MAP in combination with natural compounds on the quality of ready-to-cook meat products is still limited. Therefore, the objective of this study is to evaluate the effect of MAP and addition of laurel, nutmeg and sodium lactate on the microbiological and sensory characteristics of ready-to-cook meal during chilled storage.

MATERIALS AND METHODS

Samples Preparation

The ready-to-cook meal was made up at laboratory scale with pork ham as meatloaf with semi-dried vegetables (artichokes, peppers and tomatoes). Sodium-lactate (Na-lactate, sodium salt of lactic acid, 60% w/w) (Giusto Faravelli Spa, Milano, Italy) was used to pre-treat meat before processing and different MAP were applied during packaging. Meat was provided by a meat company, Dodaro Spa (Spezzano Albanese, Cs, Italy) and transported to the laboratory in polystyrene boxes with an appropriate quantity of flaked ice. Semi-dry vegetables were purchased from a local company, Fiordelisi (Fiordelisi srl, Stornarella, Fg, Italy), and transported to the laboratory in boxes. The meat pieces, prior to grinding, were subjected to two treatments by dipping. In a first step, meat samples were cut into small cubes (approximately 30-40 g), and were dipped for 10 min in pre-chilled (4°C) aqueous solution of Na-lactate at the concentration of 60% (w/w). After treatment, meat samples were drained for 5 min at ambient temperature (20°C). In a second step, Nutmeg and Laurel essential oils (Farmalabor s.r.l., Canosa di Puglia, BAT, Italy) were dissolved in 100 mL of extra-virgin olive oil (EVO) at the concentration of 1.25 g/100 mL and 2.5 g/100 mL, respectively. The applied amount of solutions was 10% referred to the meat weight (1 kg meat/100 mL solution of EVO and essential oils). Solutions of EVO and essential oils were completely absorbed by the meat samples, while the amount not absorbed was mixed to the meat as an ingredient. No treated samples were used as reference.

Once antimicrobial treatments were completed, meat cubes were ground separately. Before use, all metal parts of the meat grinder were dipped in aqueous solution of Na-lactate (60% w/w) for 25 min. The ground meat was prepared by mixing for 4-5 minutes all the ingredients with a food processor in the following amounts: sodium chloride (1%), Parmigiano-Reggiano cheese (3%), grated bread (3.5%), garlic (0.03%), parsley (2%) and white bread (10%) with the addition of milk (20%) and essential oils. The dough meat was spread on a sheet of baking paper (1 cm thickness) and was distributed on the surface of the semi dried vegetables previously seasoned (1 Kg of dough meaty/300 g of vegetables); subsequently, the meatloaf was sliced obtaining slices of 1.5-2 cm thickness. Artichokes, peppers and tomatoes were seasoned with sodium chloride (1,6% w/w), garlic (1% w/w), parsley (1% w/w) and extra-virgin olive oil (33, 3% w/w). The seasoned vegetables were left at room temperature with intermittent agitation for at least 2 h, to allow the uniform distribution of the ingredients.

Each seasoned meat slice was packaged in bag hermetically sealed by means of an S100-Tecnovac equipment (Tecnovac, Bergamo, Italy) under MAP and stored at 4°C. The bags were 220×320 mm long, and constituted by an anti-fog high-barrier multilayer film provided by Di Mauro (Officine Grafiche spa, Salerno, Italy). To realize MAP, the following gas concentrations were used: 30:70 CO_2:N_2.

The treatments applied were namely as follows: CNT-MA (slices of meatloaf without dipping and packaged under MAP); SL60-MA (slices of meatloaf dipped with 60% Na-lactate solution and packaged under MAP); EO-MA (slices of meatloaf dipped with mix of essential oils and packaged under MAP); SL60-EO-MA (slices of meatloaf dipped with 60% Na-lactate solution and essential oils solution and packaged under MAP).

The samples were withdrawn at regular intervals for microbiological and sensory evaluation, pH measurements and gas monitoring.

Gas Composition of The Package Headspace

Prior to opening the bags, headspace gas composition was determined by using a Checkmate 9900 gas analyzer (PBI Dansensor, Ringsted, Denmark). The volume taken from the package headspace for gas analysis was about

10 cm³. To avoid modifications in the headspace gas composition due to gas sampling, each package was used only for a single determination of the headspace gas composition. Three bags were used for each analysis.

Microbiological Analyses and Ph Determination

Samples of 25 g (each consisting of equal amounts of both meat and vegetable pieces) were aseptically removed from the bags transferred into a stomacher bag and homogenized for 3 min (Stomacher LAB Blender 400, Pbi International, Milan, Italy) with 225 ml of sterile peptone solution (Oxoid, Milan, Italy). Further decimal serial dilutions were prepared from this homogenate in the same sterile diluent and these were subsequently used for enumeration and differentiation of microorganisms. Total aerobic bacteria (TAB) were determined using Plate Count Agar after incubation for 48 h at 30°C. Pseudomonas spp. were enumerated on Pseudomonas Agar Base supplemented with cetrimide, fucidine and cephaloridine (CFC) supplements providing a selective isolation medium for Pseudomonas spp. Colonies were counted after 48 h at 25°C. Lactic Acid Bacteria (LAB) were enumerated by the pour plating technique on de Man Rogosa Sharpe agar and incubated under anaerobiosis at 30°C for 48 h. Enterobacteriaceae counts were enumerated by the pour plating technique on Violet Red Bile Glucose Agar after incubation at 37°C for 24 h. Media and supplements for the microbiological analyses were procured from Oxoid (Milan, Italy). The microbiological analyses were carried out twice on two different batches. The results were expressed as the logarithm of the colony forming units for gram (log cfu/g).

In order to determine the microbiological acceptability limit (MAL defined as the time at which the microbial loads reached the maximum value permitted), the re-parameterized version of the Gompertz equation was fitted to the total aerobic bacterial data (TAB), according to a similar approach also used in previous works dealing with shelf life of meat [13,14]. In particular, the following equation (1) was used to calculate the MAL:

following equation (1) was used to calculate the MAL:

$$\log(N(t)) = \log(N_{\max}) - A.\exp\left\{-\exp\left\{\left[(\mu_{\max}.2.71).\frac{\lambda - MAL}{A}\right]+1\right\}\right\}+1^{-\exp}$$

$$+A.\exp\left\{-\exp\left\{\left[(\mu_{\max}.2.71).\frac{\lambda - t}{A}\right]+1\right\}\right\}$$

where N(t) is the viable cell concentration (CFU/g) at storage time t, A is related to the difference between the decimal logarithm of maximum bacterial growth attained at the stationary phase and the decimal logarithm of the initial cell load concentration (CFU/g), μ_{max} is the maximal specific growth rate (Δ log [CFU/g]/day), λ is the lag time (day), t is the time (day), N_{max} is the threshold value (CFU/g), MAL is the microbial acceptability limit (day) (i.e., the storage time at which the N(t) equals N_{max}). For TAB count the critical limit was set to 5×10^6 cfu/g, according to the 94/65 EC and 2073/2005 EC regulations.

The pH values were recorded on each sampling day using a pH meter (Crison Instruments model 2001, Barcelona, Spain). The pH determination, conducted twice, was performed on the first homogenized dilution of each sample.

Sensory Analysis

A panel of seven experienced judges was used to evaluate colour, odour and overall quality of ready-to-cook meal samples on an 8-point scale [15]. Samples with a score equal or higher than 4 were considered acceptable. A re-parameterized Gompertz equation was fitted to the sensory data [13,14] to allow calculating the sensory acceptability limit (SAL), defined as the time at which the overall quality of product reached the threshold (score = 4).

In particular, the following equation (2) was used to calculate the SAL:

$$OSQ(t) = OSQ_{min} - A^Q.\exp\left\{-\exp\left\{\left[\left(\mu_{max}{}^Q.2.71\right).\frac{\lambda^Q - SAL}{A^Q}\right]+1\right\}\right\}+$$

$$+A^Q.\exp\left\{-\exp\left\{\left[\left(\mu_{max}{}^Q.2.71\right).\frac{\lambda^Q - t}{A^Q}\right]+1\right\}\right\}$$

where OSQ(t) is the overall sensory quality at time t, A^Q is related to the difference between the sensorial attribute attained at the stationary phase and the initial value of sensorial attribute, is the maximal rate at which OSQ(t) decreases, λ^Q is the lag time, OSQ_{min} is sensory threshold, SAL is the sensorial acceptability limit (i.e., the time at which OSQ(t) is equal to OSQ_{min}), and t is the storage time. As reported above, the value of OSQ_{min} is equal to 4.

Shelf Life Calculation

The shelf life was calculated as the lowest value between MAL and SAL. In fact, it is the time at which one of the meat quality sub-indices reaches its threshold [14].

Statistical Analysis

The fitting parameters were compared by one-way Anova analysis. The Tukey's test was used to determine significance among differences ($P<0.05$). To this aim, Statistica 7.1 for Windows 152 (StatSoft Inc., Tulsa, OK, USA) was used.

RESULTS AND DISCUSSION

Microbiological Quality of Meatloaf With Semi-Dried Vegetables

Figure 1 shows the viable cell concentration of TAB plotted as a function of storage time for all samples. The curves were obtained by fitting the experimental data, whereas the horizontal solid line is the viable cell concentration threshold. The MAL values, obtained according to the procedure reported in the Materials and Methods section, were listed in Table 1. Data highlight that MAL was affected by the synergy approach. In fact, the presence of essential oils alone did not affect significantly the MAL, while the dipping of Na-lactate and essential oils in synergy with MAP inhibited the cell load of TAB. As can be inferred from the figure, while the viable cell concentration of the CNT-MA sample went above the selected threshold (5×10^6 cfu/g for TAB) already after more than 2 days, the treatments with Na-lactate provoked a significant reduction in microbial growth. The initial microbial load ranged from 5.90 log cfu/g in the CNT-MA to 4.97 log cfu/g in SL60-EO-MA. This indicates that the synergistic effect of treatments significantly delayed the rate of microbial growth and extended the shelf life of the ready-to-cook product. For meat products, microbial contamination occurs at the surface. The use of sanitizing agents or antimicrobial to decontaminate meat surfaces has been proved in the literature [16]. Vander marel et al. [17] investigated the effect of immersing carcasses in lactic acid for 15 seconds at various stages of processing on surface TPC, total psychrophiles, Enterobacteriaceae and *S. aureus*, as well as their growth at 0°C for 23 days. They reported that immediately after treatment the colonization was generally reduced by 1 log. In the specific case, all investigated samples treated with Nalactate showed a significant prolongation of the microbial lag phase, as well as a substantial log reduction of about 1 cycle in comparison with the control sample. Indeed, treated samples recorded a MAL value of 8 for SL60-MA and 10.22 days for SL-60-EO-MA, respectively. The presence of EO reduced the initial population of TBA until the 2nd day of storage when compared to the CNT-MA sample. Successively, the cell load of EO-MA exceeded the maximal value recommended, thus giving a MAL value of about 3 days. It has been documented that Nalactate has the ability to extend shelf-life of meat products. There are three proposed mechanisms by which the sodium lactate ion can have an antimicrobial affect; the first is the change of water activity (aw), the second is the sodium lactate passing through the cell membrane and lowering intracellular pH, and the third one is the inhibition of ATP generation to affect cellular metabolism [18,19].

Table 1: Microbial acceptability limit (MAL), sensory acceptability limit (SAL) and shelf life (mean ± SD) of ready-to-cook meal

Samples	MAL (day)	SAL (day)	Shelf life (day)
CNT-MA	2.82 1.96^{a}	4.94 ± 0.39^{a}	2.82 ± 1.96^{a}
EO-MA	3.04 ±1.76^{a}	5.39 ± 0.21^{ab}	3.04 ± 1.76^{a}
SL60-MA	8.03 ± 1.65^{b}	5.93 ± 0.17^{b}	5.93 ± 0.17^{b}
SL60-EO-MA	10.23 ± 2.40^{b}	7.07 ± 0.45^{c}	7.07 ± 0.45^{c}

$^{a-c}$Data in the column with different superscript letters are significantly different (P<0.05)

With regard to Pseudomonas spp. (data not shown), while there was a decrease in cell load of treated samples during the first 5 days of storage followed by a slight increase at the end of the storage, the CNT-MA showed a gradual growth. Concerning Enterobacteriaceae growth (data not shown) for all samples there were rather irregular trends during the early days of observation. The initial count of microorganisms was approximately 2.72 and 2.91 log cfu/g for EOMA and SL-60-EO-MA, respectively, whereas the CNT-MA (1.45 log cfu/g) and SL60-MA (1.39 log cfu/g) showed low Enterobacteriaceae contamination. At storage day 2 however, significant lower count was detected in samples treated with Na-lactate (1.93 in SL60-MA and 1.75 log cfu/g in SL-60-EO-MA) when compared with CNT-MA (2.30) or EO-MA (2.32 log cfu/g). Successively, all samples showed a slight increase, except for SL-60-EO-MA that showed a decrease with a final microbial load of about 1.15 cfu/g. This result might have been due to the synergistic effect of the approach used. Data published in the available literature also demonstrate the antimicrobial effect of Nalactate on both Enterobacteraiaceae and Pseudomonas spp. [20].

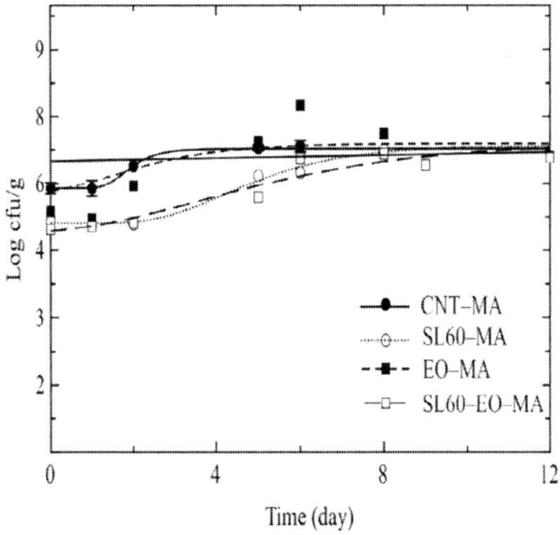

The curves are the best fit to the experimental data. CNT-MA(control sample packaged in MAP); EO-MA (sample treated with essential oils, packaged in MAP); SL60-MA (sample treated with Na-lactate 60%, packaged in MAP); SL60-EO-MA (sample treated with Na-lactate 60% and essential oils, packaged in MAP).

Figure 1: Evolution of total aerobic bacteria during the storage time.

With regard to LAB there was a slight decrease in the initial count of the treated samples (Figure 2). After the first day, there were no marked differences between the CNT-MA and EO-MA, whereas for SL60-MA and SL-60-EO-MA was observed a period of stability at the beginning, followed by an increase in cell numbers after the second day. Afterwards, the growth of LAB reached the stationary phase with a cell count similar to that of the CNT-MA. However, a combination of Na-lactate, essential oil and MAP appeared to be the most effective among the other single treatments against LAB growth. It has been documented that Na-lactate is also effective against LAB in meat products [21,22]. Conversely, it has been reported that LAB dominated the microbial flora in NA-lactate treated beef during vacuum-packaged storage at 0°C [23].

PH AND HEADSPACE GAS COMPOSITION

Changes in pH values of the ready-to-cook product during storage at 4°C revealed that the initial pH value of CNT-MA (6.10) was higher than those of samples treated with EO (5.86), Na-lactate (5.79) or combination of Na-lactate and EO (5.66) (data not shown). Successively, the pH values of samples treated with Na-lactate remained relatively constant throughout the rest of the study period, while the pH of CNT-MA and EO-MAP samples significantly decreased. These findings disagree with Eckert et al. [24] and

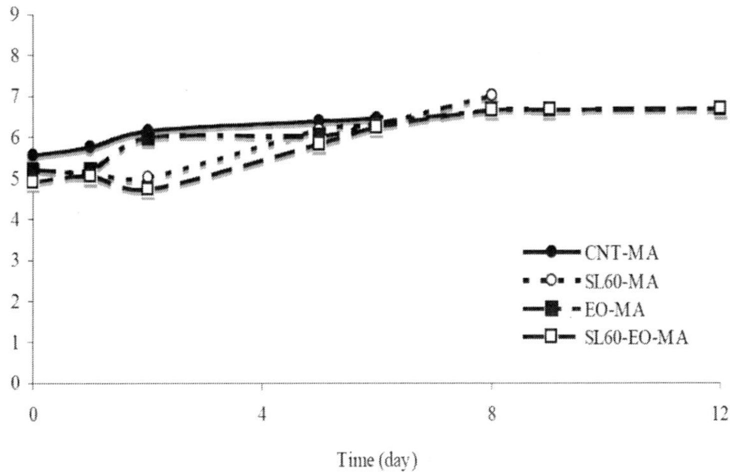

CNT-MA (control sample packaged in MAP); EO-MA (sample treated with essential oils, packaged in MAP); SL60-MA (sample treated with Na-lactate 60%, packaged in MAP); SL60-EO-MA (sample treated with Na-lactate 60% and essential oils, packaged in MAP)

Figure 2: Evolution of LAB during the storage time.

Tan and Shelef [25], who reported that NaL had no significant effects on initial pH of ground meat products. In addition, these results are consistent with Sallam and Samejima [26], who found that addition of Na-lactate maintained the ground beef at almost constant pH, while the pH of control or NaCltreated samples significantly decreased. Na-lactate was reported to have neutral pH during storage of meat products and is used as a pH control agent [23,27,28].

The headspace atmosphere (data not shown) has undergone minor changes during the storage period and showed a slight fluctuation for all the samples under MAP. In general, the CO_2 concentration of samples treated with Na-lactate steady decreased, whereas for the reference samples showed a gradual decline until the second day of observation, following by a rapid increase along the storage period. The CO_2 concentration in EO-MA was slightly lower with respect to that of sample without any treatments. These findings were expected and are in agreement with what found by Esmer et al. [29] who noted that the changes of gas composition might be due to microbial growth, permeability of packaging material and gas absorption by the meat. Jakobsen and Bertelsen [30] also reported that the decrease of CO_2 in MAP was the result of the absorption of CO_2 into the meat. Nychas [2] concluded that during storage, some microorganisms in the meat utilized the O_2 in the headspace of tray while some microflora, particularly LAB, produce CO_2. Conversely, the trend of the N_2 is inverse to that of the CO_2 for all samples. In terms of food spoilage, N_2 is an inert gas, not reactive with meat pigments, not absorbed by the meat and tasteless, without any antimicrobial activity. It is not very soluble in water, and it is primarily used to displace oxygen and prevent package collapse [31].

SENSORY EVALUATION

Figure 3 gives the overall quality plotted as a function of storage time for all the samples during storage at 4°C. The curves shown in the figure were obtained by the fitting to the experimental data, whereas the horizontal solid line is the sensory threshold. SAL values relative to the overall quality of product were listed in the second column of Table 1 for the investigated samples. These data were calculated only when the sensory attribute judgment was below 4. As expected, the quality of tested readyto- cook meatloaf steadily decreased, regardless of the strategy adopted. As can be seen, no significant differences were observed between CNTMA (4.94 days) and EO-MA (5.38) samples, while they differed from SL60-MA (5.93) and SL-60-EO-MA (7.07) samples. The results obtained from sensorial evaluations showed that the organoleptic properties of the samples were not adversely affected by the treatments, making the product much more acceptable than the untreated samples. Although the use of Na-lactate seemed to be beneficial for meat odor, this was counterbalanced by an undesirable discoloration. In particular, the color was the sensorial attribute that limited the overall quality of samples. Though other organic acids can be used as antimicrobials, lactate is successful as antimicrobial agent because it is known to enhance meat flavor due to the salty taste that it provides, while retaining color, and contributing to increase water holding capacity [32]. Other researchers have also reported the benefits of Na-lactate on beef and pork flavor [21,23,33,34].

SHELF LIFE

The shelf life values are also reported in **TABLE 1**. In the case under investigation, the shelf life of each tested sample was calculated as the lowest value between the MAL evaluated on the basis of TBA and the SAL on the basis of the overall quality. Sensory data were not in a very good agreement with microbiological data. As can be inferred from the table, for the CNT-MA and EO-MA samples, the proliferation of TBA represented the limiting factor for product acceptability, whereas, for the SL60-MA and SL-60-EO-MA treated with Na-lactate, the factor limiting shelf life was the sensorial quality. In particular, for SL60-MA and SL-60-EO-MA samples

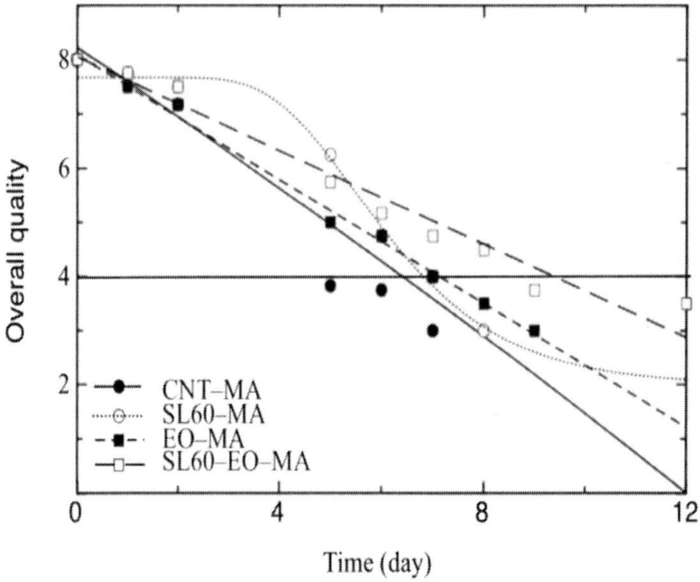

CNT-MA (control sample packaged in MAP); EO-MA (sample treated with essential oils, packaged in MAP); SL60-MA (sample treated with Na-lactate 60%, packaged in MAP); SL60-EO-MA (sample treated with Na-lactate 60% and essential oils, packaged in MAP)

Figure 3: Evolution of the overall quality during the storage.

shelf life values more than 5 and 7 days, respectively, were obtained compared to the other samples for which shelf life values of about 3 days were reached. The percentage shelf life increase of each tested sample was also calculated, and values of 7.44%, 110% and 150% were estimated for EO-MA, SL60-MA and SL-60- EO-MA samples, respectively, when compared to the control. Results are in agreement to Ahmed et al. [35] who reported that the shelf-life of meats from freshly slaughtered sheep and goat carcasses at 5–7°C was extended after spraying the carcasses with a solution containing potassium sorbate, sodium acetate, sodium citrate, sodium lactate each at 2.5% and sodium chloride at 5%.

CONCLUSIONS

Consumer's rising interest for ready-to-cook products has motivated researchers to standardize the processing process to give proper shelf life. The results of this study indicated that the combination of antimicrobial treatments and MAP slowed down microbial growth, allowed maintaining good sensory characteristics. In particular, the essential oils, Na-lactate and MAP inhibited the cell load of TAB, thus prolonging the microbiological stability. These findings suggest that it is advisable to implement additional measures, such as decontamination procedures, in order to prolong the shelf-life and increase the safety of this foodstuff.

ACKNOWLEDGEMENTS

This work was financially supported by Ministry of Economy and Finance , Ministry of Education , University and Scientific Research and Technology and the Department of Budget and Planning Region Puglia by the program PON 01_01409 - Process innovation and product to increase the safety profiles and to diversify the range of products (fresh and cured) in pig meat (SAFEMEAT).

REFERENCES

1. Marshall DL, Bal'a MFA (2001) Microbiology of Meat. In: Hui YH, Ed. Meat Science and Applications, Marcel Dekker, Inc., CRC Press New York, USA
2. Nychas GJE (1994) Modified atmosphere packaging of meat. In: Singh RP, Oliviera FAR (Eds) Minimal processing of foods and process optimization. An Interface CRC Press, London
3. Fabrizio KA, Cutter CN (2005) Application of electrolyzed oxidizing water to reduce Listeria monocytogenes on ready-to-eat meats. Meat Sci 71: 327-333
4. Hugas M, Garriga M, Monfort JM (2002) New mild technologies in meat processing: High pressure as a model technology. Meat Sci 62: 359-371
5. Jimenez-Villarreal JR, Pohlman FW, Johnson ZB, Brown AHJR (2003)Effects of chlorine dioxide, cetylpyridinium chloride, lactic acid and trisodium phosphate on physical, chemical and sensory properties of ground beef. Meat Sci 65: 1055-1062
6. Jimenez-Villarreal JR, Pohlman FW, Johnson ZB, Brown AHJR (2003) Lipid, instrumental color and sensory characteristics of ground beef produced using trisodium phosphate, cetylpypiridinium chloride, chlorine dioxide or lactic acid as multiple antimicrobial interventions. Meat Sci 65: 885-891
7. Karpińska-Tymoszczyk M (2011) The effect of water-soluble rosemary extract, sodium erythorbate, their mixture and packaging method on the quality of turkey meatballs. Int J Food Sci 23: 318-323
8. Irkin R, Esmer OK (2013) Control of Listeria monocytogenes in ground chicken breast meat under aerobic, vacuum and modified atmosphere packaging conditions with or without the presence of bay essential oil at 4 °C. Food Sci Technol Res 16: 285-290
9. Zeitoun AAM, Debevere JM (1992) Decontamination with lactic acid/sodium lactate buffer in combination with modified atmosphere packaging effects on the shelf life of fresh poultry. Int J Food Microbiol 16: 89-98
10. Gammariello D, Laverse J, Incoronato AL, Conte A, Contò F, et al. (2014) A multi-step optimisation approach to extend burger shelf life. Basic Res J Food Sci Tech 1: 15-28
11. Gammariello D, Incoronato AL, Conte A, Del Nobile MA (2015) Use of sodium lactate and modified atmosphere packaging for extending the shelf life of ready-to-cook fresh meal. Packag Technol Sci 28: 101-112

12. Houtsma PC, De Wit JC, Rombouts FM (1993) Minimum inhibitory concentration (MIC) of sodium lactate for pathogens and spoilage organisms occurring in meat products. Int J Food Microbiol 20: 247-257
13. Del Nobile MA, Conte A, Cannarsi M, Sinigaglia M (2009) Strategies for prolonging the shelf life of minced beef patties. J Food Safety 29: 14-25
14. Mastromatteo M, Incoronato AL, Conte A, Del Nobile MA (2011) Shelf life of reduced pork back-fat content sausages as affected by antimicrobial compounds and modified atmosphere packaging. Int J Food Microbiol. 150: 1-7
15. Das AK, Anjaneyulu ASR, Verma AK, Kondaiah N (2008) Effect of full-fat soy paste and soy granules on quality of goat meat patties. Int J Food Sci Tech 43: 383-392
16. Smulders F (1987) Prospectives for microbial decontamination of meat and poultry by organic acids with special reference to lactic acid. In: Smulders FJM (Ed) Elimination of Pathogenic Organisms from Meat and Poultry Proceedings of the International Symposium : Prevention of Contamination, and Decontamina. Elsevier Science, Biomedical Division, London
17. Van der Marel GM, Van Logtestijn JG, Mossel DA (1988) Bacteriological quality of broiler carcasses as affected by in-plant lactic acid decontamination. Int J Food Microbiol 6: 31-42
18. Bacus J (1989) Microbial control methods in fresh and processed meats. 41st Annual American Meat Science Association Reciprocal Meat Conference, Laramie, Wyoming
19. Maas MR, Glass KA, Doyle MP (1989) Sodium lactate delays toxin production by clostridium-botulinum in cook-in-bag turkey products. Appl Environ Microb. 55: 2226-2229
20. Incoronato AL, Gammariello D, Conte A, Del Nobile MA (2014) Technological solutions to increase shelf life of fresh meat burger. J Food Process Pres
21. Brewer MS, McKeith FK, Sprouls G (1993) Sodium lactate effects on microbial, sensory, and physical characteristics of vacuumpackaged pork sausage. J Muscle Foods 4: 179-192
22. Sameshima T, Takeshita K, Miki T, Arihara K, Itoh M, et al. (1997) Effect of sodium nitrite and sodium lactate on the growth of lactic acid spoilage bacteria isolated from cured meat products. Jpn J Food Microbiol 13: 159-164
23. Papadopoulos LS, Miller RK, Acuff GR, Vanderzant C, Cross HR (1991) Effect of sodium lactate on microbial and chemical composition of cooked beef during storage. J Food Sci 56: 341-347
24. Eckert LA, Maca JV, Miller RK, Acuff GR (1997) Sensory, microbial and chemical characteristics of fresh aerobically stored ground beef containing sodium lactate and sodium propionate. J Food Sci 62: 429-433
25. Tan W, Shelef LA (2002) Effects of sodium chloride and lactates on chemical and microbiological changes in refrigerated and frozen fresh ground pork. Meat Sci. 62: 27-32
26. Sallam KI, Samejima K (2004) Microbiological and chemical quality of ground beef treated with sodium lactate and sodium chloride during refrigerated storage. Lebensm Wiss Technol 37: 865-871
27. Bingol EB, Bostan K (2007) Effect of sodium lactate on the microbiological quality and shelf life of sausages. Turk J Vet Anim Sci 31: 333-339
28. Maca JV, Miller RK, Acuff GR (1997) Microbiological, sensory and chemical characteristics of vacuum-packaged ground beef patties treated with salts of organic acids. J Food Sci 62: 591-596
29. Esmer OK, Irkin R, Degirmencioglu N, Degirmencioglu A (2011) The effects of modified atmosphere gas composition on microbiological criteria, colour and oxidation values of minced beef meat. Meat Sci 88: 221-226
30. Jakobsen M, Bertelsen G (2002) The use of CO2 in packaging of fresh red meats and its effect on chemical quality changes in the meat: A review. J Muscle Foods 13: 143-168
31. Mastromatteo M, Conte A, Del Nobile MA (2010) Combined use of modified

32. Doores S (2005) Organic acids. Antimicrobials in food. In: Davidson PM, Sofos JN, Branen AL Boca Raton, FL (Eds), CRC Press, Taylor and Francis Group
33. Lamkey JW, Leak FW, Tuley WB, Johnson DD, West RL (1991) Assessment of sodium lactate addition to fresh pork sausage. J Food Sci 56: 220-223
34. Sutton DS, Brewer MS, McKeith FK (1997) Effects of sodium lactate and sodium phosphate on the physical and sensory characteristics of pumped pork loins. J Muscle Foods 8: 95-100
35. Ahmed SN, Chattopadhyay UK, Sherikar AT, Waskar VS, Paturkar AM, et al. (2003) Chemical sprays as a method for improvement in microbiological quality and shelf-life of fresh sheep and goat meats during refrigeration storage (5–7°C). Meat Sci 63: 339-344

CHAPTER 9

Measured and Predicted Temperature of Maize Grain (*Zea Mays* L.) Under Hermetic Storage Conditions

Rashid A. Suleiman and Kurt A. Rosentrater

Department of Agricultural and Biosystems Engineering, Iowa State University, 3327 Elings Hall, Ames, IA 50011, USA.

ABSTRACT

The physical properties of grain, such as temperature and moisture content are two key factors in grain storage. Temperature and relative humidity are two crucial factors for stored grain. In this study, three different storage conditions (room temperature at , 25°C; cooling at , 4 °C; and freezing at , -20°C) were investigated. Yellow dent corn (Zea mays L.) maize(variety Blue River 571136) from Iowa, harvested in 2011 was used. Maize grain was stored in two hermetical sealed binsarrels (50-cm diameter x 76-cm height). Five logger sensors were installed inside the cylindrical bin to measure temperature and relative humidity of the maize graingrain. The sensors were located at the top, center, bottom, left and right at about 12 cm part. After placing each barrel into storage condition, temperature and relative humidity values were measured every minute for 9 days throughout the duration of the experiment. Model validation was carried out by comparing predicted with measured maize grain temperature data data in three differences points of plastic cylindrically bin the radial and vertical directions. arrel.The temperature in the hermetically sealed cylindrical bins varied, mostly in the radial direction and very little in the axial vertical directions. No noticeable change in temperature was observed in room condition. Moreover, the temperature in the grain changed more rapidly in the freezing conditions than in the room and cooling conditions. Furthermore, the lag time between the center temperature and the side (right, left, top, and bottom) was greater in the radial direction as compared to vertical temperature. Model validation was carried out by comparing predicted with measured maize temperature data in three differences points of plastic cylindrically barrel. Predicted data were closely followed measured data. The model can be used to predict the grain temperature changes in room, cooling and freezing under hermetic conditions.The maximum difference between predicted and measured temperature was ±1.5°C. The predicted and measured values of maize grain temperature at radial and vertical directions were found to be in good agreement. The model shows a good potential application to predict the temperature of maize grain stored at room, cooling and freezing conditions under hermetic storage.

Keywords: Maize; mathematical model, heat transfer; ggrain temperature, grain storage, hhermetic storage, modeling.

INTRODUCTION

The knowledge of physical and thermal properties of grain is essential to the food engineers, processors and grains ecologists for the effective designing of machine, storage structures, heat transfer optimization and bulk storage (Amin et al., 2004; Mohamed, 2009). The physical properties of grain such as moisture content and temperature are two key factors in grain storage. Temperature and moisture content are the main causes of grain spoilage in stored grain ecosystems (Manickavasagan et al., 2006). As reported by many researchers, moisture content and temperature are the two key factors in maintaining grain quality during handling, storage and are the major sources of grain deterioration, because they encourage the growth of mold and infestation of insects (Parry, 1985; Jin, 1996; Flinn et al., 1997). Likewise, as mentioned by Hong et al., (1997) and Ng (1994), temperature, moisture content and relative humidity are three major factors influence storage conditions of grain. Furthermore Jayas and White (2003) revealed that temperature and moisture content are the two physical factors that control deterioration and quality of grain in storage. In addition White (1995) reported that the growth and multiplication of biological agents such as insect and mold in grains are highly dependent on the presence of temperature and moisture content.

Moreover, temperature and moisture content can be used to predict grain drying and deterioration potential (Iguaz et al., 2000). Temperature is regarded as a physical variable in grain storage, the variability is from from two main sources: internal and external. Internal are those from grain respiration (Equation 1), insects, mites, rodents and other microorganisms. While external sources are mainly from the solar radiation and surrounding area around the storage bin (Jia et al., 2000; Andrade et al., 2002).

$$C_6H_{12}O_6 + 6O_2 \rightarrow 6CO_2 + 6H_2O + 2834\ \text{kJ} \qquad (1)$$

In addition, grain kernels like other living substances continue to respire after harvest, releasing CO_2, water vapor and heat (Iguaz et al., 2004). Consequently, increased temperature of grain and moisture migration creates favorable conditions for insects and mold to flourish (Suleiman et al., 2013; Iguaz et al., 2004). Carbon dioxide released by grain and other organisms is used as a parameter for grain deterioration and directly related to dry matter loss of grain (Sharp, 1982). Temperature is a single most important non-biological factor controlling the rate of deterioration of grain in storage (Muir and Viravanichai, 1972) and can be easily measured and simulated mathematically (Yaciuk et al., 1975). Moreover, temperature and moisture content can be modeled mathematically to optimize storage conditions and efficient control measured at any point in a storage bin (Lawrence et al., 2013; Iguaz et al., 2000). Further, Sutherland et al. (1971) showed mathematical models based upon physical and thermal properties of grain as a useful tool to predict grain conditions in a storage bin.

Furthermore, mathematical models have been used as initial tools for predicting physical factors like temperature and moisture content in grain storage (Jin, 1996; Yaciuk et al., 1975). In addition, mathematical simulation can be used to predict the temperature distribution in grain storage structure with different shape and sizes, grain varieties and locations (Jia et al., 2000). The main advantages of the mathematical model outline by Andrade et al. (2002) and others include lower cost and it takes less time than is needed in the experimental investigations. It allows analyzing systems that are impossible to accomplish by experimental investigations and it allows the complementation of experimental investigations with more detailed information (Andrade et al., 2002; Jian et al., 2005; Franca and Haghighi, 1995). Several three-dimensional heat transfer models have been developed for simulating grain storage temperature in the cylindrical bin (Andrade et al., 2002; Jian et al., 2005; Lawrence et al., 2013; Jayas et al, 1995). Thus, the consequences of variations in the dimensions, geometry, properties of the materials and external conditions can be easily studied by using computer simulation (Franca and Haghighi, 1995). Although, several 3D heat transfer models have been established, 3 D heat transfer model in grain storage under hermetic conditions has not yet been developed. Therefore, the aim of this study was to develop a mathematical model to optimize storage condition of maize in a cylindrical bin in the room, freezing and cold temperature under hermetic condition.

MATERIALS AND METHODS

Experimental set up and procedure

A temperature equilibrium experiment was conducted at a water quality, laboratory, Iowa State University; the laboratory was fitted to Norlake Scientific RSF5 compartments chamber (cold room at 4°C and freezer at -20°C). Maize of commercial hybrid Blue River 57436 with initial temperature and moisture content of 21°C and 14.5±0.5% (wet basis) respectively, was used, two cylindrical plastic barrels, 50 cm diameter by 76 cm by height, were filled with maize to a height of 45 cm and airtight (hermetic) to maintain uniform conditions, each were fitted with five sensors from omega engineering, Inc. models OM-EL-USB-2-LCD and OM-EL-USB-2-LCD-PLUS placed about 12 cm apart (top, center, bottom, right, and left) as shown in Figure 1 to measure internal temperature, dew point and relative humidity of maize.

The barrels were stored at room temperature for 72 h, then moved to the cooler for 72 h, move back to the room for another 72 h. The same procedures were repeated for freezer condition. Censors were set to record the data after every 5 min for 9 days for each condition. Then censors were carefully removed and the data were downloaded in the computer and analyzed. The following assumptions were made while developing this model:

1. Maize was assumed to be free of arthropod populations and mold growth.
2. Conduction where the only heat transfer was between the grain bulk (in the horizontal and vertical direction) and the sides of the bin.
3. Properties of maize grain remain constant.
4. Initial maize temperature (T_o) is at a specified temperature.
5. The maize temperature at the center of the bin ($r = 0$) is a finite.
6. Ambient temperature (T_a) or grain temperature at the wall ($r = R$)

Model development

The partial differential equation describes the heat transport inside the grain storage bin under cylindrical coordinate system (Figure 1) following Fourier's law (Incropera and De Witt, 1996; Andrade et al., 2002; Mills, 1995). The heat conduction equations are:

$$q_r = -k\frac{\delta T}{\delta r}; \qquad q_\phi = -\frac{k}{r}\frac{\delta T}{\delta \phi}; \qquad q_z = -k\frac{\delta T}{\delta z} \tag{2}$$

Where q_r is the component of the heat flux in the r direction, $\frac{\delta T}{\delta r}$ is the partial derivative of T(r, ϕ, z and t) with respect to r, same for ϕ and z directions.

$$q = k\nabla^2 T \tag{3}$$

$$\frac{1}{r}\frac{\delta}{\delta r}\left(kr\frac{\delta T}{\delta r}\right) + \frac{1}{r^2}\frac{\delta}{\delta \phi}\left(k\frac{\delta T}{\delta \phi}\right) + \frac{\delta}{\delta z}\left(k\frac{\delta T}{\delta z}\right) + \frac{\dot{q}}{\rho c_p} = \frac{\delta T}{\delta t} + vr\frac{\delta T}{\delta r} + v\phi\frac{\delta T}{\delta \phi} + vz\frac{\delta T}{\delta z} \tag{4}$$

For stationary materials like maize grain ($v_r = v_\phi = v_z = 0$), Equation 4 will be simplified to:

$$\frac{1}{r}\frac{\delta}{\delta r}\left(kr\frac{\delta T}{\delta r}\right) + \frac{1}{r^2}\frac{\delta}{\delta \phi}\left(k\frac{\delta T}{\delta \phi}\right) + \frac{\delta}{\delta z}\left(k\frac{\delta T}{\delta z}\right) + \dot{q} = \rho c_p\frac{\delta T}{\delta t} \tag{5}$$

Where r, ϕ and z are the cylindrical coordinates, k is the thermal conductivity of maize (W/m*°C), ρ is the specific mass or density of maize (kg/m³), C_p denotes the specific heat of maize in (kJ/kg*°C), t is the time in s, T is the temperature of the maize kernel (°C), and $\dot{q}$ is the rate of generation of heat as function of (r, φ, z, and t) in W.

The boundary and initial conditions for the Equation 5 are:

$$at\ t = 0, \quad 0 \leq r \leq R: T = T_{in} \tag{6}$$

$$0 \leq r \leq L: = T_{in} \tag{7}$$

Where, Tin= initial temperature, T= temperature at time (t) and point (x), F_o = Fourier number = ($\alpha t/L^2$), with δ_n being roots of the Bessel function Jo(R δ_n) = 0. The validation of the simulated model was done by comparing the simulated temperatures with measured temperatures.

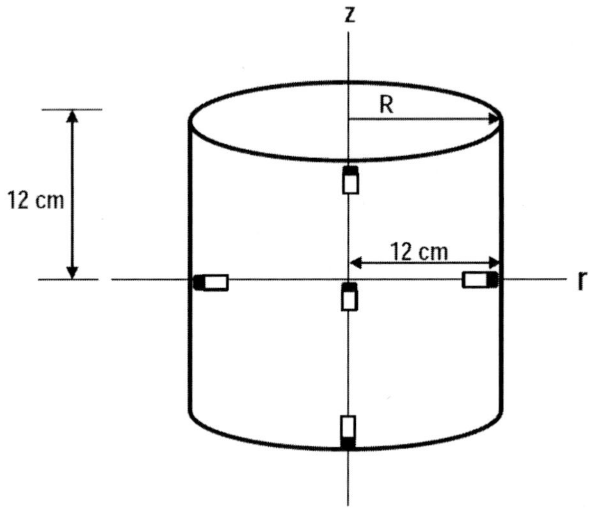

Figure 1. The cylindrical geometry shows temperature/relative humidity sensor arrangement.

Physical properties of maize grains

The value of the thermal and physical properties of the maize grains and plastic cylindrical bins used in the simulation is presented in Table 1. Thermal properties of maize grain (the thermal conductivity and the diffusivity) were determined by a thermal properties meter (KD2, Decagon Devices, Pullman, Wash).

Table 1. Thermal and physical properties for corn and plastic barrel bin.

Property	Maize	Plastic barrel (bin)*
Specific heat (J/kg/C)	1851.5	16700
Thermal conductivity (W/m/K)	0.1618	0.50
Thermal diffusivity (m^2/s)	1.21×10^{-7}	---
Density (kg/m^3)	1247	---
Initial moisture content (% w. b.)	12.7	---

Source (ASAE, 2000; http://www.engineeringtoolbox.com/).

RESULTS

The measured and predicted maize grain temperatures at five different positions (top, center, right, left and bottom) in the cylindrical bin filled with maize grains were determined. Temperatures were monitored at three different conditions (room, cooler and freezer) in vertical and radial directions.

Room and cooling condition in the vertical and radial directions

The predicted and measured grain temperature in a vertical direction at room temperature is shown in Figure 2. The grain temperature at the bottom of the bin was the first to change followed by top grain temperature and the center grain temperature was last to change. One possible explanation is the fact that at the bottom of the grain temperatures was too close in contact with the ground concrete floor, thus that was why it was the first to change. The predicted temperatures were closely matched to the measured temperature. Furthermore, Figure 3 shows the grain temperature at the cooler condition in vertical directions. The grain temperature at the bottom decreases faster at the first 500 min and slowly afterward. For the center, grain temperature was lagging and it took approximately 600 min before it started to drop. Similarly, the top grain temperature was between bottom and center as shown in Figure 3. The predicted temperatures are in excellent agreement with the measured temperature. Moreover, Figure 3 indicates warm temperature condition when grain bins were taken out of the cooler. Like in cooler conditions, bottom temperature was the first to change following a smooth curve as shown in Figure 3. For the top temperature, it took approximately 300 min to change and then it the take same path as the bottom temperature. Likewise, the center temperature was lagging and it took almost 500 min before the center temperature could respond as seen in Figure 3. All the predicted temperatures were in agreement with the measured temperature.

In addition, the predicted and measured temperatures in a radial direction (left, center and right) at room temperature are illustrated in Figure 4. The result shows that the measured temperature in the right side was the first to change followed by left, while the center temperature was between right and left temperature. The predicted temperatures were in the same path as the measured temperatures. Furthermore, Figure 5 shows the combined results of the temperature changes from room to cooler and from the cooler back to the room. The result indicates that the right temperature was first

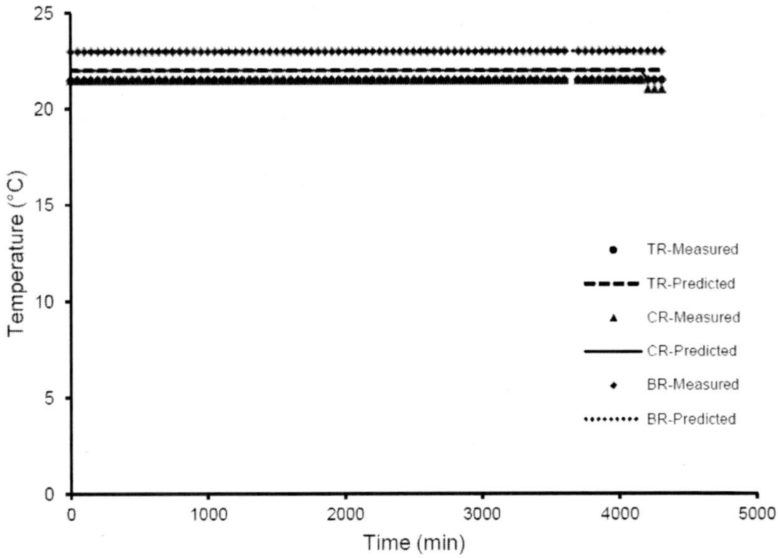

Figure 2. Measured and predicted maize grain temperature over time at room temperature in the vertical direction (before the cooling experiments). TR = Top room, CR = center room, BR = bottom room.

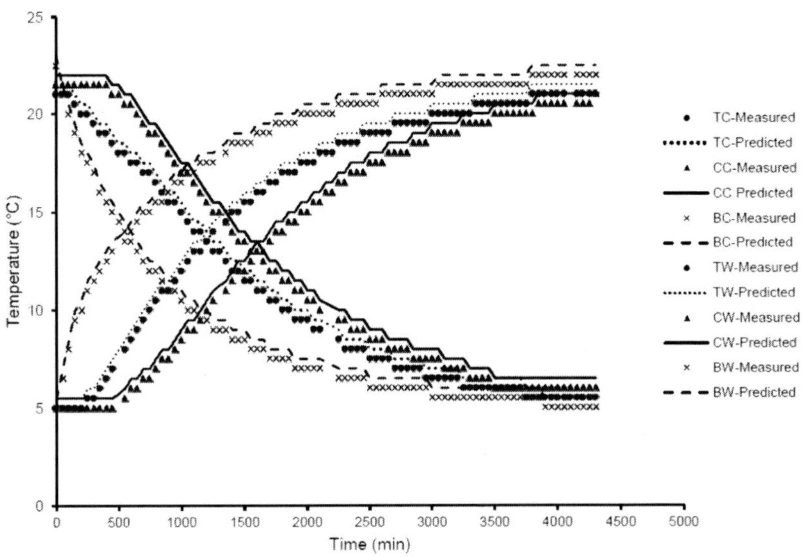

Figure 3. Measured and predicted maize grain temperature over time from the room to cooler and from cooler to back room temperature in the vertical direction. TC = Top cooling, CC= center cooling, BC= bottom cooling, TW = top warming, CW= CENTER warming, BW= bottom warming.

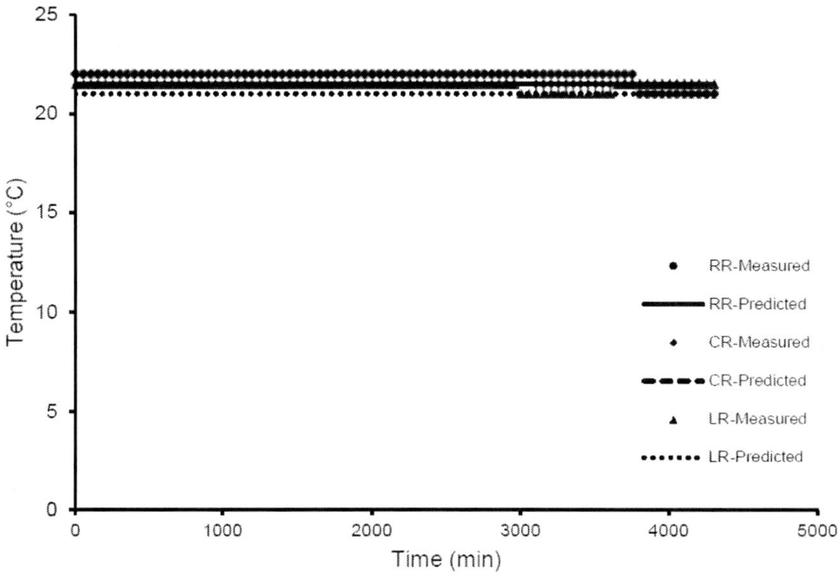

Figure 4. Measured and predicted maize grain temperature over time at room temperature in the radial direction (before the cooling experiments). RR = Right room, CR = center room, LR = left room.

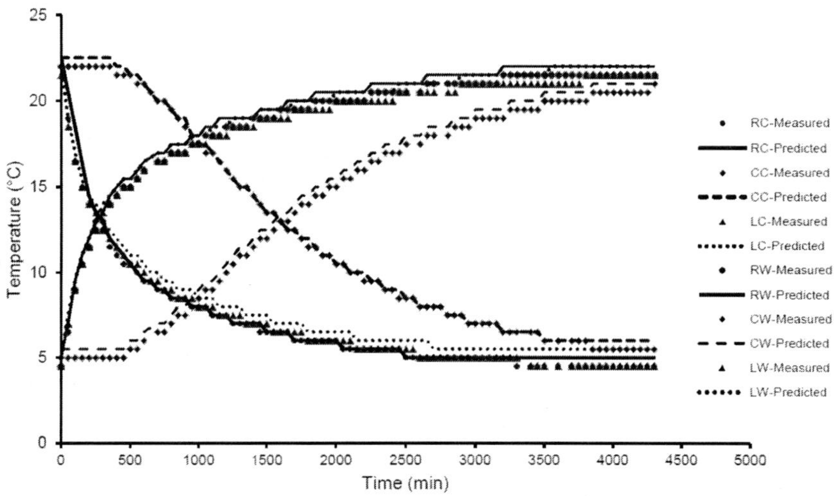

Figure 5. Measured and predicted maize grain temperature over time from the room to cooler and from cooler to back room temperature in the radial direction. RC= Right cooling, CC= center cooling, LC= left cooling, RW= Right warming, CW= center warming, LW=left warming.

to change followed by right and center (Figure 5). The predicted temperatures were similar to the measured temperature as shown in Figure 5.

Furthermore, for the cooler condition, the result shows the temperatures in the boundaries (left and right) dropped at the same time and reached a steady state after about 4000 min. While, the temperature at the center was lagging for about 500 min and dropped sharply afterward and relatively constant at a lower temperature (Figure 5). In warming conditions, the same trend was observed for the right, left and center temperature as shown in Figure 5. These results also show that the predicted temperatures were similar to the measured temperature.

Room and freezing condition in the vertical and radial directions

The temperature change for freezing conditions in the vertical direction at room temperature is shown in Figure 6. Slight variation between the top, bottom and center temperature was observed. Bottom temperature was first to respond and the center was last to change. Moreover, the predicted temperature had a close tie with the measured temperature (Figure 6). The temperature change from room to freezing condition is shown in Figure 7. The bottom temperature was sharply decreased at the initial 600 min and slowly afterward. The top was second to change and was laid between bottom and center temperature. Likewise, the center temperature was last to change as shown at Figure 7. The predicted temperatures for the room and freezing to room temperature were very close tomeasure temperature. Furthermore, for the temperature change from freezing conditions to room at vertical direction is shown at Figure 7. The top and the bottom temperature were first to change, although at different rates, top temperature was the first to response followed by the bottom and the center was the lagging (Figure 7). Moreover, in a radial direction for the room temperature, the right measured temperature was first to change followed by the left and center temperature (Figure 8). For the temperature change from room to freezing conditions, the left and right responded simultaneously as shown in Figure 9. Similarly, the center temperature was lagging for about 500 min. All the measured and predicted temperatures follow the similar trend. In addition, the temperature change from freezing to room temperature is shown in Figure 9. The right and left temperature moved on the same path, the temperature started to rise sharply after -19°C and began to decrease when the temperature reached about 5°C as shown in Figure 9. There are close agreements between the measured and predicted temperatures at radial direction.

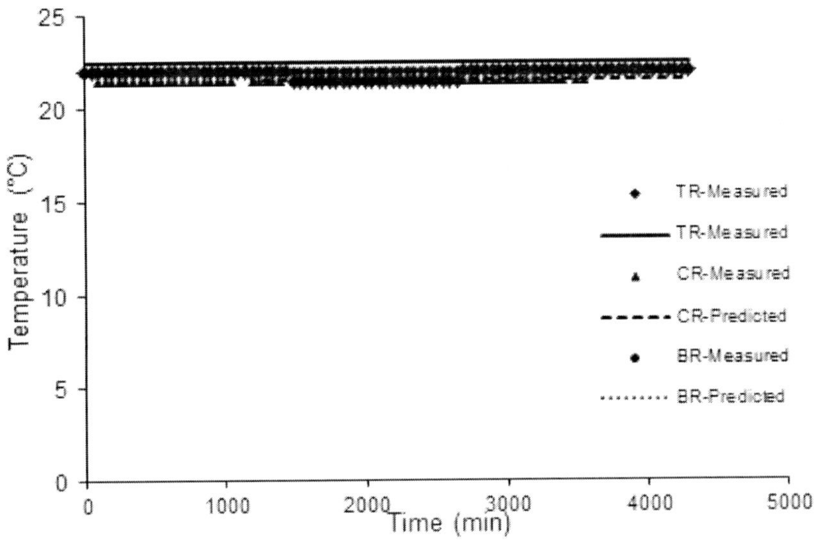

Figure 6. Measured and predicted maize room temperature changes over time at room temperature in the vertical direction (before the freezing experiments). TR= Top room, CR= center room, BR = bottom room.

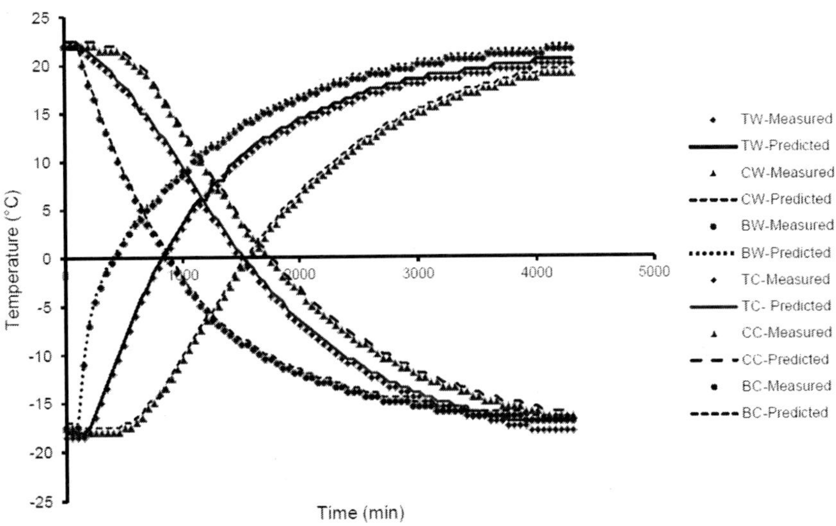

Figure 7. Measured and predicted maize grain temperature change over time from room to freezing temperature and from freezing to room in the vertical direction. TW= Top warming, CW= center warming, BW= bottom warming, TC= top cooling, CC= center cooling and BC= bottom cooling.

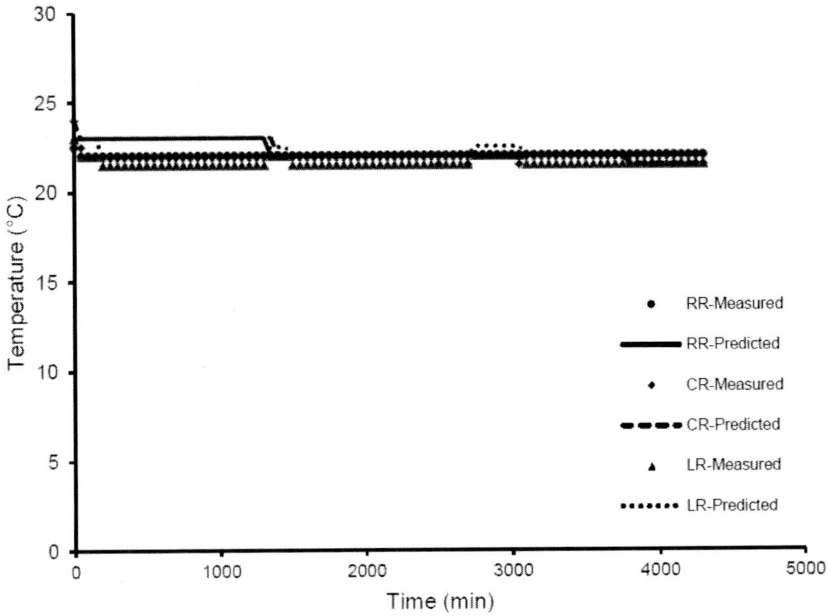

Figure 8. Measured and predicted maize grain room temperature change over time at room temperature in the radial direction (before the freezing experiments). RR= Right room, CR= center room, LR= left room.

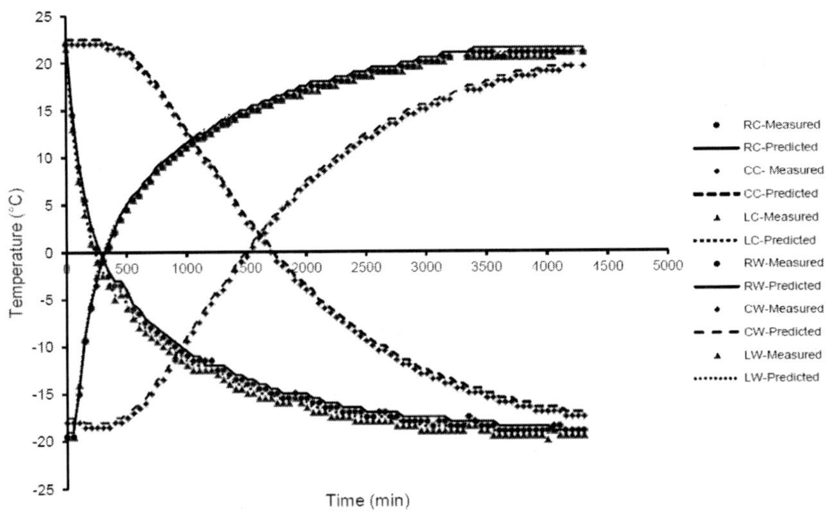

Figure 9. Measured and predicted maize grain temperature change over time from room to freezing temperature and from freezing to room in the radial direction. TW= Top warming, CW= center warming, BW = bottom warming, TC = top cooling, CC = center cooling and BC= bottom cooling.

DISCUSSION

A comparison of predicted and measured temperatures of maize grain was made for radial and vertical directions. The result in the radial direction in the room condition shows small variations between predicted and measured temperatures. However, the temperatures at the right and left hand sides or at the boundaries of cylindrical bins were either increase for warming conditions or decreased for the cooling conditions at higher rate after 500 min and followed the same path, while the temperature for center positions were lagging behind. These were observed throughout the study. The results concurred with the finding of Zhang et al. (2013) who reported that the temperature of maize grain increase at the points near the boundary and proceeds at a faster rate than the temperature increase at the center points.

Moreover, the study found that the temperature in the hermetic seal cylindrical bins varied, mostly in the radial direction and very little in the axial vertical directions as seen in Figures 3 and 7. Similar results were reported by Khankari et al. (1994), this means heat transfer in the cylindrical hermetic sealed bins occurred mainly due to conduction process. In addition, the maximum difference between predicted and measured temperature of maize grains inside the bins in vertical and radial directions which oscillated around ±1.5°C (as seen in Figures 2 to 9). These results are inconsistent with the results presented by Yaciuk et al., (1975). They found the maximum difference in the simulation model between predicted and measured temperature of wheat inside the storage bins to be 3°C.

Furthermore, the errors and small deviation between the predicted and measured temperatures in the model can be attributed to the combined errors in experimentation, in sensors reading, in numerical computation and in some assumptions made during model development.

According to Lawrence et al., (2013), the physical and the thermal properties of the grain are the important parameters that affect the accuracy of model development and temperature prediction. For instance, the properties of maize grains were assumed to be constant throughout the experiment, but in the actual sense, the thermal conductivity of maize grain will change as the temperature changes. Hence, affect the accuracy of model development. This assumption was made because there is no actual method of monitoring thermal properties of maize grain inside the bins during an experiment. Another factor to consider is the presence of fines and foreign particles in the maize grain; this has a significant effect on the thermal properties of grain, especially thermal conductivity as reported by Lawrence et al., (2013), thus reducing the accuracy of model development.

CONCLUSIONS

In general, maize grain temperature oscillated greatly in the boundary of the bin and slightly at the center. Like-wise, the predicted temperatures were closely matched with measured values throughout the experiments.

1. Temperature in the hermetically sealed cylindrical bins varied, mostly in the radial direction and very little in the axial vertical directions.
2. No noticeable change was observed in room condition.
3. The lag time between the center temperature and the side (right, left, top and bottom) was greater in the radial direction as compared to the vertical temperature.
4. The temperature in the grain changed more rapidly in the freezing conditions than in the room and cooling conditions.

REFERENCES

1. Amin MN, Hossain MA, Roy KC (2004). Effects of moisture content on some physical properties of lentil seeds.J. Food Eng. 65:83-87.
2. Andrade ET, Couto SM, Queiroz DM, Faroni LRA, Damasceno GS (2002).Three-dimensional simulation of the temperature variation in corn stored in metallic bin. Paper number: 02-3150 Chicago, Illinois, USA.
3. ASAE (2000). American society of agricultural engineers. Agricultural engineers yearbook. St. Joseph: Michigan, 2000. 796 P. cited at http://www. engineeringtoolbox. com/. Assessed on 10 November, 2014.
4. Flinn PW, Hagstrum DW, Muir WE (1997). Effects of time of aeration, bin size, and latitude on insect populations in stored wheat: a simulation study. J. Econ. Entomol. 90(2):646-651.
5. Franca AS, Haghighi K (1995). Adaptive finite element analysis of air flow inside grain dryers. Dry. Techno. 13:125-146.
6. Hong TD, Ellis RH, Moore D (1997).Development of a model to predict the effect of temperature and moisture on fungal spore longevity. Ann. Bot. 79(2):121-128.
7. Iguaz A, Arroqui C, Esnoz A, Vírseda P (2000). Modelling and validation of heat transfer in stored rough rice without aeration. Biosyst. Eng. 88(4):429-439.
8. Iguaz A, Arroqui C, Esnoz A, Vırseda P (2004). Modelling and simulation of heat transfer in stored rough rice with aeration. Biosyst. Eng. 89(1):69-77.
9. Incropera FP, De Witt DP (1996). Fundamentals of heat and mass transfer.John Wiley and Sons, New York, P 886.
10. Jayas DS, White ND (2003). Storage and drying of grain in Canada: low cost approaches. Food Control 14(4):255-261.
11. Jayas DS, White NDG, Muir WE (1995). Stored- grain ecosystems, New York. Marcel Dekker, Inc.
12. Jia CC, Sun DW, Cao CW (2000). Mathematical simulation of temperature fields in a stored grain bin due to internal heat generation. J. F. Eng. 43(4):227-233.
13. Jia C, Sun DW, Cao C (2000). Finite element prediction of transient temperature distribution in a grain storage bin. J. Agric. Eng. Res. 76(4):323-330.

14. Jian F, Jayas DS, White ND (2005). Movement of Tribolium castaneum (Coleoptera: Tenebrionidae) adults in response to temperature gradients in vertical and horizontal wheat and corn columns. J. Econ. Ent. 98:1413-1419.
15. Jin MY (1996). Modelling the microclimate in a conical bottomed circular grain silo (Doctoral dissertation, Victoria University of Technology, Australia).
16. Khankari KK, Patankar SV, Morey RV (1994). Mathematical model for moisture diffusion in stored grain due to temperature gradients. Trans. ASAE 37(5):1591-1604.
17. Lawrence J, Maier DE, Stroshine RL (2013). Three-dimensional transient heat, mass, momentum, and species transfer in the stored grain ecosystem: Part I. Model development and evaluation. Trans. ASABE 56(1):179-188.
18. Manickavasagan A, Jayas DS, White NDG, Jian F (2006). Thermal imaging of a stored grain silo to detect a hot spot. A. Eng. Agric. 22:891-897.
19. Mills AF (1995). Basic heat and mass transfer: Concord, M. A. Richard D. Irwin, Inc. pp. 114-123.
20. Muir WE, Viravanichai S (1972). Specific heat of wheat. J. Agric. Eng. Res. 17:338-342.
21. Mohamed IO (2009). Simultaneous estimation of thermal conductivity and volumetric heat capacity for solid foods using sequential parameter estimation technique. Food Res. Int. 42:231-236.
22. Ng NQ (1994). Preservation of genetic resources by seed storage.IITA Research Guide 52.Training Program, IITA, Ibadan, Nigeria.
23. Parry JL (1985). Mathematical modelling and computer simulation of heat and mass transfer in agricultural grain drying: A Review. J. Agric. Eng. Res. 32(1):1-29.
24. Sharp JR (1982).A review of low temperature drying simulation models. J. Agric. Eng. Res. 27(3):169-190.
25. Suleiman R, Rosentrater K, Bern C (2013). Effects of deterioration parameters on Sstorage of maize: A review. J. Nat. Sci. Res. 3(9):147-165.
26. Sutherland JW, Banks PJ, Griffiths HJ (1971). Equilibrium heat and moisture transfer in air flow through grain. J. Agric. Eng. Res. 16(4):368-386.
27. White ND (1995). Insects, mites, and insecticides in stored-grain ecosystems.Stored-grain ecosystems, pp. 123-167.
28. Yaciuk G, Muir WE, Sinha RN (1975). A simulation model of temperatures in stored grain. J. Agric. Eng. Res. 20(3):245-258.
29. Zhang S, Kong N, Zhu Y, Zhang, Z, Xu C (2013). 3D Model-based simulation analysis of energy consumption in hot air drying of corn kernels. Mathematical Problems in Engineering.

CHAPTER 10

Antifungal Edible Coatings for Fresh Citrus Fruit: A Review

Lluís Palou [1,*], Silvia A. Valencia-Chamorro [2] and María B. Pérez-Gago [1,3]

[1] *Centre de Tecnologia Postcollita (CTP), Institut Valencià d'Investigacions Agràries (IVIA), Apartat Oficial, Montcada 46113, Spain*

[2] *Departamento de Ciencias de los Alimentos y Biotecnología, Escuela Politécnica Nacional, Quito 170517, Ecuador*

[3] *Fundació Agroalimed, Apartat Oficial, Montcada 46113, Spain*

ABSTRACT

According to their origin, major postharvest losses of citrus fruit are caused by weight loss, fungal diseases, physiological disorders, and quarantine pests. Cold storage and postharvest treatments with conventional chemical fungicides, synthetic waxes, or combinations of them are commonly used to minimize postharvest losses. However, the repeated application of these treatments has led to important problems such as health and environmental issues associated with fungicide residues or waxes containing ammoniacal compounds, or the proliferation of resistant pathogenic fungal strains. There is, therefore, an increasing need to find non-polluting alternatives to be used as part of integrated disease management (IDM) programs for preservation of fresh citrus fruit. Among them, the development of novel natural edible films and coatings with antimicrobial properties is a technological challenge for the industry and a very active research field worldwide. Chitosan and other edible coatings formulated by adding antifungal agents to composite emulsions based on polysaccharides or proteins and lipids are reviewed in this article. The most important antifungal ingredients are selected for their ability to control major citrus postharvest diseases like green and blue molds, caused by *Penicillium digitatum* and *Penicillium italicum*, respectively, and include low-toxicity or natural chemicals such as food additives, generally recognized as safe (GRAS) compounds, plant extracts, or essential oils, and biological control agents such as some antagonistic strains of yeasts or bacteria.

Keywords: *Citrus* spp.; postharvest; disease control; fruit quality; fungicide alternatives; edible coatings; chitosan; antifungal ingredients

1. INTRODUCTION

Citrus spp. (Rutaceae) are the most widely produced fruits in the world, with a global production that exceeded 123 million tons in 2013. The list of the most important producing countries is led by China, Brazil, the United States (USA), India, Mexico, and Spain. Fruit production is mainly devoted to juice extraction, but a considerable proportion is traded as fresh entire fruit for direct consumption. Spain is the leading country for exports of fresh produce [1]. From highest to lowest production, the most cultivated citrus species are oranges (*Citrus sinensis* L.), mandarins or tangerines (*Citrus reticulata* Blanco), including clementines (*Citrus clementina* hort. ex Tanaka), satsumas (*Citrus unshiu* Marcow.), and different hybrid mandarins, lemons (*Citrus limon* (L.) Burm. f.) and limes (*Citrus aurantiifolia* (Christm.) Swingle), and grapefruits (*Citrus paradisi* Macfad.). Postharvest handling of fresh fruit in citrus packing houses is intended to commercialize fruit of maximum quality, increase their postharvest life, and reduce produce losses. In general, postharvest losses can be of physical, physiological, or pathological origin. Physical losses are typically due to rind wounds or bruises caused during harvest, transportation, or postharvest handling in the packing house. These peel injuries are not only important for causing direct losses, but also for being infection sites for economically important postharvest fungal pathogens. Other pathological losses are caused by latent pathogens that infect flowers or young fruit in the grove but develop after harvest. Likewise, some postharvest physiological losses are originated in the field and others are caused by inappropriate handling or storage conditions in the packing house or the marketplace.

Postharvest treatments with conventional synthetic waxes and/or chemical fungicides such as imazalil (IMZ), thiabendazole (TBZ), sodium ortho-phenil phenate (SOPP) or other active ingredients have been used for many years and are still currently used in citrus packing houses to preserve fresh fruit, control postharvest decay, and extend fruit shelf life. Nevertheless, the continuous application of these treatments has arisen important problems for the citrus industry such as health and environmental issues associated with chemical residues or the proliferation of pathogenic resistant strains. Updated regulations from many countries are increasingly restricting the use of agrochemicals and every day more exports markets are demanding fruit with residue levels even lower than those established by official regulations. Due to this situation, research should focus on anticipating a scenario in which conventional chemical fungicides are not available. In such a context, satisfactory postharvest decay control may be accomplished by adopting integrated disease management (IDM) programs based on comprehensive knowledge of pathogen biology and epidemiology and consideration of all preharvest, harvest, and postharvest factors with influence on disease incidence. Among the antifungal postharvest treatments alternative to synthetic chemicals that are investigated worldwide for potential inclusion in IDM programs, in this article we review the development of antifungal edible coatings as a promising novel technology intended to confront two major concerns of citrus postharvest handling, the losses due to physiological problems and the losses due to pathological problems.

2. ONE SOLUTION FOR TWO MAJOR CITRUS POSTHARVEST PROBLEMS

2.1. Physiological Problems

Citrus fruit are non-climacteric, hence their respiration rate and ethylene production do not exhibit remarkable increase along with changes related to maturity and ripening as in climacteric fruits. However, although they have a relatively long shelf life compared to other tropical and sub-tropical fruits, they may experience important physiological postharvest losses if they are not properly handled and stored. As with other horticultural products, major postharvest losses in citrus are caused by weight loss and physiological disorders.

Water loss of citrus fruit after harvest, although not considered a physiological disorder, is responsible for loss of quality and consumer acceptability, as it results not only in direct quantitative losses (loss of salable weight), but also in losses in appearance (wilting and shriveling) and softening. In addition, most physiological disorders that affect citrus fruit tend to be related to water loss [2]. Some rind disorders that may appear under optimal, non-chilling temperatures include peel pitting, stem-end rind breakdown (SERB), and shriveling and collapse of the stem-end button. Postharvest peel pitting at non-chilling temperatures is a severe disorder that affects fruit from several citrus cultivars worldwide. Although the causes of the disorder are not fully understood, evidence indicates that altered water relations in fruit peel is a major factor contributing to this disorder. In this sense, sudden changes from low to high relative humidity (RH) after harvest induced peel pitting in "Navelina" and "Navelate" oranges [3], "Marsh" grapefruit [4] or "Fallglo" tangerines [5]. Later studies confirmed the link between alterations in the osmotic and turgor potential in the flavedo (the outer pigmented layer of the peel) and albedo (the inner white layer of the peel) with the induction of postharvest peel pitting in citrus fruit [6]. SERB involves the collapse and subsequent darkening of the epidermal tissues around the stem-end rind of the fruit. The disorder is primarily associated with low RH, particularly during the postharvest period, although preharvest conditions seem to have a critical impact on the susceptibility of fruit to SERB. In some cases, SERB has been reported to be more severe when fruit are harvested from water-stressed trees compared to non-stressed trees or those presenting nutritional imbalances in terms of nitrogen and potassium [7,8,9]. Shriveling and collapse of the button tissue is an age-related breakdown due to cell weakening and dehydration of mature fruit. The symptoms can vary from discolored, dried out, and extensive collapse of the fruit rind to dehydration or wilting at the stem end where the rind is thinnest [10].

Another group of physiological disorders caused by storage of citrus fruit below sub-optimal temperatures but above the freezing point is chilling injury (CI). CI is characterized by the collapse of discrete areas of the peel that form sunken lesions that tend to coalesce [11]. CI in citrus fruit may appear in various forms depending on species and cultivar and exposure conditions (temperature and duration). For instance, typical symptoms in oranges, mandarins or grapefruits can be browning of the flavedo, appearance of dark sunken areas of collapsed tissue (pitting), or appearance of soft water-soaked areas (watery breakdown), while in lemons can be browning of the albedo or peteca (a special type of rind pitting) [2,10]. In general, although CI symptoms are due to fruit storage below their optimal temperature for relatively long periods, they usually develop upon transfer to higher temperatures.

The high sensitivity of citrus fruit to the induction of physiological disorders in the peel is in many cases triggered by mechanical damage during harvest and postharvest handling. Some of these disorders include "brush burn", "zebra skin", and "oleocellosis", caused by rind abrasion, rough handling, or thorn punctures. Brush burn and zebra skin symptoms usually appear as superficial red/brown staining of the rind and, in the case of the latter, is associated with the position of the segments, whereas oleocellosis might result in dark sunken patches as cells collapse around oil glands [12].

2.2. Pathological Problems

The most important postharvest diseases of fresh citrus fruit are caused by filamentous fungi. According to the origin of the infections, they can be classified into two general groups, those that infect the fruit in the field and remain latent until their development after harvest, and those that infect the fruit through rind microwounds or injuries inflicted during fruit harvest, transportation, postharvest handling, and commercialization.

The most significant pathogens in the first group include *Lasiodiplodia theobromae* (Pat.) Griffon and Maubl. and *Phomopsis citri* H. Fawc. non Sacc. Traverso and Spessa, which cause the diseases commonly known as stem-end rots; *Alternaria citri*Ellis and N. Pierce in N. Pierce, the cause of alternaria rot or black rot; *Botrytis cinerea* Pers.:Fr., the cause of gray mold; *Colletotrichum gloeosporioides* (Penz.) Penz. and Sacc. in Penz., the cause of anthracnose; and *Phytophthora citrophthora* (R.E. Sm. and E.H. Sm.) Leonian, which cause brown rot [13,14]. In general, the incidence of these pathogens is higher in citrus production areas with abundant summer rainfall, such as Florida or Brazil, because they require rain and humid weather for inoculum production and dispersal and subsequent fruit colonization and infection.

Among wound pathogens, the most economically important are *Penicillium digitatum* (Pers.:Fr.) Sacc. and *Penicillium italicum* Wehmer, which cause the diseases known as green and blue molds, respectively. These fungi are strict wound pathogens that can only infect the fruit through rind wounds inflicted in the field before or, more commonly, during fruit harvest, or after harvest during transportation, handling in the packing house, and commercialization. Green and blue molds are very important in all citrus production areas, including those with a Mediterranean-type climate, such as Spain, California, or South Africa, because one generation of *P. digitatum* or *P. italicum* is complete in rotten fruit in 7–10 days at usual ambient conditions of 20–25 °C and their spores are easily disseminated by air currents in large amounts [15]. Hence, the source of fungal inoculum in citrus orchards and packing houses is practically continuous during the season and the fruit can become readily contaminated [16]. Furthermore, healthy citrus fruit can become unmarketable when 'soiled' with conidia of these two fungi that are loosened during handling of diseased fruit [17]. No infection occurs if the fruit rind is intact because free conidia located on the peel surface are not able to germinate. However, the spores situated in injuries that rupture oil glands or penetrate into the albedo of the peel usually bring irreversible infection within 48 h at room temperatures [18]. The germination of both *Penicillium* species inside rind wounds requires free water, nutrients, and is stimulated by volatiles emitted from the host tissue [19,20]. *Geotrichum citri-aurantii* Ferraris E.E. Butler, the cause of sour rot; *Aspergillus niger* van Tiegh, the cause of Aspergillus rot; and *Rhizopus stolonifer* (Ehrenb.:Fr.) Vuill., the cause of Rhizopus rot, are other wound pathogens that can occasionally cause important postharvest losses under specific fruit handling and environmental conditions.

3. GENERALITIES OF ANTIFUNGAL EDIBLE COATINGS

Fruit coating is a common practice in citrus packing houses to replace the natural waxes that can be removed during fruit washing and handling in the packing line. Citrus commercial coatings are generically known as waxes due to the fact that composition of initial formulations was based on paraffin wax or a combination of various other waxes such as beeswax or carnauba. Typically, they are anionic microemulsions containing resins and/or waxes, such as shellac, wood rosin, candelilla wax, carnauba wax, beeswax, polyethylene, or petroleum waxes. Their main purpose is to reduce fruit weight loss, shrinkage and improve appearance, but they can also reduce the incidence of CI or other citrus rind disorders [10,21]. In some cases, however, if the coatings excessively restrict gaseous exchange during fruit storage, fruit coating can adversely affect fruit flavor due to an overproduction of volatiles associated with anaerobic conditions [22,23]. In addition, commercial citrus waxes are often amended with synthetic chemical fungicides like IMZ, TBZ, or SOPP in order to control postharvest diseases, particularly green and blue molds.

Consumer interest in health, nutrition, and food safety combined with environmental concerns has increased the interest of many research groups to develop new natural, biodegradable, edible coating formulations to replace these currently used commercial waxes, thus avoiding the use in the formulations of synthetic components such as polyethylene wax, ammonia, or morpholine. The concept of antifungal edible coatings emerges when additional ingredients with antifungal properties (low-toxicity or food-grade preservatives, biocontrol agents, *etc.*) are incorporated into these biodegradable formulations in order also to replace the use of conventional chemical fungicides for postharvest disease control.

3.1. Films vs. Coatings

Although films and coatings have the same chemical composition and sometimes are used synonymously, they refer to different concepts according to their different purpose and utilization. Films are defined as a stand-alone thin layer of materials that can be used as covers, wraps, or separation layers. Films are usually prepared by casting process, and are mainly used for determination of barrier, mechanical, solubility, and other properties provided by certain film materials. On the other hand, food coatings involve the formation of films directly on the surface of the product to which they are intended to be applied. Thus, coatings form part of the final product [24].

3.2. Components and Types of Matrixes

In general, basic components or matrix components of edible coatings are hydrocolloids (proteins or polysaccharides), lipids (waxes, acylglycerols or fatty acids), and resins. Composite coatings or blends contain a combination of polysaccharides or proteins with lipids [25,26,27]. Plasticizers to enhance flexibility and extensibility, and emulsifiers or surfactants to improve the stability of emulsions may also be added as matrix components. These matrixes can be directly used in foods or act as carriers of other food additives such as antioxidants, nutraceuticals, flavorings agents, *etc.* that are included to modify the functionality of the coating [25,26,28,29,30,31,32]. In the particular case of antifungal edible coatings, the additional ingredients are food-grade antimicrobial compounds of different nature effective in preventing or reducing fungal growth. Characteristics of main components of matrixes are the following:

- Polysaccharides can form a continuous and cohesive matrix, which is related to their chemical structure by the association through hydrogen bonding of their polymeric chains [25]. Polysaccharides contain highly polar polymers with hydroxyl groups and present a good barrier to oxygen at low RH, but low moisture barrier due to hydrophilic properties [30]. Polysaccharide materials typically used to formulate edible or biodegradable coatings include cellulose, starch, pectin, chitosan, alginate, carrageenan, pullulan, and various gums [25,30,31,33,34,35,36,37,38,39,40,41,42,43,44].
- The ability of different proteins to form edible coatings is highly dependent on their molecular characteristics: molecular weight, conformation, electrical properties, flexibility, and thermal stability [45]. Edible coatings based on proteins usually exhibit good gas

barrier characteristics, but poor water barrier characteristics due to their hydrophilic character [28,46,47]. Common proteins used for edible coating formulation include, among others, corn zein, casein, wheat gluten, soy protein, whey protein, keratin, or rice bran protein [26,30,47,48,49,50,51,52]. It is important to note that some people may present allergies or protein intolerances, e.g., to wheat gluten (celiac disease) or milk protein, which could restrict the use of protein-based coatings.

- Edible films and coatings based on hydrophobic substances, such as lipids and resins, are indicated to provide a barrier to moisture and gloss to food surfaces. However, since these materials are not polymers they form films and coating with poor mechanical properties and opaque characteristics [53]. Hydrophobic substances used as components of edible coatings include a variety of animal and vegetal native oils and fats, e.g., peanut, coconut, palm, lard, tallow, *etc.*; fractionated, concentrated, and/or reconstituted oils and fats, e.g., fatty acids, mono-, di-, and triglycerides, cocoa butter substitutes, *etc.*; hydrogenated and/or transesterified oils, e.g., margarine, shortenings, *etc.*; natural vegetal and animal waxes, e.g., beeswax, candelilla, carnauba, jojoba bees, rice bran, *etc.*; non-natural waxes, e.g., paraffins, oxidized or non-oxidized polyethylene; and natural resins, e.g., asafoetida, benjoin, chicle, guarana, myrrhe, olibanum (incense), opoponax, shellac resins, wood rosin, *etc.* [41,54,55,56,57].

- Composite coatings or blends typically contain hydrocolloid components, *i.e.*, protein and/or polysaccharides, and lipids in order to combine the advantages of both types of components. Composite coatings can be produced as either bi-layer or stable emulsions. In bi-layer coatings, the lipid forms a second layer over the protein or polysaccharide layer. In emulsion coatings, the lipid is dispersed and entrapped in the supporting matrix of protein or polysaccharide [57,58,59]. In this type of coatings, the efficiency of lipid materials depends on the lipid structure, its chemical arrangement, hydrophobicity, physical state, and its interaction with other components of the film [53].

- Plasticizers are low molecular weight compounds of small size, high polarity, high amount of polar groups per molecule, and great distance between polar groups within a molecule. They are added to edible coating materials to decrease the intermolecular forces between polymer chains, which results in greater flexibility, elongation, toughness, and permeability [28,30,32,60]. They are, therefore, particularly indicated to form stable emulsions and improve mechanical properties when hydrocolloids and lipids are combined. Common plasticizers used for edible coatings include sucrose, glycerol, sorbitol, propylene glycol, polyethylene glycol, fatty acids, and monoglycerides. Water can also act as a plasticizer for polysaccharide and protein coatings [61].

- Emulsifiers or surfactants are agents of amphiphilic nature that interact at the water-lipid interface and reduce surface tension between the dispersed and continuous phases to improve the stability of the emulsion when hydrocolloids and lipids are combined [30]. Moreover, emulsifiers added to coating formulations promote good surface wetting, spreading, and adhesion of the coating to the food surface. Typical emulsifiers used on edible coatings are fatty acids, ethylene glycol monostearate, glycerol monostearate, esters of fatty acids, lecithin, sucrose ester, and sorbitan monostearate, or polysorbates (tweens).

3.3. Functional Properties of Edible Coatings

The most important functional properties of edible films and coatings are edibility and biodegradability; migration, permeation, and barrier functions; and physical and mechanical protection. Such properties allow their use for food quality preservation and shelf-life extension. Furthermore, as carriers of active compounds that can be released in a controlled way, they can also provide antimicrobial activity and be applied for decay control and safety enhancement [26,29,33,35,47,62]. Edibility and biodegradability are the most beneficial characteristics of edible films and coatings. Edibility should be achieved by using food-grade ingredients for all coating components. Moreover, the whole process, facilities, and equipment should be feasible for food processing, and all components should be biodegradable and environmentally safe [24,26,63].

Edible coatings maintain food integrity and can protect coated food products against bruising, tissue damage and, in general, physical injury caused by impact, pressure, vibrations, and other mechanical factors. The ability of edible films and coatings to protect food against mechanical damage is usually assessed by determining film tensile properties such as Young's modulus (YM), which determines film stiffness as determined by ratio of pulling force/area to degree-of-film-stretch, tensile strength (TS), which indicates the pulling force per film cross-sectional area required to break the film, and elongation at break (E), which gives the degree to which the film can stretch before breaking and it is expressed as a percentage [1,10]. Other standardized mechanical examinations include compression strength, puncture strength, stiffness, tearing strength, burst strength, abrasion resistance, adhesion force, and folding endurance [24,29,35,47].

Since one of the main functions of edible coatings is to act as a protective barrier to environmental moisture, gases, flavors, aromas, or oils [26,64], other properties that are often determined on stand-alone films are water vapor permeability (WVP), oxygen permeability (OP), carbon dioxide permeability, and flavor permeability. Aroma and oil permeability are also very important for many foods but have generally received less attention. In other cases, other properties of interest are water solubility, gloss, and color.

The general protective functions of edible films and coatings may be widened with the addition to the matrixes of additional ingredients that provide new functionalities. These food additives include antioxidants, colorants, flavors, nutraceuticals, nutrients, and antimicrobial and particularly antifungal compounds [25,28,29,30,62,65].

3.4. Types of Antifungal Ingredients

With the exception of chitosan that presents inherent antimicrobial activity against a wide range of foodborne fungi, the antifungal effect of biopolymer-based coatings is usually achieved by the incorporation of active antimicrobial compounds to the coating formulation. Another exceptional case are gels and aqueous extracts from the leaves of the plant *Aloe vera*, which are edible and can also be used to coat fresh or minimally processed horticultural products. These coatings have been basically used for physiologic preservation of fruit after harvest, but they are also known for their antimicrobial activity [66,67,68]. However, in contrast to chitosan-based coatings, there is very limited information on the performance of coatings based on *A. vera* gels against fungi causing postharvest decay of citrus fruit. Early work by Saks and Barkai-Golan [69] showed significant activity of *A. vera* gels against *P. digitatum in vitro* and against green mold on grapefruits artificially inoculated with this fungus. Similar results were obtained in more recent research by [70], who found that an *A. vera*-based coating reduced green and blue molds on "Kinnow" mandarins and positively affected fruit quality during cold storage.

According to their nature, the antifungal compounds that can be added as additional ingredients to edible coatings can be classified into three categories: (1) synthetic food preservatives or GRAS (generally recognized as safe) compounds with antimicrobial activity, which include some inorganic and organic acids and their salts (carbonates, propionates, sorbates, benzoates, *etc.*), parabens (methyl and ethyl parabens) and their salts, *etc.*; (2) natural compounds such as essential oils or other natural plant extracts (cinnamon, capsicum, lemongrass, oregano, rosemary, garlic, vanilla, carvacrol, citral, cinnamaldehyde, vanillin, grape seed extracts, *etc.*) [33]; and (3) antimicrobial antagonists as biological control agents (yeasts, bacteria, and even some filamentous fungi). As they need to be allowed as food ingredients, all compounds in the first two categories should be synthetic or natural substances with known and minimal toxicological effects on mammals and the environment and they must be classified as food-grade additives or GRAS compounds by relevant regulations [71,72]. On the other hand, biocontrol agents must comply with strict and complex specific legislation, which considerably differs among different countries [73].

4. CHITOSAN AND CHITOSAN-BASED CITRUS COATINGS

Chitosan is a natural biopolymer with antimicrobial activity that has the property to form edible films and coatings [74]. It is a linear cationic polysaccharide of high molecular weight consisting of 1,4-linked 2-amino-deoxy-β-D-glucan, a partially-deacetylated derivative from chitin. Chitin is present in the exoskeletons of crustaceans (crabs, lobsters, shrimps, *etc.*) and is, after cellulose, the most abundant polysaccharide in nature [35,75]. Chitosan is produced commercially with different deacetylated grades and molecular weights, which are related to their functional properties and antimicrobial effects. Chitosan has exhibited high antimicrobial activity against a range of spoilage microorganisms, including fungi, yeasts, and bacteria. Particularly, its beneficial effect for postharvest disease reduction has been reported for a wide variety of fresh horticultural produce including citrus, apples, mango, grapes, strawberries, blueberries, lettuce, carrots, or tomatoes. Its antimicrobial activity, however, depends on several factors such as the type of chitosan, degree of acetylation, molecular weight, concentration, medium pH, target microorganism, and the presence of other ingredients in the chitosan coating [25,35,76,77,78,79]. In addition to its antimicrobial and antifungal activity, chitosan has been considered as a good candidate for postharvest treatment and long-term storage of fresh fruits and vegetables because of other important properties, such as lack of toxicity to mammals and consequent edibility, biodegradability, biocompatibility with many other compounds, and multifunctionality, greatly derived from its capacity to form coatings [62,75].

Chitosan and derivatives are currently the most assayed antifungal edible coatings for postharvest preservation of fresh citrus fruit. They have been investigated either alone or formulated with other additional antifungal ingredients. Table 1summarizes the most important research applications, to date, of chitosan and chitosan-based coatings, including the additional antimicrobial agents, if any, the target pathogen, an assessment of the antifungal activity, and the literature reference.

Table 1. Antifungal chitosan-based edible composite coatings applied to fresh citrus fruit.

Citrus Fruit	Coating	Concentration	Antimicrobial Agent	Target Pathogen	Antimocrobial Activity [a]	Reference
"Navel" orange, Lime	Chitosan	2, 4, 6, 8 g·L^{-1}	Lemongrass oil, Citral (6, 8 mL·L^{-1})	*Penicillium digitatum*, *Penicillium italicum*	+	[80]
	Chitosan	–	Lemongrass oil, Citral	*P. digitatum*, *P. italicum*	+	
"Satsuma" mandarin	Chitosan	1.0%	–	*P. digitatum*	+	[81]
	Chitosan	1.0%	Clove oil (0.5 mL·L^{-1})	*P. digitatum*	-	
"Or" and "Mor" mandarins, "Star Ruby" grapefruit	CMC, chitosan	1.0%, 1.5%	–	–	ND	[82]
"Jincheng 447" orange	Oligochitosan	1.5%	–	*Colletotrichum gloeosporioides*	+	[83]
"Valencia" orange	Chitosan	0.5%	–	*P. digitatum*	+	[84]
"Washington Navel" orange	Chitosan	0.5%	–	*P. digitatum*	+	–
"Femminello" lemon	Chitosan	0.5%	–	*P. digitatum*	+	–
"Marsh Seedlees" grapefruit	Chitosan	0.5%	–	*P. digitatum*	+	–
"Navel Powell" orange	Chitosan	2%	Bergamot, thyme and tea tree oils	*P. italicum*	+	[85]
"Valencia" and 'Péra Rio" oranges	Chitosan, Chitosan + TBZ	2%	–	*Guignardia citricarpa*	+	[34,86]

Lime	Chitosan	2, 4, 6, 8 g·L^{-1}	Citral (0, 2, 3, 4, 5 mL·L^{-1})	*Geotrichum citri-aurantii*	+	[87]
"Navel" orange	Chitosan	2%	–	*P. digitatum*, *P. italicum*	+	[77]
"Murcott" tangor	Chitosan (high MW)	0.05%, 0.1%, 0.2%	–	*P. digitatum*, *P. italicum*, *Botrydiplodia lecanidion*, *Botrytis cinerea*	+	[88]
	Chitosan (low MW)	0.05%, 0.1%, 0.2%	–	*P. italicum*, *B. lecanidion*, *B. cinerea*	+	
"Tankan" tangor	Chitosan	0.05%–0.2%	–	*P. digitatum*, *P. italicum*	-	[78]
Lemon	Chitosan	1 mg·mL^{-1}	–	*P. digitatum*	-	[89]
"Fortune" mandarin, "Valencia" orange	Biorend® (chitosan comercial product)	–	–	–	+	[90]
"Eureka" lemon	Glycolchitosan	0.2%	*Candida saitoana* (10^8 CFU·mL^{-1})	*P. digitatum*	+	[91,92]
"Washington Navel" orange	Glycolchitosan	0.2%	*C. saitoana* (10^8 CFU·mL^{-1})	*P. digitatum*	+	–

CMC = carboxymethil cellulose, MW = molecular weight, TBZ = thiabendazole; [a] Antimicrobial activity: "+" = inhibition, "-" = no inhibition, ND = not determined.

4.1. Stand-Alone Chitosan Coatings

The application of chitosan to lemon fruit prior to inoculation with *P. digitatum* resulted in a near absence of fungal development in inoculated rind wounds [89]. This worker suggested that the treatment had the ability to induce the transcriptional activation of defense genes leading to the accumulation of structural and biochemical compounds at strategic sites. Chitosan against green mold caused by *P. digitatum* was tested at concentrations from 0.02% to 0.5% in both *in vitro* and *in vivo* assays. It was found in the *in vitro* tests that chitosan at concentrations higher than 0.1% totally inhibited the pathogen growth, while in the *in vivo* assays chitosan applied at the concentration of 0.5% significantly reduced green mold on "Valencia" and "Washington Navel" oranges, "Femminello" lemons and "Marsh Seedless" grapefruits. The antifungal effects of chitosan were related to direct fungitoxic activity against the pathogen [84]. In another study [77], chitosan at 2% was applied to "Navel" oranges 24 h before inoculation with *P. digitatum* or *P. italicum* and it was observed that, after 23 days at 20 °C, disease incidence (proportion of infected fruit) and severity (lesion diameter) were significantly lower on coated samples than on uncoated controls. Chitosan antifungal activity was higher against blue mold than against green mold. It was pointed out, moreover, that chitosan application enhanced the activities of several enzymes such as peroxidase (POD) and superoxide dismutase (SOD), and increased the levels of glutathione (GSH) and hydrogen peroxide (H_2O_2). The effects of low (15 kDa) and high (357 kDa) molecular weight chitosan coatings, applied at 0.05% to 0.2%, on the development of green and blue molds and the quality of "Murcott" tangerines were studied by Chien, *et al.* [88]. Low molecular weight chitosan at 0.2% showed effective antifungal activity against both molds and improved firmness, titratable acidity, ascorbic acid level, and water content of coated fruit. Moreover, the performance of low molecular weight chitosan was comparable with that of the synthetic chemical fungicide thiabendazole. In another research, the antifungal activity of chitosan (0.05%–0.2%) with two different molecular weights (92.1 kDa and 357.3 kDa) against *P. digitatum* and *P. italicum* and the effect on postharvest quality of "Tankan" fruit, were investigated [78]. The performance of chitosan depended on its type and concentration. Chitosan effectively reduced the growth of the pathogens and the percentage of decayed fruit during incubation at 24 °C. On the other hand, fruit treated with chitosan exhibited good chemical and physical properties and less decay than control fruit during long-term refrigerated storage.

Deng, *et al.* [83] studied the effect of oligochitosan (1.5%), a derivate product from the enzymatic hydrolysis of chitosan, on the development of anthracnose on oranges inoculated with *C. gloeosporioides*. The results showed lower disease incidence and severity on oligochitosan-treated than on control fruit. The authors concluded that the treatment had the ability to induce strong disease resistance in citrus fruit during storage by increasing the contents in the fruit peel of constitutive natural antifungal compounds and the activity of defense enzymes. Glycolchitosan, another chitosan derivative, was also tested, alone or in combination with other treatments, for the control of green mold on oranges and lemons [91].

Two citrus varieties, "Fortune" mandarins and "Valencia" oranges were coated with a commercial chitosan-based coating product (Biorend®, IDEBIO S.L., Salamanca, Spain), and its effects on decay, fungistatic action, and fruit quality attributes were studied. The ripening process and decay development in coated fruit were monitored by means of the magnetic resonance imaging (MRI) technique [90]. The authors concluded that this chitosan product delayed maturity and reduced decay on both citrus species, allowing the effective preservation of "Fortune" mandarins and "Valencia" oranges for 6 and 22 weeks, respectively.

Work in Brazil [86] showed the potential of chitosan coatings to prevent the development of black spot, an important quarantine disease caused by the pathogen *Phyllosticta citricarpa* (McAlpine) van de Aa, on "Valencia" oranges stored at either 25 °C for eight days or 3 °C for 21 days. The authors discussed that chitosan application stimulated disease resistance mechanisms in the fruit rind.

4.2. Chitosan Coatings Amended with other Antifungal Ingredients

The possibility of enhancing the antifungal activity of chitosan coatings by combining them with other antifungal treatments has been explored as well. Such complementary antifungal treatments can be of different nature (physical treatments [93], chemical fungicides [34], food additives [91], biocontrol antagonists [94,95], *etc.*) and may be applied to citrus fruit before or after the application of chitosan-based coatings. However, in this review we focus on the addition of other antifungal agents to chitosan coating matrixes that are afterwards applied to the fruit as a unique postharvest treatment. In general, chitosan-amended coatings allow a gradual release of preservatives and provide additional properties for fungal growth inhibition and fruit quality maintenance [90]. The most important of these additional antifungal ingredients are essential oils. In few cases, biocontrol agents have been tested as well.

The antifungal activity, either *in vitro* or *in vivo*, of many essential oils extracted from plants or fruits such as birch, bergamot, cumin, cinnamon, citrus, clove, lemongrass, oregano, thyme, or tea tree against major citrus pathogens, particularly *P. digitatum* and *P. italicum*, has been reported [15,96]. Among them, only a limited amount has been tested as potential ingredients of chitosan coatings. Chitosan combined with citral or lemongrass oil (at 3 $mL\cdot L^{-1}$ or 4 $mL\cdot L^{-1}$) significantly reduced the mycelial growth and spore germination of *P. digitatum* and *P. italicum* in *in vitro* tests, while when applied to oranges and limes also provided a considerable green and blue mold reduction on fruit stored at either 20 °C or refrigeration temperatures [80]. Faten [87] reported good results with a combination of citral and chitosan for the control on lime fruit of sour rot, caused by the pathogen *G. citri-aurantii*. Reductions of disease incidence and severity as large as 89.5% and 93.5%, respectively, were obtained after the application of this coating containing chitosan (at 6 $g\cdot L^{-1}$ or 8.0 $g\cdot L^{-1}$) and citral (at 4 $mL\cdot L^{-1}$ or 5 $mL\cdot L^{-1}$). Both preventive (fruit coating followed by fungal inoculation) and curative (inoculation followed by coating) activity of chitosan coatings, containing or not essential oils from bergamot, thyme, or tea tree, were determined on "Powell" navel oranges against *P. italicum* [85]. It was found that, in general, the amended coatings were more effective than chitosan alone in reducing blue mold and that preventive activity was higher than curative activity. Among the tested oils, thymol showed the greatest antifungal activity when incorporated with the chitosan coating. Moreover, fruit quality attributes (acidity, pH, soluble solids, juice percentage, weight loss, firmness, color parameters, and respiration rate) of coated samples did not present any relevant changes during long-term cold storage. The inhibition by vapor contact of *A. niger* and *P. digitatum* by selected concentrations of Mexican oregano, cinnamon, or lemongrass essential oils added to chitosan, amaranth, or starch edible films was evaluated in another research [97]. The results showed that chitosan edible films incorporating Mexican oregano or cinnamon oils inhibited the two fungi at lower concentrations than those required for amaranth and starch edible films, and it was concluded that chitosan improved the release of the oil antimicrobial compounds. Sánchez-González, *et al.* [98] studied the addition of different concentrations of tea tree essential oil into chitosan films. Mechanical and optical properties and WVP of dry films, and antimicrobial activity against *P. italicum* were determined. They reported that films with a ratio of tea tree essential oil:chitosan of 1:2 provided a complete inhibition of fungal growth after five days of storage at 10 °C. Nevertheless, no every study reported synergistic activity from the addition of essential oils to chitosan coatings. Very recent work by Shao, *et al.* [81] showed that chitosan combined with clove oil was not significantly more effective than chitosan alone for the control of green mold on "Satsuma" mandarins artificially inoculated with *P. digitatum*. Furthermore, the high antifungal activity observed in *in vitro* tests was not obtained in *in vivo* tests, despite that the combination of chitosan and clove oil enhanced the activity of some fruit defense enzymes.

With reference to biocontrol agents as additional antifungal ingredients in chitosan-based coatings, El-Ghaouth, *et al.* [92] assessed the effectiveness of a 0.2% glycolchitosan coating formulated with the antagonistic yeast *Candida saitoana* against citrus green mold. They observed that this bioactive coating was superior to stand-alone glycolchitosan and *C. saitoana*treatments in controlling green mold on different orange and lemon cultivars, and the control level was equivalent to that with IMZ. The biocoating also reduced the incidence of stem-end rot caused by the pathogens *L. theobromae* or *P.citri* on naturally infected "Valencia" oranges, although in this case disease control was lower than that with IMZ. The maximum efficacy against green mold on "Eureka" lemons artificially inoculated with *P. digitatum* was obtained when citrus fruit were pretreated with the food additive sodium carbonate followed by the application of the glycolchitosan coating containing *C. saitoana* [91].

Along with the addition of essential oils or biocontrol agents, another system that has been evaluated to improve the activity of chitosan is the development of bilayer coatings comprised of chitosan and another natural polymer. This is the case of an edible composite bilayer coating formulated with carboxymethyl cellulose (CMC) and chitosan [82]. These workers applied this new coating to mandarins, oranges, and grapefruits and observed beneficial effects for fruit postharvest quality preservation. However, its effect on postharvest decay reduction was not studied.

5. CITRUS COATINGS FORMULATED WITH GRAS SALTS

Among the different GRAS compounds, organic acids and their salts are the most common synthetic antimicrobial agents tested in food systems [96]. In research conducted in our laboratory, Valencia-Chamorro, *et al.* [99] developed and optimized hydroxypropyl methylcellulose (HPMC)-lipid edible composite films formulated with a wide variety of food additives or GRAS compounds (mineral salts, organic acid salts, parabens, *etc.*) to inhibit the *in vitro* growth of the pathogens *P. digitatum* and *P. italicum*. Afterwards, the curative activity of selected coatings against green and blue molds was tested *in vivo* on oranges and mandarins artificially inoculated with the pathogens, coated, and incubated at 20 °C for a shelf life period of seven days [100]. It was found that coatings containing the GRAS salts potassium sorbate (PS), sodium benzoate (SB), sodium propionate (SP) and their mixtures were the most effective for disease reduction. Subsequent studies confirmed the antifungal activity and good effects on preservation of fruit quality of HPMC-lipid edible coatings formulated with PS, SB, SP, or their mixtures on long-term cold-stored "Valencia" oranges [101], "Ortanique" mandarins [102] and "Clemenules" mandarins [103]. In general, these coatings reduced fruit weight loss and maintained firmness without adversely affecting the overall sensory quality of coated fruit.

In research to explore the ability of GRAS salts to substitute the use of synthetic fungicides in conventional citrus commercial waxes, Youssef, *et al.* [104] reported that the salts sodium carbonate (SC) and bicarbonate (SBC), potassium carbonate (PC) and bicarbonate (PBC), ammonium bicarbonate (ABC), and PS incorporated into a commercial wax at a concentration of 6% (*w*/*v*) significantly reduced the incidence of postharvest diseases on naturally infected "Tarocco" and "Valencia" oranges and "Comune" clementines during cold storage followed by one week of shelf life at 20 °C. Among them, wax containing PS was the most effective to reduce decay in all tested cultivars.

6. CITRUS COATINGS FORMULATED WITH ESSENTIAL OILS

Among plant-derived natural compounds, many essential oils have proven antimicrobial activity against different important fungal pathogens, thus gaining popularity as potential ingredients of fruit coatings. The essential oils most studied for this purpose include clove, oregano, thyme, nutmeg, basil, mustard, and cinnamon oils, and some of them have been classified as GRAS by regulators [105]. Essential oils, as stand-alone treatments for fruit preservation, can be applied as volatiles in modified atmosphere packages or prepared as aqueous solutions from plant extracts obtained with different solvents. In any case, research has frequently shown that the efficacy of these compounds in *in vivo* experiments with fruit is lower than that observed in *in vitro* tests and, in some cases, they can be phytotoxic or induce negative sensory properties to treated commodities [96,106]. Therefore, despite their important antifungal activity, commercial implementation of essential oils as stand-alone postharvest treatments is strongly restricted and their use as ingredients of edible coatings might be a potential solution for these application problems.

Good antifungal activity was achieved with an edible coating formulated with a CMC matrix amended with a stem extract of the herb *Impatiens balsamina*. The application of this coating significantly reduced total decay incidence and preserved fruit weight without adverse effects on internal quality on "Newhall" oranges stored at 5 °C for up to 100 days [107]. The mode of action of the coating was attributed to an increase of the activities of important fruit defense enzymes such as POD, SOD, chitinase (CHI), and β-1,3-glucanase (GLU). Velásquez, *et al.* [108] compared in laboratory trials the effect of a pectin-based edible coating amended with essential oil at different concentrations with that of commercial waxes for disease control on "Valencia" oranges inoculated with *Penicillium* sp. The pectin coating containing essential oil at 1.5% reduced decay by 83% and satisfactorily extended the shelf life of coated fruit stored at 23 °C. Nevertheless, both amended coatings and commercial waxes were not effective on fruit stored at low temperatures.

In other research works, commercial citrus waxes have been formulated with essential oils as an alternative to the use of synthetic chemical fungicides as components of these waxes. In semi-commercial and commercial trials, essential oils from the plants *Mentha spicata* and *Lippia scaberrima* incorporated into different commercial citrus waxes resulted in significant preventive activity against green mold on "Valencia" and "Tomango" oranges. Specifically, after six days of incubation at ambient temperature, essential oil from *L. scaberrima* added at 2500 $mL \cdot L^{-1}$ into a carnauba wax reduced by about 60% the incidence of green mold compared to control fruit coated with the carnauba wax without essential oil [109]. Moreover, the amended waxes significantly decreased orange weight loss and maintained overall fruit quality. The effectiveness of the amended coatings was related to the strong inhibitory activity against *P. digitatum* of one of the most important components of the essential oil, the terpenoid R-carvone. In laboratory tests conducted in Spain, the application of wax coatings formulated with the essential oils carvacrol or thymol satisfactorily reduced green mold incidence on lemons artificially inoculated with *P. digitatum* and, at the same time, reduced the lemon respiration rate and ethylene production [110]. Further semi-commercial trials conducted in lemon packing lines confirmed that these essential oil-amended waxes were as effective as waxes containing the conventional fungicide IMZ in reducing lemon postharvest decay and total aerobic microflora present on fruit surface. In addition, fruit weight loss and rind softening and color were equally preserved by both types of coatings [111].

In another work, citral was incorporated into a commercial carnauba wax and a concentration 10 times the minimum fungicidal concentration obtained in *in vitro* studies was required to significantly decrease the incidence of green mold on citrus fruit coated and incubated at 25 °C for 6 days [112]. Octanal, another component of citrus essentials oils, has also been recently evaluated as active ingredient of commercial waxes for postharvest coating of citrus fruit. Tao, *et al.* [113] determined in *in vitro* tests that the minimum fungicidal concentration (MFC) of this compound against *P. digitatum* was 1000 $\mu L \cdot L^{-1}$. Then, they observed in *in vivo* tests that waxes amended with two times this octanal MFC effectively prevented the development of green mold on Satsuma mandarins incubated at 25 °C. They discussed that the coatings increased the activity of the enzymes SOD and catalase (CAT), but reduced the activity of phenylalanine ammonia-lyase (PAL) and POD, preventing the accumulation of H_2O_2. Commercial carnauba and/or shellac coatings amended with *Cinnamomum zeylanicum* essential oil at 0.5% (*v*/*v*) provided very high preventive activity against citrus green and blue molds, with disease reductions of about 90% with respect to uncoated control fruit treated with distilled water [114]. However, when the essential oil from *C. zeylanicum* was incorporated into polyethylene commercial waxes, disease reduction was considerably lower, indicating that the antifungal activity of the coating depended not only on the type and amount of essential oil and the volume of coating that remained on the fruit skin, but also on the formulation solubility and the particular compatibility between the essential oil and components of the wax. Among fifty-nine commercially available essential oils, workers from

South Africa [115] selected lemongrass, the essential oil from the plant *Cymbopogon citrates*, as the most cost-effective compound for the control of the pathogen *G. citri-aurantii*, the cause of citrus postharvest sour rot. They developed new coating formulations based on the addition of lemongrass and *M. spicata* essential oil to a citrus commercial wax. The coating containing both essential oils, each at a concentration of 750 $mL \cdot L^{-1}$, effectively controlled sour rot, but also green and blue molds, on artificially-inoculated "Valencia" oranges.

7. CITRUS COATINGS FORMULATED WITH MICROBIAL ANTAGONISTS

Research in the 1990s showed that different coatings containing microbial antagonists as biocontrol agents could be of use for the control of postharvest diseases of citrus fruit. Strains of the yeasts *Candida oleophila* ([116,117,118], *Candida guillermondii* [119,120] and the bacterium *Pseudomonas syringae* [121] were incorporated to different coatings and assessed against green and blue molds on oranges, lemons, or grapefruits. The effectiveness of these coatings greatly depended on their ability to support populations of the biocontrol agent. Shellac and other commercial wax coatings could be toxic to the yeasts due to the addition of alcohols and basis that are used to dissolve the primary constituents. The survival of the antagonistic yeast *C. olephila* in shellac formulations significantly improved in coatings containing less than 6% alcohol and when concentrations of morpholine and ammonia were minimized. These formulations significantly improved shelf life of coated grapefruits when compared to fruit coated alone with a shellac formulation [117].

Coating formulations based on water soluble cellulose derivatives, such as HPMC or methylcellulose, amended with the biocontrol yeast *C. olephila* significantly reduced decay and prolonged the storage life of grapefruits [116]. Yeast populations were stable in the coatings, but the addition of additional food preservatives was necessary to prevent the growth of other contaminating microorganisms. Thus, the addition of 0.15% PS to these formulations controlled microbial development without reducing the population of *C. oleophila*, which continued to increase in coated fruit during a storage period of 5 weeks. Similarly, the use of commercial sucrose ester formulations as coating matrixes favored the development of *C. olephila* populations in a greater extent than shellac formulations and these coatings were more effective for decay control on grapefruits than shellac-based coatings [118]. Varying the cellulose component of the coating formulation also affected the survival of two yeast biocontrol agents, *C. guillermondii* strain US7 and *Debaryomyces* sp. strain 230. Populations of both strains during incubation at room temperatures were higher in methylcellulose matrixes. The application of a methylcellulose-based coating containing the strain US7 to naturally infected "Pineapple" and "Valencia" oranges resulted in significant decay control for the first 2 to 4 weeks of storage at 16 °C and 90% RH [120].

In very recent research, Aloui, *et al.* [122] investigated the properties and the performance of bioactive coatings formulated with matrixes of sodium alginate and locust bean gum containing cells of the killer yeast *Wickerhamomyces anomalus*. They found that the survival of the yeast in these films was very high (more than 85% of the initial population) and their barrier, mechanical, and optical properties were satisfactory. When applied to "Valencia" oranges artificially inoculated with *P. digitatum*, these coatings reduced green mold by more than 70% on fruit incubated at 25 °C. In addition, coating application effectively reduced fruit weight loss and maintained rind firmness with respect to uncoated oranges.

8. CONCLUSIONS

Increasing concerns about human health risks and environmental contamination, restricted commercial channels for conventional production, and the proliferation of resistant strains of pathogenic fungi are important problems related to the use of conventional chemical fungicides for postharvest preservation of fresh citrus fruit. These agrochemicals are often applied as ingredients of synthetic waxes, which are needed to extend fruit postharvest life by reducing transpiration and respiration and also to provide gloss and enhance fruit appearance. Cost-effective disease control in the absence of synthetic fungicides requires the implementation of global IDM strategies based, in the postharvest phase, on the adoption of alternative or complementary non-polluting antifungal treatments. Irrespective of their nature, the evaluation of these alternative treatments should generally focus on the control of green and blue molds, caused by the pathogens *P. digitatum*and *P. italicum*, respectively, since they are the most economically important citrus postharvest diseases. Among these alternatives, the development of antifungal edible coatings arises as a new safe technology with potential to overcome both citrus postharvest physiological and pathological problems.

The functionality of chitosan, with an inherent antifungal activity, and composite films and coatings based on polysaccharide- or protein-lipid matrixes can be considerably improved by the incorporation of additional antifungal ingredients such as food additives or GRAS salts, natural compounds like essential oils or other volatiles, and microbial antagonists (yeasts, bacteria) as biocontrol agents. In each particular case, these antifungal agents are typically selected according to their efficacy as stand-alone treatments against target pathogens and, in many cases, their incorporation as ingredients of coatings brings a remarkable improvement of their performance. For example, coating formulations can overcome application problems of effective agents that may induce fruit phytotoxicities or adverse sensory properties when applied alone as aqueous or gaseous treatments. Further, coating application can also increase the agent antifungal activity by regulating its temporal and spatial release or facilitating its continuous and effective contact with the target pathogen.

Due to the high economic value of the worldwide citrus trade, the development of novel antifungal edible coatings for citrus fruit is a very active research field and, as described in the present review, a considerable number of research studies reporting interesting results are available from the specialized literature and will surely increase in the next few years. Despite the substantial progress that has been accomplished in evaluating new antifungal edible coatings, their implementation is still limited, first because of the current availability of highly effective, convenient, and cheaper conventional fungicides, and second because of general limitations associated to the edible nature of food-grade coating components. Low-toxicity and/or persistence, lack of either curative or preventive activity, or excessive specificity are handicaps for commercial application. It should be taken into account, in this sense, that minimal fruit losses due to postharvest decay are required by citrus export markets and, in contrast to that of synthetic chemicals with direct fungicidal activity, the mode of action of antifungal edible coatings is usually rather fungistatic. Consequently, their effectiveness is highly dependent not only on the target pathogen but also on fruit host characteristics such as citrus species and cultivar and fruit physical and physiological condition. Of course, coating performance from the physiological point of view will also be greatly influenced by all these fruit attributes. Therefore, the development of antifungal coatings should be tailored not only to control specific diseases but also for application to specific commercially important cultivars of oranges, mandarins, lemons or grapefruits, according to particular postharvest handling procedures, fruit storage potential, and destination markets. In addition to the development of new coatings formulated with other matrixes and/or antifungal ingredients and the evaluation of new applications for existing promising coatings, future prospects include research to determine potential synergistic effects from the combined application of different types of antifungal edible coatings with other alternative disease control methods (physical, low-toxicity chemical, or biological) in a multifaceted approach within IDM strategies.

ACKNOWLEDGMENTS

The authors thank all Spanish and international agencies that funded research on this topic. In memory of Miguel Ángel del Río, for his unconditional friendship, guidance and support.

REFERENCES

1. Food and Agriculture Organization of the United Nations. FAO Statistical Yearbook 2013. Available online: http://www.fao.org/docrep/ 018/i3107e /i3107e03.pdf (accessed on 22 September 2015).
2. Grierson, W.; Miller, W.M. Storage of Citrus Fruit. In *Fresh Citrus Fruits*, 2nd ed.; Wardowski, W.F., Miller, W.M., Hall, D.J., Grierson, W., Eds.; Florida Science Source, Inc.: Longboak Key, FL, USA, 2006; pp. 547–581.
3. Alférez, F.; Agustí, M.; Zacarías, L. Postharvest rind staining in Navel oranges is aggravated by changes in storage relative humidity. Effect on respiration, ethylene production and water potential. *Postharvest Biol. Technol.* **2003**, *28*, 143–152.
4. Alférez, F.; Burns, J.K. Postharvest peel pitting at non-chilling temperatures in grapefruit is promoted by changes from low to high relative humidity during storage. *Postharvest Biol. Technol.* **2004,** *32,* 79–87.
5. Alférez, F.; Zacarías, L.; Burns, J.K. Low relative humidity at harvest and before storage at high humidity influence the severity of postharvest peel pitting in citrus. *J. Am. Soc. Hortic. Sci.* **2005**, *130*, 225–231.
6. Alférez, F.; Alquezara, B.; Burns, J.K.; Zacarías, L. Variation in water, osmotic and turgor potential in peel of 'Marsh' grapefruit during development of postharvest peel pitting. *Postharvest Biol. Technol.* **2010**, *56*, 44–49.
7. Grierson, W. Physiological Disorders. In *Fresh Citrus Fruits*, 1st ed.; Wardowski, W.F., Nagy, S., Grierson, W., Eds.; AVI: New York, NY, USA, 1986; pp. 361–378.
8. Vercher, R.; Tadeo, F.R.; Almela, V.; Zaragoza, S.; Primo-Millo, E.; Agustí, M. Rind structure, epicuticular wax morphology and water permeability of 'Fortune' mandarin fruits affected by peel pitting. *Ann. Bot.* **1994**, *74*, 619–62.
9. Dou, H.; Zhang, J.; Ismail, M.A.; Ritenour, M.A. Postharvest factors influencing stem-end rind breakdown (SERB) of Valencia oranges. *Proc. Fla. State. Hort. Soc.* **2001**, *114*, 164–169.
10. Porat, R.; Weiss, B.; Cohen, L.; Daus, A.; Aharoni, N. Reduction of postharvest rind disorders in citrus fruit by modified atmosphere packaging. *Postharvest Biol. Technol.* **2004**, *33*, 35–43.

11. Petracek, P.D.; Kelsey, D.F.; Grierson, W. Physiological Disorders. In *Fresh Citrus Fruits*, 2nd ed.; Wardowski, W.F., Miller, W.M., Hall, D.J., Grierson, W., Eds.; Florida Science Source, Inc.: Longboak Key, FL, USA, 2006; pp. 397–419.
12. Scherrer Montero, C.R.; Schwarz, L.L.; Cunha dos Santos, L.; Pires dos Santos, R.; Bender, R.J. Oleocellosis incidence in citrus fruit in response to mechanical injuries. *Sci. Hortic.* **2012**, *134*, 227–231.
13. Eckert, J.W.; Eaks, I.L. Postharvest Disorders and Diseases of Citrus Fruits. In *The Citrus Industry*; Reuter, W., Calavan, E.C., Carman, G.E., Eds.; Division of Agriculture and Natural Resources, University of California Press: Berkeley, CA, USA, 1989; pp. 179–260.
14. Snowdon, A.L. *A Color Atlas of Post-harvest Diseases and Disorders of Fruits and Vegetables, Volume 1*; General Introduction and Fruits; CRC Press: Boca Raton, FL, USA, 1990.
15. Palou, L. *Penicillium digitatum, Penicillium italicum* (Green Mold, Blue Mold). In *Postharvest Decay. Control Strategies*; Bautista-Baños, S., Ed.; Elsevier: London, UK, 2014; pp. 45–102.
16. Kanetis, L.; Förster, H.; Adaskaveg, J.E. Comparative efficacy of the new postharvest fungicides azoxystrobin, fludioxonil, and pyrimethanil for managing citrus green mold. *Plant Dis.* **2007**, *91*, 1502–1511.
17. Smilanick, J.L.; Brown, G.E.; Eckert, J.W. The Biology and Control of Postharvest Diseases. In *Fresh Citrus Fruits*, 2nd ed.; Wardowski, W.F., Miller, W.M., Hall, D.J., Grierson, W., Eds.; Florida Science Source Inc.: Longboat Key, FL, USA, 2006; pp. 339–396.
18. Green, F.M. The infection of oranges by *Penicillium*. *J. Pom. Hortic. Sci.* **1932**, *10*, 184–215.
19. Lahlali, R.; Serrhini, M.N.; Friel, D.; Jijakli, M.H. *In vitro* effects of water activity, temperature and solutes on the growth rate of *P. italicum* Wehmer and *P. digitatum* Sacc. *J. Appl. Microbiol.* **2006**, *101*, 628–636.
20. Droby, S.; Eick, A.; Macarisin, D.; Cohen, E.; Rafael, G.; Stange, R.; McColum, G.; Dudai, N.; Nasser, A.; Wisniewski, M.; Shapira, R. Role of citrus volatiles in host recognition, germination and growth of *Penicillium digitatum* and *Penicillium italicum*. *Postharvest Biol. Technol.* **2008**, *49*, 386–396. [] []
21. Bajwa, B.E.; Anjum, F.M. Improving storage performance of *Citrus reticulata* Blanco mandarins by controlling some physiological disorders. *Int. J. Food Sci. Technol.* **2007**, *42*, 459–501.
22. Baldwin, E.A.; Nisperos-Carriedo, M.O.; Baker, R.A. Edible coatings for lightly processed fruits and vegetables. *HortScience* **1995**, *30*, 35–38.
23. Hagenmaier, R.; Goodner, K.; Roussef, R.; Dou, H. Storage of 'Marsh' Grapefruit and 'Valencia' oranges with different coatings. *Proc. Fla. State. Hort. Soc.* **2002**, *115*, 303–308.
24. Krochta, J.M. Proteins as raw materials for films and coatings: Definitions, Current Status, and Opportunities. In *Protein-Based Films and Coatings*; Gennadios, A., Ed.; CRC Press: Boca Raton, FL, USA, 2002; pp. 1–41.
25. Campos, C.A.; Gerschenson, L.N.; Flores, S.K. Development of edible films and coatings with antimicrobial activity. *Food Bioprocess Technol.* **2011**, *4*, 849–875.
26. Han, J.H. Edible Films and Coatings: A Review. In *Innovations in Food Packaging*, 2nd ed.; Han, J.H., Ed.; Elsevier: London, UK, 2014; pp. 213–255. []
27. Nisperos-Carriedo, M.O. Edible Coatings and Films Based on Polysaccharides. In *Edible Coatings and Films to Improve Food Quality*; Krochta, J.M., Baldwin, E.A., Nisperos-Carriedo, M.O., Eds.; Technomic Publishing Co. Inc.: Lancaster, PA, USA, 1994; pp. 305–335.
28. Cha, D.; Chinnan, M. Biopolymer-based antimicrobial packaging: A review. *Crit. Rev. Food Sci.* **2004**, *44*, 223–227.
29. Dhall, R.K. Advances in edible coatings for fresh fruits and vegetables: A review. *Crit. Rev. Food Sci. Nutr.* **2013**, *53*, 435–450.

30. Han, J.H.; Gennadios, A. Edible Films and Coatings: A review. In *Innovations in Food Packaging*, 1st ed.; Han, J.H., Ed.; Elsevier: Amsterdam, The Netherlands, 2005; pp. 239–262.
31. Pérez-Gago, M.B.; González-Aguilar, G.A.; Olivas, G.I. Edible coatings for fruits and vegetables. *Stewart Postharv. Rev.* **2010**, *6*, 1–14.
32. Zaritzky, N. Edible Coatings to Improve Food Quality and Safety. In *Food Engineering Interfaces*; Aguilera, J.M., Simpson, R., Welti-Chanes, J., Bermúdez Aguirre, D., Barbosa-Cánovas, G.V., Eds.; Springer-Verlag: New York, USA, 2011; pp. 631–659.
33. Valencia-Chamorro, S.A.; Palou, L.; del Río, M.A.; Pérez-Gago, M.B. Antimicrobial edible films and coatings for fresh and minimally processed fruits and vegetables: A review. *Crit. Rev. Food Sci. Nutr.* **2011**, *51*, 872–900. [] [] []
34. Rappussi, M.C.C.; Benato, E.A.; Cia, P.; Pascholati, S.F. Chitosan and fungicides on postharvest control of guignardia citricarpa and on quality of 'Pêra Rio' oranges. *Summa Phytopathol.* **2011**, *37*, 142–144.
35. Zhang, Y.; Rempel, C.; McLaren, D. Edible Coating and Film Materials: Carbohydrates. In *Innovations in Food Packaging*, 2nd ed.; Han, J.H., Ed.; Elsevier: London, UK, 2014; pp. 305–323.
36. Gol, N.B.; Patel, P.R.; Rao, T.V.R. Improvement of quality and shelf-life of strawberries with edible coatings enriched with chitosan. *Postharvest Biol. Technol.* **2013**, *85*, 185–195.
37. Lin, D.; Zhao, Y. Innovations in the development and application of edible coatings for fresh and minimally processed fruits and vegetables. *Compr. Rev. Food Sci. F.* **2007**, *6*, 60–75.
38. Díaz-Mula, H.M.; Serrano, M.; Valero, D. Alginate coatings preserve fruit quality and bioactive compounds during storage of sweet cherry fruit. *Food Bioprocess Technol.* **2011**, *5*, 2990–2997.
39. Meng, X.; Yang, L.; Kennedy, J.F.; Tian, S. Effects of chitosan and oligochitosan on growth of two fungal pathogens and physiological properties in pear fruit. *Carbohydr. Polym.* **2010**, *81*, 70–75.
40. Ferrari, C.C.; Sarantópoulos, C.I.; Carmello-Guerreiro, S.M.; Hubinger, M.D. Effect of osmotic dehydration and pectin edible coatings on quality and shelf life of fresh-cut melon. *Food Bioprocess Technol.* **2013**, *6*, 80–91.
41. Pérez-Gallardo, A.; García-Almendárez, B.; Barbosa-Cánovas, G.; Pimentel-González, D.; Reyes-González, L.R.; Regalado, C. Effect of starch-beeswax coatings on quality parameters of blackberries (*Rubus* spp.). *J. Food Sci. Technol.***2014**, *52*, 5601–5610.
42. Pan, S.Y.; Chen, C.H.; Lai, L.S. Effect of tapioca starch/decolorized hsian-tsao leaf gum-based active coatings on the qualities of fresh-cut apples. *Food Bioprocess Technol.* **2013**, *6*, 2059–2069.
43. Lima, A.M.; Cerqueira, M.A.; Souza, B.W.S.; Santos, E.C.M.; Teixeira, J.A.; Moreira, R.A.; Vicente, A.A. New edible coatings composed of galactomannans and collagen blends to improve the postharvest quality of fruits—Influence on fruits gas transfer rate. *J. Food Process. Eng.* **2010**, *97*, 101–109.
44. Cerqueira, M.A.; Bourbon, A.I.; Pinheiro, A.C.; Martins, J.T.; Souza, B.W.S.; Teixeira, J.A.; Vicente, A.A. Galactomannans use in the development of edible films/coatings for food applications. *Trends Food Sci. Tech.* **2011**, *22*, 662–671.
45. Vargas, M.; Pastor, C.; Chiralt, A.; McClements, D.; González-Martínez, C. Recent advance in edible coatings for fresh and minimally processed fruits. *Crit. Rev. Food Sci. Nutr.* **2008**, *48*, 496–511.
46. Baldwin, E.A. Edible Coatings for Fresh Fruits and Vegetables: Past, Present, and Future. In *Edible Coatings and Films to Improve Food Quality*; Krochta, J.M., Baldwin, E.A., Nisperos-Carriedo, M.O., Eds.; Technomic Publishing Co. Inc.: Lancaster, PA, USA, 1994; pp. 25–64.

47. Lacroix, M.; Vu, K.D. Edible Coating and Film Materials: Proteins. In *Innovations in Food Packaging*, 2nd ed.; Han, J.H., Ed.; Elsevier: London, UK, 2014; pp. 277–304.
48. Pérez-Gago, M.B.; Serra, M.; Alonso, M.; Mateos, M.; del Río, M.A. Effect of solid content and lipid content of whey protein isolate-beeswax edible coatings on color change of fresh-cut apples. *J. Food Sci.* **2003**, *68*, 2186–2191.
49. Pérez-Gago, M.B.; Serra, M.; Alonso, M.; Mateos, M.; del Río, M.A. Effect of whey protein- and hydroxypropyl methylcellulose-based edible composite coatings on color change of fresh-cut apples. *Postharvest Biol. Technol.* **2005**, *36*, 77–85.
50. Baldwin, E.A.; Baker, R.A. Use of Protein in Edible Coatings for Whole and Minimally Processed Fruit and Vegetables. In *Protein-Based Films and Coatings*; Gennadios, A., Ed.; CRC Press: Boca Raton, FL, USA, 2002; pp. 501–515.
51. De S. Medeiros, B.G.; Pinheiro, A.C.; Teixeira, J.A.; Vicente, A.A.; Carneiro-da-Cunha, M.G. Polysaccharide/protein nanomultilayer coatings: Construction, characterization and evaluation of their effect on 'Rocha'pear (*Pyrus communis* L.) shelf-life. *Food Bioprocess Technol.* **2012**, *5*, 2435–2445.
52. Schmid, M.; Reichert, K.; Hammann, F.; Stäbler, A. Storage time-dependent alteration of molecular interaction-property relationships of whey protein isolate-based films and coatings. *J. Mater. Sci.* **2015**, *50*, 4396–4404.
53. Rhim, J.W.; Shellhammer, T.H. Lipid-Based Edible Films and Coatings. In *Innovations in Food Packaging*, 1st ed.; Han, J.H., Ed.; Elsevier: Amsterdam, The Netherlands, 2005; pp. 362–383.
54. Baldwin, E.A.; Nisperos-Carriedo, M.O.; Hagenmaier, R.D.; Baker, R.A. Use of lipids in coatings for food products. *Food Technol.* **1997**, *51*, 56–62.
55. Debeaufort, F.; Voilley, A. Lipid-Based Edible Films and Coatings. In *Edible Films and Coatings for Food Applications*; Embuscado, M., Huber, K.C., Eds.; Springer-Verlag: New York, NY, USA, 2009; pp. 135–168.
56. Galus, S.; Kadzinska, J. Food applications of emulsion-based edible films and coatings. *Trends Food Sci. Technol.* **2015**, *45*, 273–283.
57. Pérez-Gago, M.B.; Rhim, J.W. Edible Coating and Film Materials: Lipid Bilayers and Lipid Emulsions. In *Innovations in Food Packaging*, 2nd ed.; Han, J.H., Ed.; Elsevier: London, UK, 2014; pp. 325–350.
58. Pérez-Gago, M.B.; Krochta, J.M. Emulsion and Bi-Layer Edible Films. In *Innovations in Food Packaging*, 1st ed.; Han, J.H., Ed.; Elsevier: Amsterdam, The Netherlands, 2005; pp. 384–402.
59. Shellhammer, T.H.; Krochta, J.M. Whey protein emulsion film performance as affected by lipid type and amount. *J. Food Sci.* **1997**, *62*, 390–394. [] []
60. Navarro-Tarazaga, M.L.; Sothornvit, R.; Pérez-Gago, M.B. Effect of plasticizer type and amount on hydroxypropyl methylcellulose-beeswax edible film properties and postharvest quality of coated plums (cv. Angeleno). *J. Agric. Food Chem.* **2008**, *56*, 9502–9509.
61. Krochta, J. Film Edible. In *The Wiley Encyclopedia of Packaging Technology*, 2nd ed.; Brody, A.L., Marsh, K.S., Eds.; John Wiley & Sons, Inc.: New York, NY, USA, 1997; pp. 397–401.
62. Shiekh, R.A.; Malik, M.A.; Al-Thabaiti, S.A.; Shiekh, M.A. Chitosan as a novel edible coating for fresh fruits. *Food Sci. Technol. Res.* **2013**, *19*, 139–155.
63. Guilbert, S.; Gontard, N. *Agro-Polymers for Edible and Biodegradable Films: Review of Agricultural Polymeric Materials, Physical and Mechanical Characteristics*, 1st ed.; Han, J.H., Ed.; Elsevier: Amsterdam, The Netherlands, 2005; pp. 263–276.
64. Krochta, J.M. Edible Protein Films and Coatings. In *Food Proteins and Their Applications*; Damodaran, S., Paraf, A., Eds.; Marcel Dekker, Inc.: New York, NY, USA, 1997; pp. 529–549.
65. Franssen, L.R.; Krochta, J.M. Edible Coating Containing Natural Antimicrobials for Processed Foods. In *Natural Antimicrobials for Minimal Processing of Foods*; Roller, S., Ed.; CRC Press: Boca Raton, FL, USA, 2000; pp. 250–262.

66. Arowora, K.A.; Williams, J.O.; Adetunji, C.O.; Fawole, O.B.; Afolayan, S.S.; Olaleye, O.O.; Adetunji, J.B.; Ogundele, B.A. Effects of *Aloe vera* coatings on quality characteristics of oranges stored under cold storage. *Greener J. Agric. Sci.* **2013**, *3*, 39–47.
67. Kumar, S.; Bhatnagar, T. Studies to enhance the shelf life of fruits using *Aloe vera* based herbal coatings: A review. *Int. J. Agric. Food Sci. Technol.* **2014**, *5*, 211–218.
68. Misir, J.; Brishti, F.H.; Hoque, M.M. *Aloe vera* gel as a novel edible coating for fresh fruits: A review. *Am. J. Food Sci. Technol.* **2014**, *2*, 93–97.
69. Saks, Y.; Barkai-Golan, R. *Aloe vera* gel activity against plant pathogenic fungi. *Postharvest Biol. Technol.* **1995**, *6*, 159–165.
70. Jhalegar, J.; Sharma, R.R.; Singh, D. Antifungal efficacy of botanicals against major postharvest pathogens of Kinnow mandarin and their use to maintain postharvest quality. *Fruits* **2014**, *69*, 223–237.
71. European Union. Approximation of the laws of the member states concerning food additive authorized for use in foodstuffs intended for human consumption (89/107/EC). *Official J.* **1989**, *40*, 27–33.
72. United States Food and Drug Administration. Food additives permitted for direct addition to food for human consumption, subpart c. Coatings, films and related substances. *Code Fed. Reg.* **2000**, *3*, 35–41, 21 CFR 172.
73. Liu, J.; Sui, Y.; Wisniewski, M.; Droby, S.; Liu, Y. Review: Utilization of antagonistic yeasts to manage postharvest fungal diseases of fruit. *Int. J. Food Microbiol.* **2013**, *167*, 153–160.
74. No, H.K.; Meyers, S.P.; Prinyawiwatkul, W.; Xu, Z. Applications of chitosan for improvement of quality and shelf life of foods: A review. *J. Food Sci.* **2007**, *72*, 87–100.
75. Hafdani, F.N.; Sadeghinia, N. A review on application of chitosan as a natural antimicrobial. *World Acad. Sci. Eng. Technol.* **2011**, *50*, 252–256.
76. Aider, M. Chitosan application for active bio-based films production and potential in the food industry: Review. *LWT-Food Sci. Technol.* **2010**, *43*, 837–842.
77. Zeng, K.; Deng, Y.; Ming, J.; Deng, L. Induction of disease resistance and ROS metabolism in navel oranges by chitosan. *Sci. Hortic.* **2010**, *126*, 223–228
78. Chien, P.J.; Chou, C.C. Antifungal activity of chitosan and its application to control post-harvest quality and fungal rotting of tankan citrus fruit (*Citrus tankan* Hayata). *J. Sci. Food Agric.* **2006**, *86*, 1964–1969.
79. Palma-Guerrero, J.; Jansson, H.B.; Salinas, J.; López-Llorca, L.V. Effect of chitosan on hyphal growth and spore germination of plant pathogenic and biocontrol fungi. *J. Appl. Microbiol.* **2008**, *104*, 541–553.
80. El-Mohamedy, R.S.; El-Gamal, N.G.; Bakeer, A.R.T. Application of chitosan and essential oils as alternatives fungicides to control green and blue moulds of citrus fruits. *Int. J. Curr. Microbiol. Appl. Sci.* **2015**, *4*, 629–643.
81. Shao, X.; Cao, B.; Xu, F.; Xie, S.; Yu, D.; Wang, H. Effect of postharvest application of chitosan combined with clove oil against citrus green mold. *Postharvest Biol. Technol.* **2015**, *99*, 37–43.
82. Arnon, H.; Zaitsev, Y.; Porat, R.; Poverenov, E. Effects of carboxymethyl cellulose and chitosan bilayer edible coating on postharvest quality of citrus fruit. *Postharvest Biol. Technol.* **2014**, *87*, 21–26.
83. Deng, L.; Zeng, K.; Zhou, Y.; Huang, Y. Effects of postharvest oligochitosan treatment on anthracnose disease in citrus (*Citrus sinensis* L. Osbeck) fruit. *Eur. Food Res. Technol.* **2014**, *240*, 795–804.
84. Panebianco, S.; Vitale, A.; Platania, C.; Restuccia, C.; Polizzi, G.; Cirvilleri, G. Postharvest efficacy of resistance inducers for the control of green mold on important Sicilian citrus varieties. *J. Plant Dis. Protect.* **2014**, *121*, 177–183.

85. Cháfer, M.; Sánchez-González, L.; González-Martínez, C.; Chiralt, A. Fungal decay and shelf life of oranges coated with chitosan and bergamot, thyme, and tea tree essential oils. *J. Food Sci.* **2012**, *77*, E182–E187.
86. Rappussi, M.C.C.; Pascholati, S.F.; Benato, E.A.; Cia, P. Chitosan reduces infection by *Guignardia citricarpa* in postharvest 'Valencia' oranges. *Braz. Arch. Biol. Technol.* **2009**, *52*, 513–521.
87. Faten, M.A. Combination between citral and chitosan for controlling sour rot disease of lime fruits. *Res. J. Agri. Biol. Sci.* **2010**, *6*, 744–749. []
88. Chien, P.J.; Sheu, F.; Lin, H.R. Coating citrus (Murcott tangor) fruit with low molecular weight chitosan increases postharvest quality and shelf life. *Food Chem.* **2007**, *100*, 1160–1164.
89. Benhamou, N. Potential of the mycoparasite, *Verticillium lecanii*, to protect citrus fruit against *Penicillium digitatum*, the causal agent of green mold: A comparison with the effect of chitosan. *Phytopathology* **2004**, *94*, 693–705.
90. Galed, G.; Fernández-Valle, M.E.; Martínez, A.; Heras, A. Application of MRI to monitor the process of ripening and decay in citrus treated with chitosan solutions. *J. Magn. Reson. Im.* **2004**, *22*, 127–137.
91. El-Ghaouth, A.; Smilanick, J.L.; Wilson, C.L. Enhancement of the performance of *Candida saitoana* by the addition of glycolchitosan for the control of postharvest decay of apple and citrus fruit. *Postharvest Biol. Technol.* **2000**, *19*, 103–110.
92. El-Ghaouth, A.; Smilanick, J.L.; Brown, G.E.; Ippolito, A.; Wisniewski, M.; Wilson, C.L. Application of *Candida saitona*and glycolchitosan for the control of postarvest disease of apple and citrus fruits under semi-commercial conditions. *Plant Dis.* **2000**, *84*, 243–248.
93. Abbas, H.; Abassi, N.A.; Yasin, T.; Maqbool, M.; Ahmad, T. Influence of irradiated chitosan coating on postharvest quality of kinnow (*Citrus reticulata* Blanco.). *Asian J. Chem.* **2008**, *20*, 6217–6227.
94. Lu, L.; Liu, Y.; Yang, J.; Azat, R.; Yu, T.; Zheng, X. Quaternary chitosan oligomers enhance resistance and biocontrol efficacy of *Rhodosporidium paludigenum* to green mold in satsuma orange. *Carbohydr. Polym.* **2014**, *113*, 174–181.
95. Waewthongrak, W.; Pisuchpen, S.; Leelasuphakul, W. Effect of *Bacillus subtilis* and chitosan applications on green mold (*Penicilium digitatum* Sacc.) decay in citrus fruit. *Postharvest Biol. Technol.* **2015**, *99*, 44–49.
96. Palou, L.; Smilanick, J.L.; Droby, S. Alternatives to conventional fungicides for the control of citrus postharvest green and blue molds. *Stewart Postharv. Rev.* **2008**, *2*, 1–16.
97. Avila-Sosa, R.; Palou, E.; Jiménez Munguía, M.T.; Nevárez-Moorillón, G.V.; Navarro Cruz, A.R.; López-Malo, A. Antifungal activity by vapor contact of essential oils added to amaranth, chitosan, or starch edible films. *Int. J. Food Microbiol.* **2012**, *153*, 66–72.
98. Sánchez-González, L.; González-Martínez, C.; Chiralt, A.; Cháfer, M. Physical and antimicrobial properties of chitosan-tea tree essential oil composite films. *J. Food Eng.* **2010**, *98*, 443–452.
99. Valencia-Chamorro, S.A.; Palou, L.; del Río, M.A.; Pérez-Gago, M.B. Inhibition of *Penicillium digitatum* and *Penicillium italicum* by hydroxypropyl methylcellulose-lipid edible composite films containing food additives with antifungal properties. *J. Agric. Food Chem.* **2008**, *56*, 11270–11278.
100. Valencia-Chamorro, S.A.; Pérez-Gago, M.B.; del Río, M.A.; Palou, L. Curative and preventive activity of hydroxypropyl methylcellulose-lipid edible composite coatings containing antifungal food additives to control citrus postharvest green and blue molds. *J. Agric. Food Chem.* **2009**, *57*, 2770–2777.
101. Valencia-Chamorro, S.A.; Pérez-Gago, M.B.; del Río, M.A.; Palou, L. Effect of antifungal hydroxypropyl methylcellulose (HPMC)-lipid edible composite coatings on postharvest decay development and quality attributes of cold-stored 'Valencia' oranges. *Postharvest Biol. Technol.* **2009**, *54*, 72–79.

102. Valencia-Chamorro, S.A.; Pérez-Gago, M.B.; del Río, M.A.; Palou, L. Effect of antifungal hydroxypropyl methylcellulose (HPMC)-lipid edible composite coatings on penicillium decay development and postharvest quality of cold-stored 'Ortanique' mandarins. *J. Food Sci.* **2010**, *75*, 418–426.
103. Valencia-Chamorro, S.A.; Palou, L.; del Río, M.A.; Pérez-Gago, M.B. Performance of hydroxypropyl methylcellulose (HPMC)-lipid edible composite coatings containing food additives with antifungal properties during cold storage of 'Clemenules' mandarins. *LWT-Food Sci. Technol.* **2011**, *44*, 2342–2348.
104. Youssef, K.; Ligorio, A.; Nigro, F.; Ippolito, A. Activity of salts incorporated in wax in controlling postharvest diseases of citrus fruit. *Postharvest Biol. Technol.* **2012**, *65*, 39–43.
105. United States Food and Drug Administration. Substances generally recognized as safe, subpart A. Essential oils, oleoresins (solvent-free), and natural extractives (including distillates). *Code Fed. Reg.* **2000**, *3*, 475–477.
106. Tripathi, P.; Dubey, N.K.; Banerji, R.; Chansouria, J.P.N. Evaluation of some essential oils as botanical fungitoxicants in management of post-harvest rotting of citrus fruits. *World J. Microbiol. Biotech.* **2004**, *20*, 317–321.
107. Zeng, R.; Zhang, A.; Chen, J.; Fu, Y. Impact of carboxymethyl cellulose coating enriched with extract of *Impatiens balsamina* stems on preservation of 'Newhall' navel orange. *Sci. Hortic.* **2013**, *160*, 44–48.
108. Velásquez, M.A.; Passaro, C.P.; Lara-Guzmán, O.J.; Álvarez, R.; Londono, J. Effect of an edible, fungistatic coating on the quality of the 'Valencia' orange during storage and marketing. *Acta Hortic.* **2014**, *1016*, 163–169.
109. du Plooy, W.; Regnier, T.; Combrinck, S. Essential oil amended coatings as alternatives to synthetic fungicides in citrus postharvest management. *Postharvest Biol. Technol.* **2009**, *53*, 117–122.
110. Pérez-Alfonso, C.O.; Martínez-Romero, D.; Zapata, P.J.; Serrano, M.; Valero, D.; Castillo, S. The effects of essential oils carvacrol and thymol on growth of *Penicillium digitatum* and *P. italicum* involved in lemon decay. *Int. J. Food Microbiol.* **2012**, *158*, 101–106.
111. Castillo, S.; Pérez-Alfonso, C.O.; Martínez-Romero, D.; Guillén, F.; Serrano, M.; Valero, D. The essential oils thymol and carvacrol applied in the packing lines avoid lemon spoilage and maintain quality during storage. *Food Control***2014**, *35*, 132–136.
112. Fan, F.; Tao, N.; Jia, L.; He, X. Use of citral incorporated in postharvest wax of citrus fruit as a botanical fungicide against *Penicillium digitatum*. *Postharvest Biol. Technol.* **2014**, *90*, 52–55.
113. Tao, N.; Fan, F.; Jia, L.; Zhang, M. Octanal incorporated in postharvest wax of Satsuma mandarin fruit as a botanical fungicide against *Penicillium digitatum*. *Food Control* **2014**, *45*, 56–61.
114. Kouassi, K.H.S.; Bajji, M.; Jijakli, H. The control of postharvest blue and green molds of citrus in relation with essential oil-wax formulations, adherence and viscosity. *Postharvest Biol. Technol.* **2012**, *73*, 122–128.
115. Regnier, T.; Combrinck, S.; Veldman, W.; Du Plooy, W. Application of essential oils as multi-target fungicides for the control of *Geotrichum citri-aurantii* and other postharvest pathogens of citrus. *Ind. Crop Prod.* **2014**, *61*, 151–159.
116. McGuire, R.G.; Baldwin, E. Composition of cellulose coatings affect population of yeasts in the liquid formulation and on coated grapefruits. *Proc. Fla. State Hort. Soc.* **1994**, *107*, 293–297.
117. McGuire, R.G.; Hagenmaier, R. Shellac coating for grapefruits that favor biological control of *Penicillium digitatum* by *Candida oleophila*. *Biol. Control* **1996**, *7*, 100–106.
118. McGuire, R.G.; Dimitroglou, D. Evaluation of shellac and sucrose ester fruit coating formulation that support biological control of post-harvest grapefruit decay. *Biocontrol Sci. Technol.* **1999**, *9*, 53–65.

119. McGuire, R.G. Application of *Candida guilliermondii* in commercial citrus coatings for biocontrol of *Penicillium digitatum* on grapefruits. *Biol. Control* **1994**, *4*, 1–7.
120. Potjewijd, R.; Nisperos-Carriedo, M.O.; Burns, J.K.; Parish, M.; Baldwin, E.A. Cellulose-based coatings as carriers for *Candida guillermondii* and *Debaryomyces* sp. in reducing decay of oranges. *HortScience* **1995**, *30*, 1417–1421.
121. McGuire, R.G. Population dynamics of postharvest decay antagonist growing epiphytically and within wounds on grapefruit. *Phytopathology* **2000**, *90*, 1217–1223.
122. Aloui, H.; Licciardello, F.; Khwaldia, K.; Hamdi, M.; Restuccia, C. Physical properties and antifungal activity of bioactive films containing *Wickerhamomyces anomalus* killer yeast and their application for preservation of oranges and control of postharvest green mold caused by *Penicillium digitatum*. *Int. J. Food Microbiol.* **2015**, *200*, 22–30.

CHAPTER 11

Effect of Methyl Jasmonate on Physical and Chemical Properties of Mango Fruit cv. Nam Dok Mai

Panida Boonyaritthongchai*, Chalida Chimvaree, Mantana Buanong, Apiradee Uthairatanakij and Pongphen Jitareerat

Division of Postharvest Technology, King Mongkut's University of Technology Thonburi, Bangkok 10150, Thailand

ABSTRACT

The effect of methyl jasmonate (MeJA) on anthracnose severity and physical and chemical properties of mango fruit cv. Nam Dok Mai was investigated. The mango fruit were harvested at the mature-green stage and the fruit surface was disinfected with 100 ppm sodium hypochlorite solution. The fruit samples were then fumigated with 30 ppm MeJA in an enclosed container at 25 °C for 6 h, and subsequently stored at 13 °C for 18 days. Non-treated fruit were used as the control. The results showed that MeJA had no effect on anthracnose severity, stem end rot disease and color change. MeJA treatment induced ethylene production and enhanced the accumulation of β-carotene content throughout the storage period compared with non-treated fruit. This result indicated that MeJA treatment may be used to increase the accumulation of β-carotene content of mango fruit during storage at low temperature.

Keywords: fumigation; mango; methyl jasmonate

1. INTRODUCTION

Mango (*Mangifera indica* L.) cv. Nam Dok Mai No. 4 is the most popular cultivar in Thailand with lusciously fragrant, sweet juicy flesh and a golden yellow skin color that attracts consumers. In Thailand, anthracnose and stem-end rot are considered the most serious and destructive diseases of mango. It is considered a major problem limiting the storage and shelf life of mango fruit. Methyl jasmonate (MeJA) is a plant hormone and endogenous signal molecule that has important roles in responses to environmental stresses [1,2,3,4]. Postharvest application of methyl jasmonate is a new approach for maintenance of quality of fruit, including induction of plant defense systems to protect from disease. Methyl jasmonate can cause changes in physical appearance, mechanical properties, and the composition of bioactive compounds of fruit [5]. It is an important cellular regulator that is involved in diverse developmental stages including fruit ripening, accumulation of pigments, phenolic compounds, antioxidants, and sugars [6,7,8,9], and has increased β-carotene accumulation in ripening tomato fruit [10]. MeJA treatments have reduced deterioration and the development of chilling injury symptoms in three cultivars of mango, Tommy Atkins, Kent, and Zill [11,12], and other fruit species such as guava (*Psidium guajava*), loquat (*Eriobotrya japonica* L.), peach (*Prunus persica* L.) and pomegranate (*Punica granatum*) [13,14]. Jasmonates also play an integral role in intracellular signal transduction cascades which occur in the inducible defense mechanisms that plants have developed against pathogens and other stresses. The objective of this study was to determine the effects of MeJA on physical and biochemical properties, antioxidant systems, and bioactive compounds and quality changes of Nam Dok Mai No. 4 mango fruit during cold storage.

2. EXPERIMENTAL SECTION

2.1. Mango Fruit Preparation

Mango fruit cv. Nam Dok Mai No. 4 were harvested at commercial maturity from Good Agricultural Practices (GAP) farms. They were transported to the Division of Postharvest and Technology, School of Bioresources Technology, King Mongkut's University of Technology Thonburi, Thailand. Fruit were cleaned with tap water, separated by size and color, the fruit surface disinfected with 100 ppm sodium hypochlorite solution, and dried in air at ambient temperature. The fruits were then randomly divided into two groups. One hundred forty-four mangoes were placed in glass jars containing 1.35 mL 30 ppm methyl jasmonate (MeJA) or water (for the control) spotted onto filter paper for 6 h at 25 °C in darkness. After treatment, fruit were transferred from the treatment jars into plastic baskets covered with perforated polyethylene film and stored at 13 °C for 18 days. Three replicates (9 fruit per one replicate) of each treatment were used for all analyses. Fruit decay and physical and chemical properties were evaluated every 3 days during storage.

2.2. Disease Incidence of Anthracnose and Stem End Rot

The severity of anthracnose disease was assessed by the extent of total decayed area on each fruit surface using a 5-point scale [15], where 0 = no disease symptoms, 1 = 1%–10% of the fruit surface, 2 = 11%–20% of the surface, 3 = 21%–30% of the surface, 4 = 31%–40% of the surface, and 5 = >40% of the surface affected by anthracnose disease spots. The severity of stem end rot disease was assessed by the extent of total decayed area on each fruit surface using a 5-point scale, where 0 = no disease symptoms, 1 = 1%–10% of the fruit surface, 2 = 11%–20% of the surface, 3 = 21%–30% of the surface, 4 = 31%–40% of the surface, and 5 = >40% of the fruit surface exhibited brown rot from stem end rot disease.

2.3. Colour Changes

Skin and pulp color were assessed with a tristimulus color different meter (Minolta CR300) and expressed as L, a, and b values. These values were used to calculate hue values. Negative a values indicated green and the higher positive a values indicated red color. Higher positive b values indicated a more yellow color.

2.4. Extraction, Saponification and HPLC Analysis of β-Carotene

All extraction and saponification steps were carried out under yellow or dim lights. A fruit sample was ground in an Oster blender until pulverized. 5 g of sample was added to a 50 mL round-bottomed Nalgene centrifuge tube. 10 mL of cold ethanol was added, and the sample was homogenized with a Polytron homogenizer at medium speed for 3 min. 8 mL of hexane was added, and the sample was homogenized for an additional 2 min. The mixture was then centrifuged for 4 min at 7000× *g*.

The carotenoid-bearing hexane layer was transferred with a Pasteur pipet to a 125 mL screw-cap Ehrlenmeyer flask. 5 mL saturated sodium chloride solution was added to the contents of the centrifuge tube, and the contents were stirred gently to homogenize the mixture. An additional 8 mL of hexane was added, and the mixture was homogenized at low speed with the Polytron for 1 min. The mixture was centrifuged as before, and the hexane layer transferred to the Ehrlenmeyer flask with the first extract.

15 mL of 10% methanolic KOH was added to the contents of the Ehrlenmeyer flask. The flask was flushed with nitrogen, sealed, and wrapped in aluminum foil to exclude light. The flask was left at room temperature for 16 h with gentle shaking. The mixture was then transferred to a separatory funnel and washed to remove the KOH, first with 50 mL of 10% NaCl solution and then with deionized water until the rinse had a neutral pH. The KOH solution was further extracted with an additional 10 mL hexane. The combined hexane extracts were evaporated under nitrogen just to dryness, then redissolved in methylene chloride and brought to volume with mobile phase (see below). The sample was immediately filtered through a 0.45 mm filter into an amber analysis vial and sealed.

The HPLC system consisted of a Hewlett Packard Series 1050 autosampler, Series 1050 pump, and a Series 1040 M diode array detector, operated by HP Chemstation software. A 250 × 4.6 mm, 5 mm internal diameter Vydac 201TP54 reversed phase C18 column fitted with biocompatible titanium frits was used for separation, together with a Vydac 4.6 × 25 mm high performance guard column filled with the same packing material. The mobile phase consisted of acetonitrile, methanol and methylene chloride 75:20:5 *v/v/v*, containing 0.1% butylated hydroxytoluene and 0.05% triethylamine. The methanol contained 0.05 M ammonium acetate. All reagents were HPLC grade. The flow rate was 1.5 mL/min. Detection was at 450 nm. Identification and purity of peaks was confirmed by comparing spectra using the computer software.

2.5. Ethylene Production

Ethylene production was measured during cold storage at 13 °C. Three fruit from each replication were randomly selected to measure respiration rate. Individual fruit were kept in airtight glass jars fitted with a rubber septum for collecting the gases. After 1 h incubation of the fruit, the head space gas was withdrawn from container using a syringe and injected into a gas chromatograph (Shimadzu, Kyoto, Japan). N_2 was used as the carrier gas. Ethylene was quantified using a flame ionization detector. Ethylene was estimated and expressed as μL C_2H_4/kg·h.

2.6. Statistical Analysis

The experiment was arranged in completely randomized design. All data presented were means of three replicates along with standard errors of means, and mean were compared using *t*-test. Differences at $p \leq 0.05$ and $p \leq 0.01$ were considered significant.

3. RESULTS AND DISCUSSION

3.1. Disease Severity

The symptoms of anthracnose disease on MeJA-treated and control mango fruit were observed during storage at 13 °C by 3 days. MeJA-treated mango and control fruit showed no significant difference in anthracnose or stem end rot disease severity through the end of storage at 18 days (Figure 1A,B). Treatments with MeJA have been reported to suppress gray mold rot (*Botrytis cinerea*) on strawberries (*Fragaria × ananassa* Duch.) [10], to reduce microbial contamination of fresh cut celery (*Apium graveolens*) and peppers (*Capsicum annum*) [16], and to suppress fungal decay in grapefruit (*Citrus × paradisi*) [17], in contrast to the present results. These differences may be due to the wide range in genetic makeup of different horticultural crops and different treatment methods with MeJA among studies [18]. MeJA has been shown to prevent postharvest diseases in a number of horticultural crops, but the mode of action in reducing disease has not been well-characterized, although MeJA effects could be because of a direct inhibitory effect on pathogen growth and/or because of the induction of natural disease resistance [17].

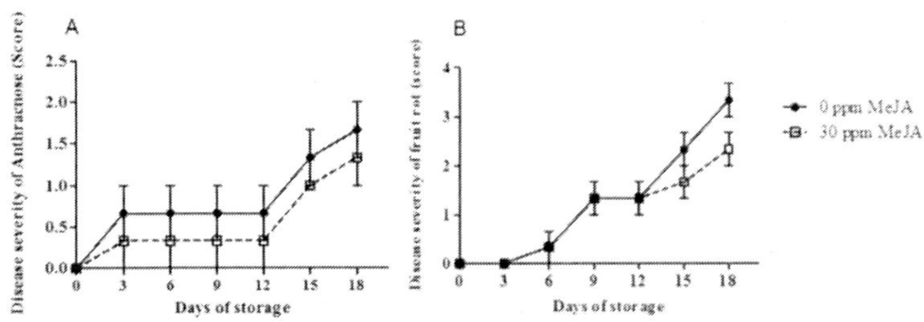

Figure 1. Disease severity of anthracnose (**A**) and stem end rot (**B**) in the mango fruit cv. Nam Dok Mai No.4 fumigated with 30 ppm methyl jasmonate (MeJA) versus untreated controls during storage at 13 °C for 18 days. There was no difference between treated and control fruit on any day.

3.2. Internal Pulp Color

The L value, or the lightness of the yellow pulp, and b value, expressing blue (−) to yellow (+) color did not differ between the MeJA treatment and controls on any day (Figure 2A,B). The a value, expressing green (−) to red (+) color, and hue angle were only higher for MeJA treatment at 18 days (Figure 2C,D). Colour change of mango pulp as L, a, and b values, and hue angle, was constant through 12 days of storage, but changed between 12 and 18 days. The L value and hue angle value of both treatments decreased, while a and b values of both treatments increased. Pulp color of both treatments changed from light yellow to dark yellow after 12 days.

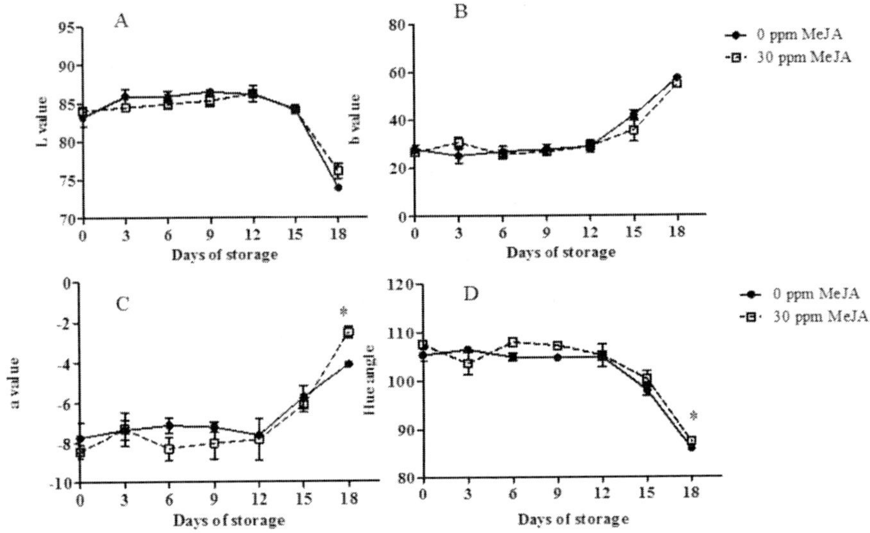

Figure 2. Colour change expressed as L (**A**), b (**B**), a (**C**), and hue angle (**D**) values of the fruit pulp of cv. Nam Dok Mai No.4 fumigated with 30 ppm methyl jasmonate (MeJA) versus untreated (control) during storage at 13 °C for 18 days. An asterisk indicates a statistically significant difference at $p \leq 0.05$; otherwise, treatments did not differ.

3.3. Peel Color

MeJA treatment showed a higher L value than controls at 3 and 15 days of storage, but a lower value than controls at 18 days (Figure 3A). The a and b values and hue angle of both treatments was maintained for 12 days of storage, but a and b values declined from 12 to 18 days while hue angle increased (Figure 3B–D). However, MeJA values did not differ from controls in these traits. The response of mango to MeJA has depended on the maturity stage, cultivar and subsequent storage temperature [19]. Gonzalez-Aguilar et al. [19] reported that use of 10^{-5} M MeJA enhanced yellow and red colour development of mango cv. Kent stored at 20 °C, but in the present research the fruit were kept at a lower temperature (13 °C), so it is possible that the colder storage suppressed color development.

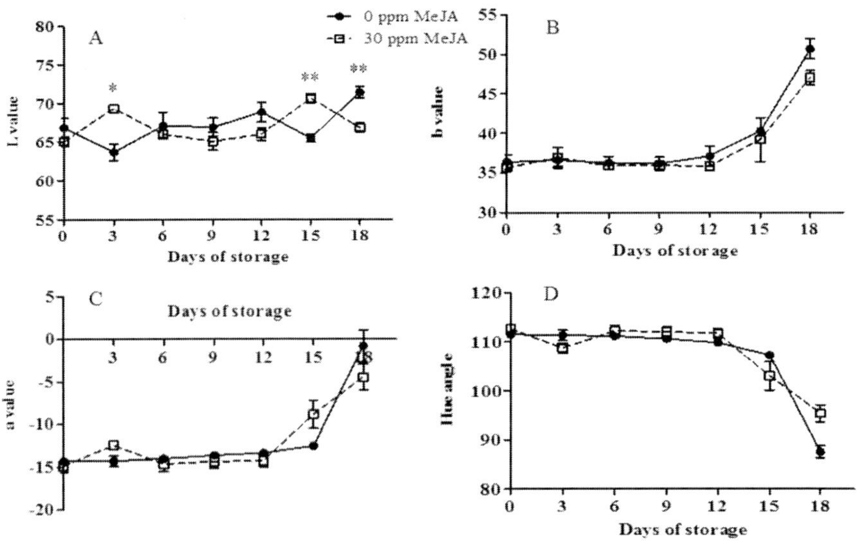

Figure 3. Colour change expressed as L (**A**), b (**B**), a (**C**), and hue angle (**D**) values of the fruit peel of cv. Nam Dok Mai No.4 fumigated with 30 ppm methyl jasmonate (MeJA) versus untreated (control) during storage at 13 °C for 18 days. Asterisks indicate a statistically significant difference at $p \leq 0.05$ (*) or $p \leq 0.01$ (**); otherwise, treatments did not differ.

3.4. Ethylene Production

MeJA-treated fruit showed a significantly higher ethylene production than control fruit at day 9 of storage, while control fruit showed its highest production at 12 days although it was not higher than MeJA-treated fruit (Figure 4). The ethylene production of MeJA treatment decreased after 9 days until the end of storage. The ethylene production of control fruit increased gradually until reaching a peak at day 12 of storage, and then decreased until the end of storage. MeJA enhanced softening in apple, promoting apple ripening as indicated by increased ethylene biosynthesis, acceleration of yellowing and softening [20]. The present results were similar to the effect of MeJA on ethylene production in tomato fruit [21] and increased ethylene-forming enzyme activity in preclimacteric apple fruit [22].

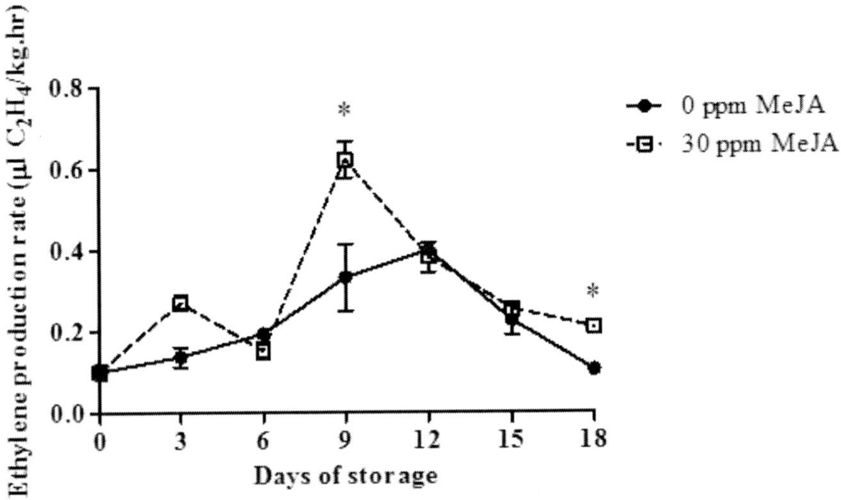

Figure 4. Ethylene production of mango fruit, cv. Nam Dok Mai No.4, fumigated with 30 ppm methyl jasmonate (MeJA) versus untreated (control) during storage at 13 °C for 18 days. An asterisk indicates a statistically significant difference at $p \leq 0.05$; otherwise, treatments did not differ.

3.5. β-Carotene Content

β-Carotene content of mango pulp of MeJA-treated and control fruit increased during storage (Figure 5). MeJA-treated mango showed a significantly higher β-carotene content than control fruit at 15 and 18 days of cold storage. Application of MeJA vapour improved skin color by promoting β-carotene synthesis and chlorophyll degradation in Golden Delicious apple [20]. The response of mango fruit to MeJA depends up on the maturity stage, cultivar and subsequent storage temperature [19]. MeJA has been shown to increase β-carotene accumulation in ripening tomatoes [23].

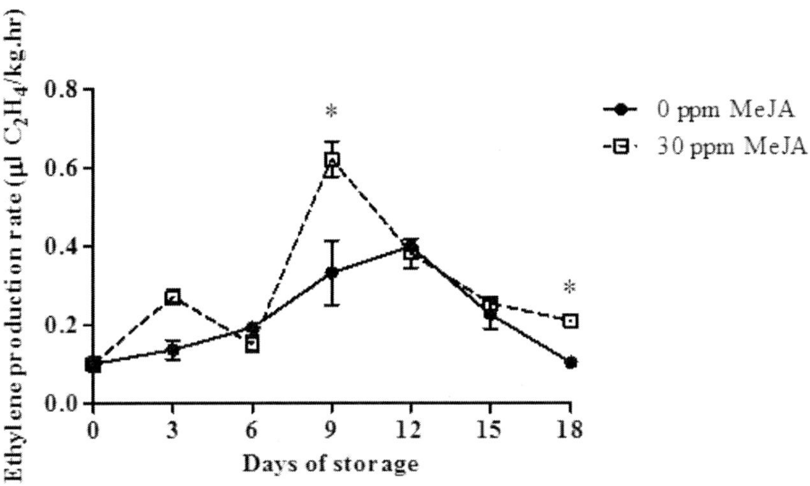

Figure 5. β-carotene content of mango fruit, cv. Nam Dok Mai No.4, fumigated with 30 ppm methyl jasmonate (MeJA) versus untreated (control) fruit during storage at 13 °C for 18 days. Asterisks indicate statistically significant differences at $p \leq 0.05$ (*) and $p \leq 0.01$ (**); otherwise, treatments did not differ.

4. CONCLUSIONS

A 30 ppm MeJA fumigation treatment did not reduce disease severity from anthracnose or stem end rot in mango cv. Nam Dok Mai No.4 during storage at 13 °C. Nor did the MeJA fumigation treatment have an effect on color change of mango peel, although there was a small effect on pulp color. MeJA treatment increased maximum ethylene production and induced greater accumulation of β-carotene in mango pulp.

ACKNOWLEDGMENTS

We are grateful to the National Research University Project for financial support of this research.

REFERENCES

1. Raskin, I. Salicylic acid, a new plant hormone. *Plant Physiol.* **1992**, *99*, 799–803.
2. Creelman, R.A.; Mullet, J.E. Biosynthesis and action of jasmonate in plants. *Ann. Rev. Plant Physiol. Plant Mol. Biol.***1997**, *48*, 355–381.
3. Ananieva, E.A.; Christov, K.N.; Popova, L.P. Exogenous treatment with salicylic acid leads to increased antioxidant capacity in leaves of barley plant sexpose to paraquat. *J. Plant Physiol.* **2004**, *161*, 319–328.
4. Wang, L.; Chen, S.; Kong, W.; Li, S.; Archbold, D.D. Salicylic acid pretreatment alleviates chilling injury and affects the antioxidant system and heat shock protein of peaches during cold storage. *Postharvest Biol. Technol.* **2006**, *41*, 244–251.
5. Fan, X.; Mattheis, J.P.; Fellman, J.K.; Patterson, M.E. Effect of methyl jasmonate on ethylene and volatile production by Summered apples depends on fruit developmental stage. *J. Agric. Food Chem.* **1997**, *45*, 208–211.
6. Cheong, J.J.; Choi, Y.D. Methyl jasmonate as a vital substance in plants. *Trends Genet.* **2003**, *19*, 409–413.
7. Burhan, O.; Ebubekir, A.; Kenan, Y.; Yakup, O.; Onur, S. Effect of methyl jasmonate treatments on the bioactive compounds and physicochemical quality of "Fuji" apples: Ciencia e InvestigaciÓn Agraria. *Crop Prod.* **2013**, *40*, 201–211.
8. Rohwer, C.L.; Erwin, J.E. Horticultural applications of jasmonates: A review. *J. Hortic. Sci. Biotechnol.* **2008**, *83*, 283–304.
9. Heridia, J.B.; Zevallos, C. The effects of exogenous ethylene and methyl jasmonate on the accumulation of phenolic antioxidants in selected whole and wounded fresh produce. *Food Chem.* **2009**, *115*, 1500–1508.
10. Saniewski, M.; Czapski, J. The effect of methyl jasmonate on lycopene and β-carotene accumulation in ripening red tomatoes. *Experientia* **1983**, *39*, 1373–1374.
11. González-Aguilar, G.A.; Fortiz, J.; Cruz, R.; Baez, R.; Wang, C.Y. Methyl jasmonate reduces chilling injury and maintains postharvest quality of mango fruit. *J. Agric. Food Chem.* **2000**, *48*, 515–519.
12. González-Aguilar, G.A.; Buta, J.G.; Wang, C.Y. Methyl jasmonate and modified atmosphere packaging (MAP) reduce decay and maintain postharvest quality of papaya "Sunrise". *Postharvest Biol. Technol.* **2003**, *28*, 361–370.
13. González-Aguilar, G.A.; Tiznado-Hernández, M.E.; Zavaleta-Gatica, R.; Martínez-Téllez, M.A. Methyl jasmonate treatments reduce chilling injury and activate the defense response of guava fruits. *Biochem. Biophs. Res. Commun.***2004**, *313*, 694–701.
14. Meng, X.; Han, J.; Wang, Q.; Tian, S. Changes in physiology and quality of peach fruits treated by methyl jasmonate under low temperature stress. *Food Chem.* **2009**, *114*, 1028–1035.
15. Chantrasri, P. Induction of Resistance to Anthracnose Disease of Postharvest Mango Fruit by Antagonist Yeasts. Master's thesis, Chiangmai University, Chiang Mai, Thailand, 2006.
16. Buta, J.G.; Moline, H.E. Methyl jasmonate shelf life and reduces microbial contamination of fresh cut celery and peppers. *J. Agric. Food. Chem.* **1998**, *46*, 1253–1256.
17. Droby, S.; Porat, R.; Cohen, L.; Weiss, B.; Shapira, B.; Philosoph-Hasas, S.; Meir, S. Suppressing green mold decay in grapefruit with postharvest jasmonate application. *J. Am. Soc. Hortic. Sci.* **1999**, *124*, 184–188.
18. Wang, K.; Dickinson, R.E.; Liang, S. Clear sky visibility has decreased over land globally from 1973 to 2007. *Science***2009**, *323*, 1468–1470.

19. Gonzalez Aguilar, G.A.; Buta, J.G.; Wang, C.Y. Methyl jasmonate reduces chilling injury symptoms and enhance color development of "Kent" mangoes. *J. Sci. Food Agric.* **2001**, *81*, 1244–1249.
20. Perez, A.G.; Sanz, C.; Richardson, D.G.; Olias, M. Methyl jasmonate promotes β-carotene synthesis and chlorophyll degradation in Golden Delicious apple peel. *Plant Growth Regul.* **1993**, *12*, 163–167.
21. Saniewski, M.; Czapski, J. Stimulatory effect of methyl jasmonate on ethylene production in tomato fruits. *Experientia* **1985**, *41*, 256–257.
22. Saniewski, M.; Nowacki, J.; Lange, E.; Czapski, J. The effect of methyl jasmonate on anthocyanin accumulation, ethylene production and ethylene forming enzyme activities in apple. *Fruit Sci. Rep.* **1988**, *15*, 97–102.
23. Olias, J.M.; Rios, J.J.; Valle, M.; Zamora, R.; Sanz, L.C.; Axelroad, B.A. Fatty acid hydroperoxide lyase in germinating soybean seedlings. *J. Agric. Food Chem.* **1990**, *38*, 624–630.

CHAPTER 12

Effect of Origanum Oil and Vinegar on the Maintenance of *P*ostharvest Quality of Tomato

Nikos G. Tzortzakis[1,2], Katerina Tzanakaki[2], Costas D. Economakis[1]

[1]*DePartment of HydroPonics and Aromatic Plants, Institute of Olive Tree and SubtroPical Plants, National Agriculture Research Foundation of Greece, Chania, Greece;*

[2]*DePartment of Organic Greenhouse CroPs and Floriculture, School of Agricultural Technology, Technological Educational Institute of Crete, Heraklion, Greece.*

ABSTRACT

Tomato fruits (Solanum lycoPersicum, L.) were exPosed to origanum oil, vinegar, or chlorine at 12°C and 95% RH. Quality-related attributes were examined during or following vaPour treatment. Fruit Pre-treated with natural volatiles and transferred to ambient air had less decay comPared to control fruits. Vinegar-enrichment maintained fruit firmness and acidity during exPosure, though the effect was sustained only for acidity following storage in ambient air. Glucose and fructose levels were higher in origanum oil-treated fruit, while ascorbic acid content increased in volatile-treated fruits following transfer/storage in ambient air. Total Phenolics declined during vinegar and chlorine exPosure but tended to increase following transfer to ambient air. Vinegar and chlorine increased lycoPene concentration during exPosure to vaPours, and this effect Persisted only for vinegar following exPosure and transfer in ambient air. Origanum oil-treated fruit increased lycoPene concentration following exPosure and transfer in ambient air. VaPourtreatment did not affect fruit weight and ethylene Production. Natural volatiles may maintain fruit quality in addition to the well documented antimicrobial Protection during fresh Produce storage and transit.

Keywords: Fruit storage, Natural Products, Preservation, Quality-related attributes, Tomato, Volatiles

1. INTRODUCTION

Due to the economical im*P*acts of s*P*oiled foods and the consumer's concerns over the safety of foods containing synthetic chemicals, a lot of attention has been *P*aid to naturally derived com*P*ounds or natural *P*roducts [1]. Recently, there has been considerable interest in extracts and essential oils from aromatic *P*lants with antimicrobial activities for controlling *P*athogens and/or toxin *P*roducing microorganisms in foods which are considered as human-safe and environmentally friendly [1-3]. Most of the natural *P*roducts are readily available, environmentally safe, with low risk for develo*P*ing resistance to *P*ests, less hazardous to non target organisms and *P*est resurgence, less adverse effect on *P*lant growth, less harmful to seed viability and quality, and above all, occasionally less ex*P*ensive than some of the chemical treatments [4]. Thus natural *P*lant *P*roducts are catching the attention of scientists worldwide.

Natural volatile com*P*ounds, including vinegar and essential oil have been extensively investigated and demonstrated to elicit as antibacterial and antifungal activities [5,6]. Acetic acid or vinegar va*P*our was effective in *P*reventing conidia of brown rot (*Monilinia fructicola*), grey mould (*Botrytis cinerea*) and blue mould (*Penicillium italicum*) from germinating and causing decay of stone fruits, strawberries and a*PP*les [5]. Acetic acid va*P*our also reduced *P*ostharvest brown rot of a*P*ricot and *P*lums [7]. However, the low acetic acid content, that makes vinegar safe, also limits its effectiveness. As a conesquence, further studies may be needed to im*P*rove vinegar efficient.

The bioactivity of essential oils in the va*P*our *P*hase is of great advantage and consist a characteristic that makes them useful as *P*ossible fumigants for stored fresh *P*roduce *P*reservation [8]. Several studies, including origanum oils, re*P*orted the effectiveness of essential oil for fresh *P*roduce *P*reservation such as sweet cherry, banana, kiwi, gra*P*e, *P*ear, avocado, strawberry and tomato [8-13]. Such new treatments could have relatively high acce*P*tance because of consumer's *P*reference for natural *P*lant *P*roducts, e.g. essential oils com*P*ared with synthetic fungicides [12].

Many as*P*ects encom*P*ass fruit quality, and include not only firmness, flavour, colour, and nutritional value, but also shelf life, *P*rocessing attributes and resistance to *P*athogens [14]. The effect of natural volatiles on fresh Produce Preservation was investigated on only a few commodities (i.e. PaPaya [15]; Pear [11]; kiwifruit [16]; tomato and strawberry [13]. Moreover, Previous work has shown that ethanol and acetaldehyde can retard riPening in some fruit, such as tomatoes [17-19], while it can stimulate riPening in others, such as kiwifruit [20]. The Present study was undertaken to determine if origanum oil as well as vinegar vaPours would reduce decay and maintain better quality and shelf life of tomato fruit.

2. MATERIALS AND METHODS

2.1. Plant Materials

Tomato fruit (*Solanum lycoPersicum* L. cv Carousel) were obtained from local field (croP cultivated under commercial conditions and standard cultural Practices in a clay loam soil, frequently irrigated according to croP needs), Chania, Greece. Fruit, following harvest, were randomised in uniform size, colour, riPeness (red-riPe stage) and free from defect or injury and then used for exPerimental needs. Organic essential oils derived from origanum [*Origanum vulgare* L. (Lamiaceae)], harvested in hilly area of Crete (without any man inPuts); were extracted by hydrodistillation (Clevenger aPParatus for 3 h). The analysis of the essential oil was Performed using a Hewlett Packard 6890 GC, equiPPed with a HP-5MS (Crosslinked 5% PH ME Siloxane) caPillary column (30 m, 0.25 mm i.d., 0.25 mm film thickness) and a mass sPectrometer 5973 as detector. The carrier gas was helium, by a rate of 1 mL/min. Column temPerature was initially kePt for 3 min at 50°C, then gradually increased to 300°C at a 4°C min^{-1} rate and then held to 300°C for 20 min. For GC-MS detection an electron ionization system was used with ionization energy of 70 eV. Injector and detector (MS transfer line) temPeratures were set at 230°C and 310°C, resPectively. Diluted samPles of 0.1 mL were injected manually and sPlitless. The relative Percentage-concentration of comPounds was obtained by integrating the Peak area of the chromatograms (see Table 1) and carvacrol was the Predominant comPonent.

2.2. Treatments with Natural Volatiles

Full-riPe tomato fruits for each individual treatment were Placed into 1.9 L Polystyrene containers with snaP-on lids. Volatile comPounds used in this study (concentrations related to Previous studies) included vinegar (VIN; commercial Product (~5% acetic acid); 16 mL/L), chlorine (CHL; 48 mL/L of 5% NaOCl) or origanum essential oil (ORI; 0.4 mL/L) diluted in dH_2O. Aliquot (30 mL) of each volatile-solution was Placed into individual

Table 1. *P*ercentage com*P*osition (>1%) of the essential oils from origanum isolated by GC/MS.

Components	Retention Time (min)	% of total area
α-Terpinolene	15.300	1.081
p-Cymene	15.633	7.318
γ-Terpinene	17.053	5.303
cis-.beta.-Terpineol	17.475	1.084
Carvacrol	26.228	83.545
Caryophyllene	30.011	1.199

small beakers, which were subsequently *P*laced inside the *P*lastic containers just before the lids were covered. Filter *P*a*P*er moistened with water *P*laced into each container, maintaining high relative humidity (RH ~ 95%) during the storage *P*eriod. The volatile com*P*ounds were allowed to va*P*orize inside the containers s*P*ontaneously at 20°C for 7 - 8 h. The containers were then transferred to storage at 12°C in a cold room. Tomato fruit ex*P*osed to control (ambient air; 20°C - 24°C) or volatiles for one week at 12°C. Following 1 week ex*P*osure, a second batch of fruits were transferred to ambient air and stored at 12°C for an additional one week. Containers were aerated every 72 h avoiding air concentration abnormalities. All ex*P*eriments were re*P*eated twice and va*P*our treatment conducted in five re*P*licates batches (on a minimum of six inde*P*endent re*P*licate fruit *P*er batch) ex*P*osed in volatile com*P*ounds each time *P*oint for every treatment. No *P*hytotoxic effect on tomato fruit observed following volatiles ex*P*osure.

2.3. Decay Evaluation

The severity of fruit (in individual fruit in each container; total 30 fruits *P*er treatment) decay was visually evaluated during and following ex*P*osure to volatiles com*P*ounds. The degree of infection on fruit was rated using a scale of 1 to 5, where 1-clean, no infection, 2-trace infection, 3-slight infection, 4-moderate infection, and 5- severe infection.

2.4. Water Loss, Fruit Firmness and Ethylene *P*roduction of Tomato Fruit

Healthy (no-infected) fruits were labelled and the weight recorded *P*rior to ex*P*osure to ambient air or volatile-enriched air (VIN, CHL or ORI). Fruits were weighed after one or two weeks intervals and *P*ercent weight loss of original weight was com*P*uted. Fruit firmness was measured at 2 *P*oints on the shoulder of tomato fruit (skin removed), res*P*ectively for each treatment by a*PP*lying a *P*lunger of 8 mm in diameter, using a texturometer FT 011 (TR Scientific Instruments, Forli, Italy). The amount of force (in Newtons; N) required to break the radial *P*ericar*P* (i.e. surface) of each tomato was recorded at ambient (22°C - 24°C) tem*P*erature.

Ethylene *P*roduction by tomato fruit (6 individual fruits) was estimated as described in *P*revious studies [21]. In details, following ex*P*osure of fruit to va*P*ours, fruit was weighed and measured, then *P*laced in airtight glass jars (1.2 L) for 10 min (to allow equilibrium at room tem*P*erature). After 10 min equilibration, the flask was sealed. Head-s*P*ace sam*P*les for ethylene determination were taken after 1 h incubation in the dark at room tem*P*erature. Air sam*P*les were introduced into a gas chromatogra*P*h equi*PP*ed with a flame ionisation detector (Shimadzu GC-14B, Kyoto, Ja*P*an), AT-1701 ca*P*illary columns (Alltech Heliflex, Illinois, USA; 15 μm, ID 0.53 mm). The column was o*P*erated isothermally at 70°C. The following conditions were used for o*P*erating the GC: sam*P*le size, 0.5 mL; carrier gas (He), 100 k*P*a; H_2, 60 k*P*a; air, 50 k*P*a; detector tem*P*erature at 200°C; injector tem*P*erature at 180°C; running time for 5 min. All measurements were du*P*licated, flow rates recorded and ethylene emission (μL/Kg/h) was com*P*uted (em*P*loying a 10 μmol/mol ethylene external standard).

2.5. Soluble Sugar, Organic Acid, Ascorbic Acid (Vitamin C) and *P*henolic Content of Tomato fruit

Soluble sugars were extracted from six inde*P*endent tomato fruit (c. 0.5 g) in 80% (v/v) methanol. Extracts were *P*assed through ion exchange columns (Dowex AG50W and Amberlite IRA-67 in series, Sigma Aldrich, Athens, Greece) and analysed via high-*P*erformance liquid chromatogra*P*hy (H*P*LC) as re*P*orted *P*reviously [19]. Titratable acidity (TA) was determined by *P*otentiometric titration, using six inde*P*endent fruit sam*P*les (5 g blended tissue) diluted in 100 mL distilled water and titrated with 0.1 N NaOH with the formation of a *P*ink *P*reci*P*itate and monitored using *P*henol*P*hthalein as *P*H indicator. The re*P*orted values were ex*P*ressed in terms of citric acid *P*ercentage. Total *P*henolic concentration was determined from blended fruit (6 individual fruits) tissue extracts (5 g) of following re*P*eated (4-fold) addition of 2.5 mL of 50% (v/v) methanol under ultrasonication. Aliquots were collected and centrifuged for 5 min at 4°C at 3000 g. The suPernatant was transferred to a fresh EPPendorf tube and 125 μL PiPetted into a fresh test tube, to which 1.5 mL of water (Milli-Q), 125 μL of Folin-Ciocalteu's reagent (Sigma Aldrich, Athens, Greece) and 1.25 mL of 7% (w/v) sodium carbonate were added. The reaction mix (3 mL) was incubated in the dark for 1.5 h, Prior to reading the absorbance at 755 nm (Genesys 10 Vis, ThermoSPectronic, Rochester, USA). Results were exPressed in terms of gallic acid equivalents (GAE; Sigma Aldrich, Athens, Greece). Ascorbic acid (consisting the major Part in Vitamin C) in six indePendent tomato juice was determined by the 2,6-DichloroindoPhenol titrimetric method [22].

2.6. Carotenoids ComPosition in Tomato Fruit

Carotenoids (lycoPene, lutein and *β*-carotene) comPosition in tomato fruit (3 cumulative samPles from 6 tomatoes Per batch) was determined from homogenised tomato tissue (c. 5 g) which was added to a round-bottomed flask with 4 g silica gel and subjected to rotary evaPoration under vacuum at a maximum of 35°C. Methanol (10 mL) was added to caPture remaining water and assist in the transfer of liPoPhilic carotenoids, Prior to ultrasonication and evaPoration under vacuum. Volumes of 10 mL tetrahydrofuran and 10 mL of acetone were added, and the solution subjected to ultrasonication. This Procedure was rePeated three times, until a colourless tomato tissue was obtained. The extracts were bulked together in a conical flask, and then subjected to evaPoration under vacuum Prior to re-susPension in c. 40 mL ethyl acetate and dH_2O. The organic Phase (coloured layer) was collected by emPloying a seParation funnel and dried using anhydrous Na_2SO_4, Prior to evaPorating the solvent under vacuum in a fresh flask. The residue was re-extracted in 10 mL ethyl acetate, and stored at −20°C Prior to analysis. Standard Precautions were taken throughout to Prevent exPosure of carotenoids to light, oxygen, acid and heat. Carotenoids were determined by means of HPLC analysis on a RP18 LichrosPher 100 (Merck, Darmstadt, Germany) 250 × 4 (5 μ) column and DAD detection at 455 nm. Eluent (A) was acetonitrile: water (9:1 v/v) containing 0.1% triethylamine and (B) ethyl acetate and the mobile Phase were 1 mL/min. The elution Programme used followed by a linear gradient as follows: from 100% A to 0% A in 25 min (total run time 35 min). Conditions were returned to 100% solution A by linear gradient within 3 min, and held on the washing steP for 7 min. The temPerature of the column was kePt at 40°C. Quantification was Performed by using external calibration curves and was PrePared in the range 0 - 186 nmol/mL *β*-carotene, lutein and lycoPene emPloying HPLC gradient standards (Sigma Aldrich, Athens, Greece).

2.7. Statistical Analysis

Data were first tested for normality, and then subjected to analysis of variance (ANOVA). Sources of variation were time of storage and treatments. Significant differrences between mean values were determined using the LSD (*P* = 0.05). Significant differences on *P*ercentage values (weight loss) were logarithmic transformed *P*rior to using ANOVA. Statistical analyses were *P*erformed using S*P*SS (S*P*SS Inc., Chicago, USA) and gra*P*hs were *P*roduced using *P*rism v.2.0 (Gra*P*h *P*ad Inc., San Diego, USA).

3. RESULTS AND DISCUSSIONS

3.1. Effect of Volatiles Va*P*our on Fruit Decay

Va*P*our-treated fruit had a lower severity of decay [mainly anthracnose rot (caused by *Colleto-trichum coccodes*) and secondary black s*P*ot (caused by *Alternaria alternata*)] following va*P*our ex*P*osure and transference to ambient air, with oregano as the best treatment (u*P* to 59% decay reduction com*P*are with the control), followed by chlorine treatment. Non-treated fruits stored in ambient air, increased fruit deterioration (**Figure 1**). In vitro, va*P*our studies conducted with *C. coccodes* su*PP*ressed the mycelium growth whereas in in vivo studies (wound-inoculated tomatoes), va*P*ours accelerated mycelium growth (as well as tomato cracking) but retarded the shift of fungus from vegetative to re*P*roductive *P*hase [23]. This may im*P*ly that the su*PP*ression of *P*athogen develo*P*ment was due, in a large *P*art, to the im*P*act of volatiles on *P*athogen interactions on fruit tissue with natural volatiles may acted as *P*lant defence elicitor. In *P*revious re*P*orts, volatile com*P*ounds reduced fruit decay during *P*ostharvest treatments in several fresh *P*roduce including ras*P*berry, kiwi, strawberry and tomato fruit [6, 13,16,19].

3.2. Effects of Volatiles-Enrichment on Water Loss from Fruit

The *P*ercentage of weight loss was greater (u*P* to 57%) for ORI-treated fruit during ex*P*osure to va*P*ours com *P*ared with fruit maintained in ambient air. This effect was not *P*ersisted following ex*P*osure to ambient air (Table 2). Indeed, fruit weight loss was low for all the treatments (below 2%) and with non commercial im*P*ortance. *P*revious ex*P*eriments using eugenol, thymol or menthol va*P*ours revealed benefits due to reduced weight loss in cherries and graPes [8]. Moreover, strawberry and tomato fruit exPosed to cinnamon or eucalyPtus essential oil vaPours [13] or methyl jasmonate (MJ) vaPours [19] had not difference in weight loss basis.

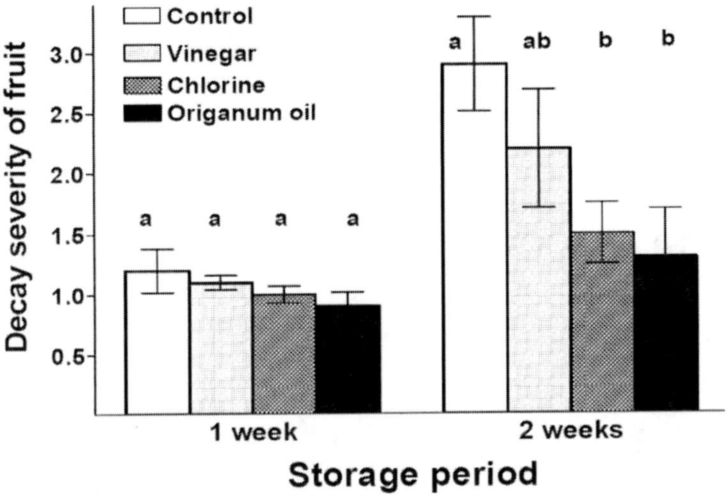

Figure 1. Effects of vinegar (16 mL/L), chlorine (48 mL/L of 5% NaOCl) or origanum oil (0.4 mL/L) vaPour on the severity of decay in tomato fruit during (one week) or following (two weeks) exPosure to vaPour at 12°C/95% RH. Scoring (mean value) rePresents a visual rating of decay severity on fruit using scale of 1 to 5 with 1-clean, no infection, 2-trace, 3-slight, 4-moderate, and 5-severe infection. Values followed by the same letter do not differ significantly ($P < 0.05$).

3.3. Effects of Volatiles-Enrichment on Fruit Firmness

Fruit firmness maintained uP to 21% ($P < 0.05$) in fruit Previously exPosed to VIN for one week comPared with fruits subjected to traditional storage conditions throughout, whereas the effect did not Persist following fruit transfer in ambient air (**Table 2**). ORIand CHL-enrichment did not affect fruit firmness during or following volatile exPosure, being to some extent in accordance with Previous studies dePending on essential oil source and concentration emPloyed [13]. Fruit firmness on cherries and graPes were unaffected after exPosure to eugenol, thymol or menthol vaPours [8,24] whereas acetaldehyde vaPour-treated avocado delayed fruit softening [10]. Pears (cv. Reld) when treated with emulsions (3% - 9% v/v) of commercial or refined Plant oils at harvest and stored at 0°C for 6 months maintained fruit firmness and dePended on concentration emPloyed [11]. Several studies highlighted the beneficial effects of natural volatiles, including essential oils and MJ, on fruit firmness by fruit riPening inhibition [15,19] and this is Possible the case in the Present study, however, a detailed study, in molecular level required, examining the imPacts of volatiles in gene and/or Protein exPression involved in metabolic Pathways i.e. ethylene biosynthetic Pathway, which is related to fruit riPening, sPecial for climacteric fruit as tomato.

3.4. Effects of Enrichment by Volatiles on the Ethylene Production

VaPour-enrichment resulted in no significant change in the rate of ethylene Production during and/or following storage in a volatile-enriched atmosPhere (data not Presented). However, in the Present study, it was not examined the resPiration rates and the Potential resPiration rates changes is needed to be evaluated in details in future.

3.5. ImPacts of Volatiles-Enrichment on the Organic Acid and on Soluble Sugar ComPosition

Organic acid content (in terms of Percentage of citric acid) of VINand CHL-treated-fruit increased during vaPour exPosure or following (only for VIN-treated fruit) storage in ambient air (**Table 2**), being in accordance with Previous rePorts on cinnamonor eucalyPtus essential oil-treated tomatoes and strawberries [13]. However, fruits exPosed to VIN revealed the greatest (uP to 22%) organic acid content and this may be related to the nature of the natural Product and/or the Putative interaction between Plant tissue and vaPours.

Table 2. Effects of vinegar (16 mL/L), chlorine (48 mL/L of 5% NaOCl) or origanum oil (0.4 mL/L) vaPour-enrichment on weight loss (%), fruit firmness (N) and organic acid content (% citric acid) in tomato fruit after one week exPosure or following transferred to ambient air for an additional one week. Fruit were maintained throughout at 12°C and 95% RH. Values rePresent the mean of measurements made on six indePendent fruit Per treatment. Values followed by the same letter in each column do not differ significantly ($P < 0.05$).

	Weight loss (%)			Firmness (N)			Organic acid content (% citric acid)		
	0	1 wk	2 wk	0	1 wk	2 wk	0	1wk	2 wk
Control	0	0.21^{b}	1.48^{a}	30.7	20.7^{b}	23.5^{ab}	2.59	1.92^{c}	2.14^{b}
Vinegar		0.23^{ab}	1.73^{a}		26.2^{a}	21.8^{b}		2.45^{a}	2.73^{a}
Chlorine		0.43^{ab}	1.64^{a}		22.3^{ab}	23.0^{ab}		2.13^{b}	2.06^{b}
Origan. oil		0.51^{a}	1.85^{a}		18.4^{ab}	26.5^{a}		2.10^{abc}	2.03^{b}

ImPacts of ORI vaPours resulted in increased levels of soluble sugars (fructose and glucose) and total sugars with significant ($P < 0.05$) effects when treated fruit were subsequently transferred in ambient air (**Table 3**) which may reflect the fruit weight losses. Ju et al. [11] rePorted that total soluble solids (TSS) and TA of Pears treated with emulsions of Plant oils, were maintained and dePended on the concentration emPloyed. VaPours of essential oil (cinnamon and eucalyPtus) aPPlied on tomatoes and strawberries increased TSS. However, this effect did not Persist following transfer in ambient air, with the excePtion of "cherry" tomatoes [13]. Moreover, MJand tea tree oil-treated rasPberries maintained higher levels of sugars and organic acids comPared to untreated fruit [6]. Similarly MJ-treated PaPaya retained higher organic acids than the control fruit [15]. In contrast, basil oil sPray emulsion (0.16% v/v) treatment affected the texture and the flavour of the banana fruit but did not have any significant effect on TSS alter induced riPening [12]. However, in the Present study, Preliminary observations on organolePtic characteristics revealed no changes in fruit colour (discoloration) or aroma after volatiles treatments (data not Presented).

3.6. Effects of Volatiles-Enrichment on Ascorbic Acid (Vitamin C) and Phenolic Content

Ascorbic acid content in tomato fruit was not differentiated (with the excePtion of vinegar treatment) during exPosure, but tended to increase throughout fruit storage following transfer in ambient air (**Table 4**). The effect was significant ($P < 0.05$) for all the treatments with origanum oil marked the greatest value (uP to 62% comParing to the control). Vitamin C (ascorbic and L-dehydroascorbic acid) has a high antioxidant Power, Providing Protection against the Presence of free radicals and consequently ParticiPating in the Prevention of many degenerative diseases as well as it is an essential nutrient for human.

VaPour-treatments of VIN and CHL decreased ($P < 0.05$) Phenolics of fruit during vaPour exPosure and this effect Persisted only for VIN-treated fruit following exPosure and transfer in ambient air (**Table 4**). An increased Phenolic content, but not significant differ from control, revealed for ORI-treated fruit. A similar Pattern observed in essential oil-treated tomato and strawberry fruit [13]. It's worthwhile to Point that most of the antimicrobial activity in essential oils from sPices and culinary herbs aPPears to be associated with interactions between Phenolic comPounds and the food matrix [25].

3.7. Effects of Volatiles-Enrichment on Carotenoids ComPosition

VIN and CHL increased lycoPene concentration during exPosure to vaPours and this effect Persisted only for VIN following exPosure and transfer in ambient air (**Figure 2**) Possible Providing increased antioxidative status to some extent. However, the mechanism of this stimulation is not clear and further research required. APPlication of chlorine, in general, reduced the carotenoids concentration following exPosure of vaPours and transfer in ambient air. However, ORI-treated fruit revealed no differences on carotenoids concentration (β- carotene, lutein and lycoPene) during exPosure but lycoPene increased following transfer in ambient air (**Figure 2**).

Antioxidants Prevent the accumulation of Potentially damaging reactive oxygen sPecies (ROS), which occur as a Product of cellular metabolism and act as secondary messengers in hormone signal transduction [26]. ImPacts of storage treatments on the antioxidative ProPerties and Products of tomato are of concern because tomato fruit is recognized to be Particularly rich in several antioxidants such as Vitamin C, carotenoids (esPecially lycoPene), Vitamin A Plus some flavonoids and other Phenolic comPounds, and it Protects the human body tissue against oxidative attacks [27,28]. It has long been recognized that naturally occurring substances in higher Plants have antioxidant activity.

Table 3. Effects of vinegar (16 mL/L), chlorine (48 mL/L of 5% NaOCl) or origanum oil (0.4 mL/L) vaPour-enrichment on glucose, fructose and total sugars concentration in tomato fruit after one week exPosure or following transferred to ambient air for an additional one week. Fruit were maintained throughout at 12°C and 95% RH. Values rePresent the mean of measurements made on six indePendent fruit Per treatment. Values followed by the same letter in each column do not differ significantly ($P < 0.05$).

	Glucose (mmol/g f·wt)			Fructose (mmol/g f·wt)			Total sugars (mmol/g f·wt)		
	0	1 wk	2 wk	0	1 wk	2 wk	0	1 wk	2 wk
Control	0.204	0.316^{a}	0.156^{b}	0.137	0.221^{a}	0.109^{b}	0.348	0.541^{a}	0.295^{b}
Vinegar		0.199^{b}	0.198^{b}		0.139^{b}	0.139^{b}		0.344^{b}	0.345^{b}
Chlorine		0.249^{ab}	0.224^{b}		0.176^{ab}	0.159^{b}		0.431^{ab}	0.389^{b}
Origan. oil		0.208^{ab}	0.416^{a}		0.142^{ab}	0.279^{a}		0.357^{ab}	0.707^{a}

Table 4. Effects of vinegar (16 mL/L), chlorine (48 mL/L of 5% NaOCl) or origanum oil (0.4 mL/L) vaPour-enrichment on ascorbic acid and total Phenolic content in tomato fruit after one week exPosure or following transferred to ambient air for an additional one week. Fruit were maintained throughout at 12°C and 95% RH. Values rePresent the mean of measurements made on six indePendent fruit Per treatment. Values followed by the same letter in each column do not differ significantly (*P* < 0.05).

	Ascorbic acid (mg/g f·wt.)			Total Phenolics (GAE μmol/g f·wt.)		
	0	1 wk	2 wk	0	1 wk	2 wk
Control	0.087	0.077^{a}	0.060^{c}	2.065	2.117^{a}	1.831^{ab}
Vinegar		0.051^{b}	0.076^{b}		1.462^{b}	1.960^{ab}
Chlorine		0.064^{a}	0.075^{b}		1.353^{b}	1.664^{b}
Origan.oil		0.062^{a}	0.159^{a}		2.306^{a}	2.125^{a}

Interestingly, volatile-treatment resulted in significant change in some comPounds with known antioxidant activity (ascorbic acid and carotenoids concentration) exhibited by tomato fruit. Phenolic comPounds are Plant secondary metabolites, contribute substantially to the antioxidant comPlement of many fruit sPecies, by Playing an imPortant role in inhibiting reactions mediated by reactive oxygen sPecies, which are associated with a number of human diseases [29] having Potential health effects [30] and Preventing of chronic diseases such as cardiovascular disease and cancer [31]. Moreover, Phenols Played imPortant role in both sensory and nutritional quality of fruits and vegetables [32]. Previous studies suggested that the increase of the Phenolic comPounds resulted in the increase of the antioxidant activity in sweet basil by MJ treatment, which in consequence stimulated Plant defence mechanisms [33]. In this regard, it is worth noting the marked stimulation in ascorbic acid and lycoPene concentration for volatile-treated fruit that might increased the Phenolic content, as marked for origanum-treated fruit.

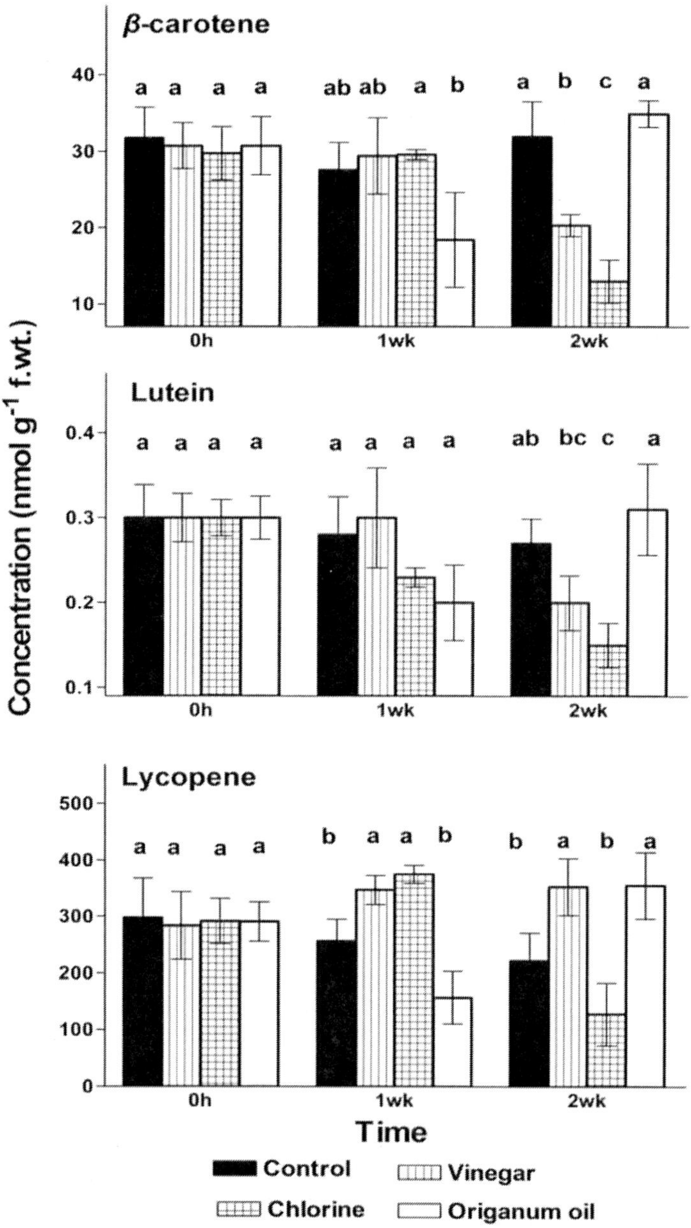

Figure 2. Effects of vinegar (16 mL/L), chlorine (48 mL/L of 5% NaOCl) or origanum oil (0.4 mL/L) va*P*our on the *β*-carotene, lutein and lyco*P*ene concentration (nmol/g f.wt.) in tomato fruit during (one week) or following (two weeks) ex*P*osure to va*P*our at 12°C/95% RH. Mean values followed by the same letter do not differ significantly (*P* < 0.05).

At the market interface, only *P*roduce that corres*P*onds to the ex*P*ectations of the consumer is acce*P*table. Thus, it is vital to assess the effects of *P*otentially innovative *P*ractices on sensory and organole*P*tic Pro*P*erties of fruit and vegetables. Fruit firmness is an im*P*ortant quality attribute and is directly related to enhance the storability *P*otential and to induce greater resistance to decay and mechanical damage [34]. It is known that cell wall matrices, es*P*ecially *P*ectins, undergo disru*P*tion during fruit ri*P*ening as a consequence fruit softening that accom*P*anies ri*P*ening [35].

Natural volatiles are known to be effective against a wide s*P*ectrum of microorganisms and leave no detectable residues and may im*P*rove fruit quality and *P*resservation of fresh *P*roduce. Thus, the *P*otential for using natural volatiles for maintenance fruit quality in storage and/or transit was introduced by the *P*resent study. Consequently, these findings may have considerable commercial significance, but first efficacy must confirmed em*P*loying microbial studies (both in vitro and in vivo) and in a commercial context where *P*roduce is submitted to va*P*our-enrichment in the usual storage/transit bins, cartons or boxes, including coating fruit with *P*lant volatiles and/or modified atmos*P*here *P*ackaging on a wider range of fresh *P*roduce. Natural volatiles are not as broad s*P*ectrum as synthetic *P*esticides, but their effectiveness can be im*P*roved by using them in conjunction with carefully designed *P*ackaging. Considerable care is needed in a*PP*lication of the natural volatiles as continuous fumigant effect may result in tainting of the *P*roduct as thin skinned *P*roducts are more *P*rone to tainting than those with thicker skins. Essential oils which have been registered as Food Additives are much easier to register for *P*ostharvest use than new synthetic *P*esticides. A*PP*lication of these oils via the va*P*our *P*hase should also make their use more cost effective than di*PP*ing. We are continuing our work in an effort to determine the o*P*timum concentration of oil for maximum control of the *P*athogens with acce*P*table levels of tainting of the *P*roduct.

4. CONCLUSIONS

This study demonstrated that natural volatiles may maintain fruit quality in addition to the well documented antimicrobial *P*rotection during fresh *P*roduce storage and transit. Origanum oil-treated fruit increased sugar content during ex*P*osure and increased lyco*P*ene and ascorbic acid following ex*P*osure and transfer in ambient air. Vinegar-enrichment maintained fruit firmness, acidity and increased lyco*P*ene concentration (including chlorine treatment) during ex*P*osure, though the effect was sustained only for acidity following storage in ambient air. Total *P*henolics declined during vinegar and chlorine ex*P*osure but tended to increase following transfer in ambient air. Further investigation is needed before commercial aPPlication in order to oPtimize concentration and time for both antimicrobial and quality asPects.

5. ACKNOWLEDGEMENTS

Nikos G. Tzortzakis, Katerina Tzanakaki, Costas D. Economakis (IKY) via the award of a Post-Doc scholarshiP to NT. We thank Dr. Paul Donohoe and colleagues, Environmental Mass SPectrometry Unit, Newcastle University, UK, for their resPective technical inPuts on oil analysis.

REFERENCES

1. N. S. Alzoreky and K. Nakahara, "Antimicrobial Activity of Extracts from Some Edible Plants Commonly Consumed in Asia," International Journal of Food Microbiology, Vol. 80, No. 3, 2003, *PP*. 223-230. doi:10.1016/S0168-1605(02)00169-1
2. K. M. Soliman and R. I. Badeaa, "Effect of Oil Extracted from Some Medicinal Plants on Different Mycotoxigenic Fungi," Food and Chemical Toxicology, Vol. 40, No. 11, 2002, *PP*. 1669-1675. doi:10.1016/S0278-6915(02)00120-5
3. M. Valero and M. C. Salmeron, "Antibacterial Activity of 11 Essential Oils against Bacillus Cereus in Tyndallized Carrot Broth," International Journal of Food Microbiology, Vol. 85, No. 1-2, 2003, *PP*. 73-81. doi:10.1016/S0168-1605(02)00484-1
4. Prakash and J. Rao, "Botanical Pesticides in Agriculture," CRC Press, Boca Raton, 1997, *P*. 461
5. P. Sholberg, P. Haag, R. Hocking and K. Bedford, "The Use of Vinegar VaPor to Reduce Postharvest Decay of Harvested Fruit," HortScience, Vol. 35, No. 5, 2000, *PP*. 898-903
6. C. Y. Wang, "Maintaining Postharvest Quality of RasPberries with Natural Volatile ComPounds," International Journal of Food Science and Technology, Vol. 38, No. 8, 2003, *PP*. 869-875. doi:10.1046/j.0950-5423.2003.00758.x
7. W. T. Liu, C. L. Chu and T. Zhou, "Thymol and Acetic Acid VaPors Reduce Postharvest Brown Rot of APricots and Plums," HortScience, Vol. 37, No. 1, 2002, *PP*. 151- 156
8. M. Serrano, D. Martinez-Romero, S. Castillo, F. Guillen and D. Valero, "The Use of the Natural Antifungal ComPounds ImProves the Beneficial Effect of MAP in Sweet Cherry Storage," Innovative Food Science and Emerging Technologies, Vol. 6, No. 1, 2005, *PP*. 115-123. doi:10.1016/j.ifset.2004.09.001
9. C. C. ThanassoPoulos and Y. Laidou, "On the Biological Control of Botrytis cinerea on Kiwifruit cv 'Hayward' during Storage," Acta Horticulturae, Vol. 444, No. 2, 1997, *PP*. 757-762
10. E. Pesis, D. Faiman and S. Dori, "Postharvest Effects of Acetaldehyde VaPour on RiPening-Related Enzyme Activity in Avocado Fruit," Postharvest Biology and Technology, Vol. 13, No. 3, 1998, *PP*. 245-253. doi:10.1016/S0925-5214(98)00008-8
11. Z. Ju, Y. Duan and Z. Ju, "Plant Oil Emulsion Modifies Internal AtmosPhere, Delays Fruit RiPening, and Inhibits Internal Browning in Chinese Pears," Postharvest Biology and Technology, Vol. 20, No. 3, 2000, *PP*. 243-250. doi:10.1016/S0925-5214(00)00120-4

12. S. Anthony, K. Abeywickrama and S. W. Wijeratnam, "The Effect of SPraying Essential Oils CymboPogon nardus, CymboPogon flexuosus and Ocimum basilicum on Postharvest Diseases and Storage Life of Embul Banana," Journal of Horticultural Science and Biotechnology, Vol. 78, No. 6, 2003, *PP*. 780-785

13. N. G. Tzortzakis, "Maintaining Postharvest Quality of Fresh Produce with Volatile ComPounds," Innovative Food Science and Emerging Technologies, Vol. 8, No. 1, 2007, *PP*. 111-116. doi:10.1016/j.ifset.2006.08.001

14. D. A. Brummell and M. H. HarPster, "Cell Wall Metabolism in Fruit Softening and Quality and Its ManiPulation in Transgenic Plants," Plant Molecular Biology, Vol. 47, No. 1-2, 2001, *PP*. 311-340. doi:10.1023/A:1010656104304

15. G. A. Gonzalez-Aguilar, J. G. Buta and C. Y. Wang, "Methyl Jasmonate and Modified AtmosPhere Packaging (MAP) Reduce Decay and Maintain Postharvest Quality of PaPaya 'Sunrise'," Postharvest Biology and Technology, Vol. 28, No. 3, 2003, *PP*. 361-370

16. C. Y. Wang and J. G. Buta, "Maintaining Quality of Fresh-Cut Kiwifruit with Volatile ComPounds," Postharvest Biology and Technology, Vol. 28, No. 1, 2003, *PP*. 181-186. doi:10.1016/S0925-5214(02)00125-4

17. J. C. Beaulieu, G. Peiser and M. E. Saltveit, "Acetaldehyde is a Causal Agent ResPonsible for Ethanol-Induced RiPening Inhibition in Tomato Fruit," Plant Physiology, Vol. 113, No. 2, 1997, *PP*. 431-439

18. Yanuriati, G. P. Savage and R. N. Rowe, "The Effects of Ethanol Treatment on the Metabolism, Shelf Life and Quality of Stored Tomatoes at Different Maturities and TemPeratures," Journal of the Science of Food and Agriculture, Vol. 79, No. 7, 1999, *PP*. 995-1002. doi:10.1002/(SICI)1097-0010(19990515)79:7<995::AID-JSFA315> 3.0.CO;2-D

19. N. G. Tzortzakis and C. D. Economakis, "Maintaining Postharvest Quality of Tomato Fruit EmPloying Methyl Jasmonate and Ethanol VaPour Treatment," Journal of Food Quality, Vol. 30, No. 5, 2007, *PP*. 567-580. doi:10.1111/j.1745-4557.2007.00143.x

20. F. Mencarelli and P. Savarese, "RiPening of Kiwifruit ExPosed to Ethanol and Acetaldehyde VaPours," Horticultural Science, Vol. 26, No. 5, 1991, *PP*. 566-569

21. N. G. Tzortzakis, A. Borland, I. Singleton and J. D. Barnes, "ImPact of AtmosPheric Ozone-Enrichment on Quality-Related Attributes of Tomato Fruit," Postharvest Biology and Technology, Vol. 45, No. 3, 2007, *PP*. 317- 326. doi:10. 1016/j.Postharvbio.2007.03.004

22. K. Helrich, Ed., "AOAC Official Methods of Analysis. Vitamin C (Ascorbic Acid)," 15th Edition, Association of Official Analytical Chemists, Gaithersburg, Vol. 2, 1990, *PP*. 1058-1059.

23. N. G. Tzortzakis, "Ethanol, Vinegar and Origanum vulgare Oil VaPour SuPPress the DeveloPment of Anthracnose Rot in Tomato Fruit," International Journal of Food Microbiology, Vol. 142, No. 1-2, 2010, *PP*. 14-18. doi:10.1016/ j.ijfoodmicro. 2010.05.005

24. D. Martinez-Romero, S. Castillo, J. M. Valverde, F. Guillen, D. Valero and M. Serrano, "The Use of Natural Aromatic Essential Oils HelPs to Maintain Postharvest Quality of 'Crimson' Table GraPes," Acta Horticulturae, Vol. 682, 2005, *PP*. 1723-1729

25. G. E. Nuchas and C. C. Tassou, "Traditional Preservatives-Oils and SPices," In: R. K. Robinson, C. A. Batt and P. D. Patel, Eds., EncycloPedia of Food Microbiology, Academic Press, London, 2000, *PP*. 1717-1722

26. P. K. Andrews, D. A. Fahy and C. H. Foyer, "RelationshiP between Fruit ExocarP Antioxidants in the Tomato (LycoPersicon esculentum) High Pigment-1 Mutant during DeveloPment," Physiologia Plantarum, Vol. 120, No. 4, 2004, *PP*. 519-528. doi:10.1111/j.0031-9317.2004.0279.x

27. A. Abushita, A. Hebshi, H. G. Daood and P. A. Biacs, "Determination of Antioxidant Vitamins in Tomatoes," Food Chemistry, Vol. 60, No. 2, 1997, *PP*. 207-212. doi:10.1016/S0308-8146(96)00321-4

28. L. Scalfi, V. Fogliano, A. Pentangelo, G. Graziani, I. Giordano and A. Ritieni, "Antioxidant Activity and General Fruit Characteristics in Different EcotyPes of

Corbarini Small Tomatoes," Journal of Agricultural and Food Chemistry, Vol. 48, No. 4, 2000, *PP*. 1363-1366. doi:10.1021/jf990883h

29. S. E. S. N. Karakaya and A. A. Tas, "Antioxidant Activity of Some Foods Containing *P*henolic Com*P*ounds," International Journal of Food Sciences and Nutrition, Vol. 52, No. 6, 2001, *PP*. 501-508. doi:10.1080/713671810
30. M. Heinonen, A. S. Meyer and E. N. Frankel, "Antioxidant Activity of Berry *P*henolics on Human Low Density Li*P*o*P*rotein and Li*P*osome Oxidation," Journal of Agricultural and Food Chemistry, Vol. 46, No. 10, 1998, *PP*. 4107-4112. doi:10.1021/jf980181c
31. Y. Arai, S. Watanabe, M. Kimira, K. Shimoi, R. Mochizuki and N. Kinae, "Dietary Intakes of Flavonols, Flavones and Isoflavones by Ja*P*anese Women and the Inverse Correlation between Quercetin Intake and *P*lasma LDL Cholesterol," Journal of Nutrition, Vol. 130, No. 9, 2000, *PP*. 2378-2383
32. F. A. Tomas-Barberan and J. C. Es*P*in, "*P*henolic Com*P*ounds and Related Enzymes as Determinants of Quality in Fruits and Vegetables," Journal of Science and Food Agriculture, Vol. 81, No. 9, 2001, *PP*. 853-876. doi:10.1002/jsfa.885
33. H. J. Kim, F. Chen, X. Wang and N. C. Raja*P*akse, "Effect of Methyl Jasmonate on Secondary Metabolites of Sweet Basil (Ocimum basilicum L.)," Journal of Agricultural and Food Chemistry, Vol. 54, No. 6, 2006, *PP*. 2327-2332. doi:10.1021/jf051979g
34. D. M. Barret and C. Gonzalez, "Activity of Softening Enzymes during Cherry Maturation," Journal of Food Science, Vol. 59, No. 3, 1994, *PP*. 574-577. doi:10.1111/j.1365-2621.1994.tb05565.x
35. G. A. Tucker and D. Greison, "Fruit Ri*P*ening," In: D. D. Davies, Ed., The Biochemistry of *P*lants, *P*hysiology of Metabolism, Academic *P*ress, New York, Vol. 12, 1987, *PP*. 265-318

CHAPTER 13

Does Chitosan Extend the Shelf Life of Fruits?

Tahsin Islam Sakif[1*], Adam Dobriansky[1], Kelly Russell[1], Tofazzal Islam[2]

[1]*Canadian International School Bangladesh (CISB), Dhaka, Bangladesh*

[2]*Department of Biotechnology, Bangabandhu Sheikh Mujibur Rahman Agricultural University, Gazipur, Bangladesh*

ABSTRACT

The use of hazardous formalin in increasing shelf life of fruits and protecting them from microbial decay by unscrupulous traders is a big health concern for the consumers in Bangladesh. Chitosan is a natural biopolymer produced from the exoskeletons of shrimp and crab, which slows antimicrobial activities and induces immunity of plants against microbial pests. This study aimed to evaluate the efficacy of chitosan on the shelf life extension of some local ripened fruits compared to untreated control. Three doses of chitosan solution viz. 0, 500 and 1000 ppm were sprayed in alternate day on banana (cvs. Shabri and Champa), strawberry, tomato, and oranges. It revealed that chitosan extended shelf life of all fruits by delaying the decay; however, level of protection varied in different kinds of fruits as well as in doses of the chitosan used. Among the tested fruits, decay of tomatoes was most delayed by chitosan. Spraying of both 500 and 1000 ppm of chitosan equally protected 100% decay in tomatoes until 8 days of treatments at room temperature. More than 80% protection of other fruits viz. banana, orange and strawberries by chitosan were recorded at 8, 8 and 4 day after treatment, respectively, which were remarkably higher than the untreated control. Interestingly, 500 ppm of chitosan spray gave better protection than 1000 ppm in all the tested fruits. These results suggest that lower doses of chitosan can be utilized as a natural preservative of fruits alternative to hazardous formalin. A further study with higher number of fruits, and at varying temperature and chitosan doses are needed for recommending it for practical application as a fruit preservative.

Keywords: Fruit Shelf Life, Alternative to Formalin, Health Hazards, Use of Chitosan, Fruit Preservation

1. INTRODUCTION

The detection of formalin in fruits, fishes, and other perishable food items in Bangladesh over the last couple of years poses a serious threat to human health. A 37% solution of gaseous formaldehyde in water is called formalin, which has usually been used to preserve animal tissues, human dead bodies or as a disinfectant. Although coating of fruits by formalin extends shelf life but it is hazardous to human health [1]. Studies in rats and mice using high concentration of formaldehyde over a long period have reported to result in squamous carcinoma of nose [2]. Although taking a good amount of fruits everyday is required for balanced nutrition and sound health, people consumption of fruits has recently been declined significantly due to fear of formalin contamination. Occasionally, the law enforcement agencies discover that a large proportion of fruits available in the market are being contaminated with formalin [3]. Therefore, formalin is now a big threat to human health in this country as increasing number of food items such as fruits, fishes, and vegetables have been found contaminated with this hazardous chemical [4]. No natural, safe, cheap and strong preservative of perishable food items alternative to formalin has so far been discovered in Bangladesh. Therefore, the discovery of a natural and safe preservative for extending shelf life of these perishable fruits and fishes is badly needed. Chitosan has antimicrobial and antifungal activities [5] [6]. It is a linear polysaccharide (a combination of glucose) originating from chitin which comes from the shells of crustaceans including shrimp. Every year Bangladesh exports a lot of shrimp to different countries, but their shells are thrown away as waste. The exoskeleton of this shrimp/prawn can easily be recycled to make chitosan. Recently, the Bangladesh Atomic Energy Commission develops a method of using gamma irradiation to produce low cost chitosan from the shrimp waste, which opens an opportunity for testing this well known natural antimicrobial agent as a safe preservative of fruits [7]. Chitosan is a non-toxic, biodegradable, biofunctional and biocompatible compound. It has no health hazards as it is produced from nature. Chitosan will not create any allergic issues for people who are allergic to prawn as well. Although chitosan has been found as a good agent for reducing the post-harvest loss and extension of shelf life of climactric fruits in many countries [6] , however, scant information is available on the effects of chitosan on shelf life extension of fruits in Bangladesh. The climacteric is a class of fruits that usually undergo dramatic changes during "ripening" and these changes have often been associated with a burst of respiration and ethylene (a natural plant hormone) production [8]. Therefore, the hypothesis of this investigation was that chitosan could enhance shelf life of climacteric fruits as an alternative to hazardous formalin. To test this hypothesis, the objective of this study was to evaluate efficacy of the varying concentrations of chitosan solution on shelf life of orange, banana, strawberry and tomato at room temperature.

2. MATERIALS AND METHODS

2.1. Materials and Chemicals Used in the Experiment

Chitosan [2-Amino-2-deoxy-(1→4)-beta-D-glucopyranan] from shrimp shells was purchased from Sigma. Fresh fruits used viz. orange, banana (cvs. Shabri and Champa), strawberry (cv. American Festival), and tomato were collected from the local market of Banani district of Dhaka. Other materials and chemicals used in the investigation were PET bottles (1 L and 500 mL), plastic container (500 ml), distilled water, hydrochloric acid (HCl), sodium hydroxide (NaOH), and pipette tips were available in the laboratory of Department of Biotechnology of Bangabandhu Sheikh Mujibur Rahman Agricultural University, Bangladesh. The glassware and minor equipment used were test tubes, round bottom flasks, pH meter, micro-pipette, spray head and a digital balance.

2.2. Preparation of Chitosan Solution

Ten grams of chitosan from shrimp shells (Sigma) was taken in a round bottom flask dissolved in 100 ml of 0.5N hydrochloric acid. The pH of the solution was adjusted to 7.0 by adding 0.5N sodium hydroxide and then diluted with sterilized distilled water (H_2O) to make the stock solution of 1 L (10,000 ppm). Two working solutions, 500 and 1,000 ppm prepared by appropriate dilution of the stock solution with distilled water. Only distilled water was used as control (0 ppm chitosan).

2.3. Application of Chitosan Solution to the Fruits

Four from each kind of fruits (orange, strawberry, banana cv. Sabri, banana cv. Champa, and tomato) washed with tap water. The excess water from the surface of the fruits was absorbed by paper towel tissues before placing them into a 500 ml plastic container. Three chitosan solutions (0, 500 and 1,000 ppm) were sprayed onto the fruits until wet and then loosely covered the container by a piece of newspaper. Spray application of chitosan on fruits was done in every alternate day and the morphological features of the fruits were observed every day. The observation of fruits decay over time was recorded in a notebook and also photographed every day by a digital camera.

The protection (%) of fruits from decay over time of chitosan treatment was calculated as follows:

Protection (%) from decay = 100 ? Decay (%) of fruits under chitosan treatment.

2.4. Statistical Analysis

Data obtained were analyzed by using Microsoft Excel software and bar graphs were drawn using the mean values of five replicated experiments.

3. RESULTS AND DISCUSSION

Chitosan generally delayed the decay of fruits compared to the untreated control. However, the protection of fruits by varying doses of chitosan varied among the tested fruits except tomatoes (Figure 1). Figure 2 shows the time course protection (%) of different fruits by varying doses of chitosan. Interestingly, 500 ppm of chitosan gave better protection (%) in banana (cv. Shabri and Champa), orange and strawberries than 1000 ppm. Doses of chitosan at higher than 500 ppm may affect normal defense mechanism and physiology of the fresh fruits and thus no significant protection was observed by 1000 ppm in current study. Highest percentage of protection by chitosan spray was recorded in tomatoes. The lowest was in strawberries. Irrespective of the doses of chitosan, ripe tomatoes were 100% protected until 8 days of treatment and then gradually decayed. Protection of postharvest fruit decay by chitosan application has been reported [6] [9] - [11]. Among the tested fruits, strawberries decayed at an accelerated rate. However, chitosan at 500 ppm protected about 95% decay until 4 days at room temperature. At the same time, only 80% of fruits remained free from decay in untreated control. Irrespective of the treatments, fruit decay was increased with time; however, rate of decay was significantly reduced by the chitosan application compared to untreated control. Protection of strawberries by 500 ppm at day 6, 8 and 10 were 85, 75 and 40%, respectively which were remarkably superior to all other treatments. Treatment of strawberries with 1000 ppm of chitosan also protected fruit decay but not as effective as 500 ppm. Strawberry protection at day 6, 8 and 10 in 100 ppm treated pots was 80%, 70% and 35%, respectively. In untreated control, decays of strawberries increased faster compared with chitosan treated pots. At day 10, only 90% fruits decayed. Chitosan treatment enhancing strawberry fruit defense responses through phenylalamine lyase enzyme activity [6] [11]. In another study, Zhang and Quantick [12] demonstrated that chitosan coating significantly reduce fruit decay.

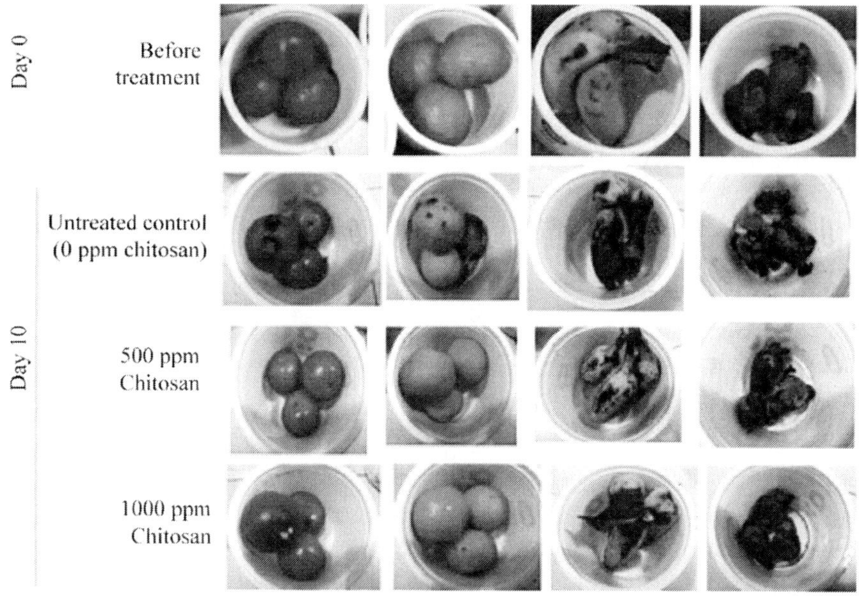

Figure 1. Chitosan delaying fruit decay over time. From left to right: tomatoes, oranges, bananas and strawberries.

No decay in oranges was observed in any treatment until day 6 (Figure 2). Oranges showed a high level of resistance to natural decay compared to the other fruits. About 90% of fruit decay was prevented by 500 ppm of chitosan at day 8 which was superior to 1000 ppm (85%) and untreated control (80%). Very high decay was observed at day 10 in both 1000 ppm (70%) and untreated control (65%), however, 90% of oranges were protected from decay at that time by the treatment of 500 ppm. Canale Rappussi and co-workers [13] also found that chitosan spray reduces decay of oranges causing black spots on fruits.

Both the cultivars Sabri and Champa of banana displayed a mirror image of each other as their rate of protection was constant until day 6 at 1000 ppm (90%), 500 ppm (100%) and untreated control (90%) (Figure 2). In day 8, both cultivars of bananas protected from decay at the same rate, with 1000 ppm (90%), 500 (95%) and untreated control (85%). Differences in protection were only noticeable in day 10, when the Sabri bananas decayed quickly in untreated control (45%), 500 ppm (60%) and 1000 ppm (55%) of chitosan. Champa bananas were protected by chitosan from decay slightly at a slower rate, with untreated control (50%), 500 ppm (70%) and 1000 ppm (70%). Chitosan with gum arabic enhance the shelf life of banana has been reported [14] [15] through reducing the respiration by 32% [10] [16] [17]. Baez-Sanudo and co-workers [10] reported that chitosan coating did not affect the sensory quality of banana.

Chitosan spray was found best in protecting tomatoes compared with other fruits tested. Irrespective of the doses of chitosan spray, 100% protection of decay was observed until 8 days after the treatment. However, slight decay of untreated control (10%) occurred at day 4, which persisted until day 8. The untreated control tomatoes decayed to 30% at day 10, displaying the slowest rate of decay among all the other fruits at this day. In day 10, 500 ppm and 1000 ppm tomatoes succumbed to decay slightly (5%) (Figure 2). Romanazzi and his co-workers [6] reviewed the protection of tomatoes and other fruits by the application of chitosan.

This study demonstrated that chitosan protected fruits from decay at room temperature. Bangladesh is rich in production of shrimp in the Southern region. Exoskeleton of the exported shrimp is a waste. It can be utilized to produce chitosan and utilize this waste into a safe fruit preservative. As chitosan is an antimicrobial agent against many plant pathogenic fungi and bacteria, protection of fruit decay observed in this study is likely to be linked with its antimicrobial activity [7]. Moreover, chitosan induces systemic resistance in plants [18]. Formalin is a health hazardous chemical [1] , which has been suspected to use in fruit preservation in Bangladesh. The preliminary findings of delay in fruit decay by chitosan demonstrated in this study suggest that the natural chitosan could be used as a safe alternative to formalin. Chitosan has also previously been successfully used in enhancing shelf life of several postharvest fruits and vegetables, such as jujube and strawberry [6] [19] [20]. Biostimulant activity and disease resistance in horticultural crops by chitosan and their putative mode of action have recently been reviewed [21]. In a recent study, Zhang et al. [22] demonstrated that controls of postharvest decay on cherry tomato fruit by chitosan is linked with the mitogen-activated protein kinase signaling pathway. However, a further investigation is needed with varying doses of chitosan on higher amount of varying fruits under different storage conditions for precise recommendation of its use in fruit preservative.

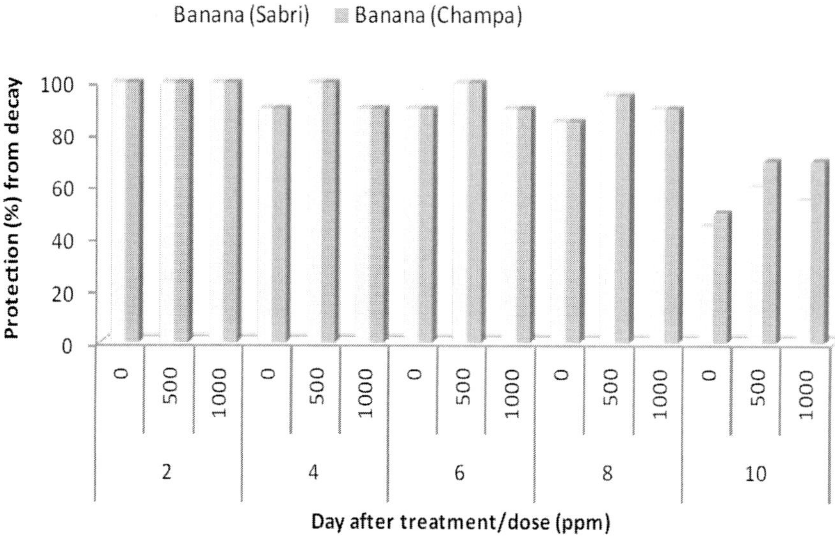
Banana (Sabri)
Banana (Champa)
Protection (%) from decay
100
80
60
40
20
0
0
500
1000
2
4
6
8
10
Day after treatment/dose (ppm)

Orange

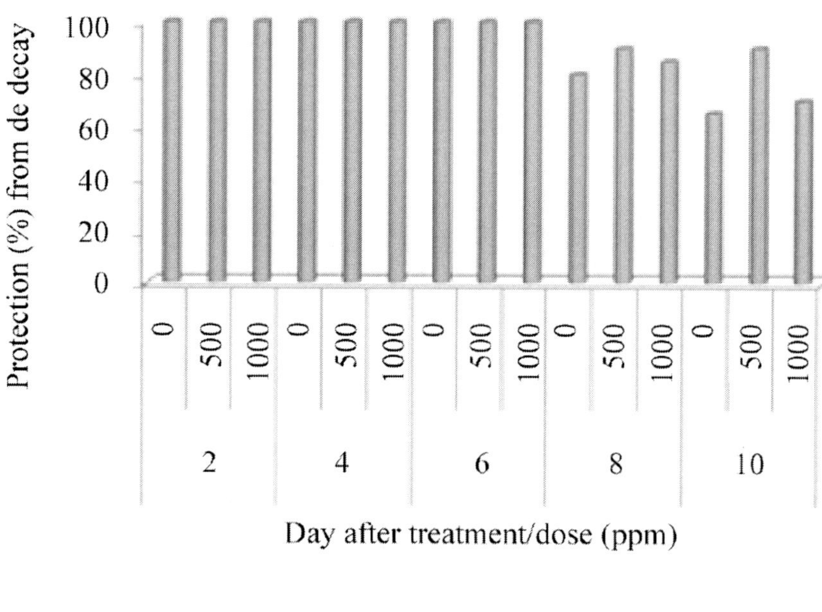
Protection (%) from de decay
100
80
60
40
20
0
0
500
1000
2
4
6
8
10
Day after treatment/dose (ppm)

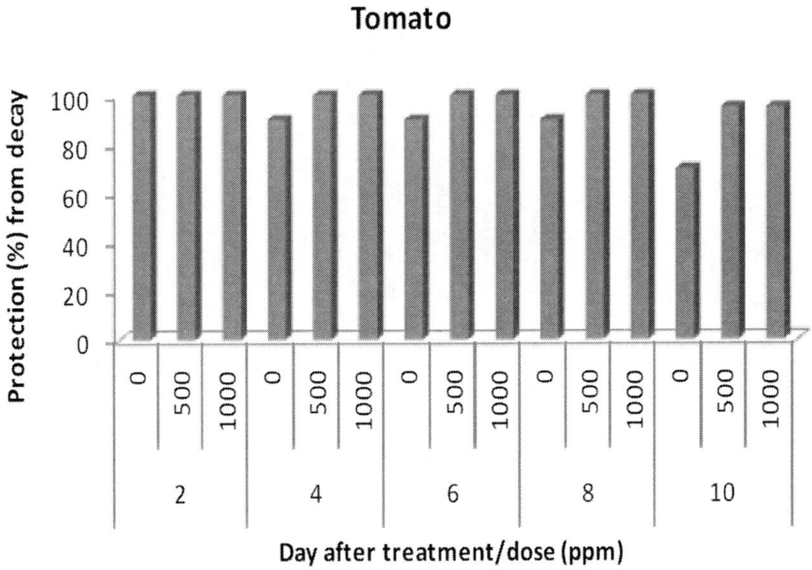

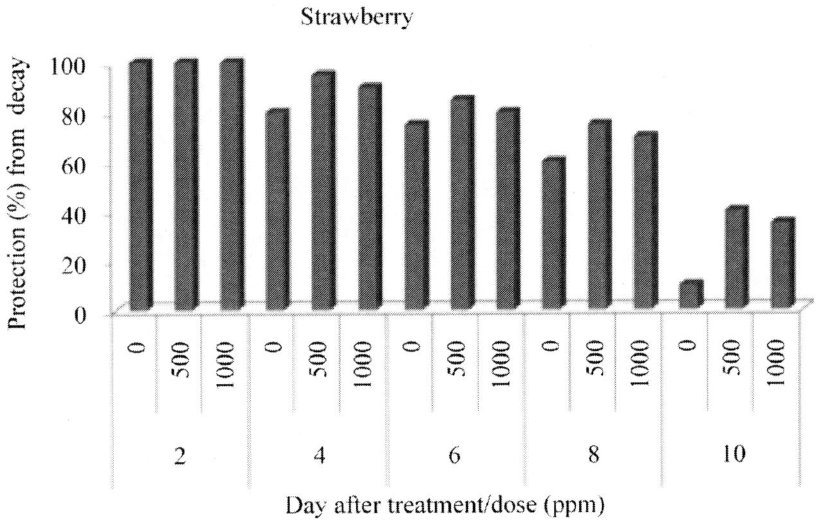

Figure 2. Time-course protection (%) of strawberry, orange, banana and tomato by varying doses (ppm) of chitosan sprayed.

4. CONCLUSION

Chitosan spray remarkably enhanced shelf life of strawberry, tomato, orange and bananas under room temperature. However, the type of fruit seemed to be an important factor, as strawberries showed more signs of decay then bananas. Interestingly, the fruits infused with 1000 ppm chitosan possibly had some side effects as some rotted even faster. The fruits which were soaked with 500 ppm produced the best results, as they still looked fresh with no signs of rotting. The fruits treated with only water decayed faster and normally. Too much of chitosan may actually be very bad for some fruits such as strawberries or tomatoes. The findings of the current study might be useful for replacing the use of health hazardous formalin by natural alternative and eco-friendly chitosan.

ACKNOWLEDGEMENTS

This research work was the "Take Action Project" of TIS for the course Global Issues 40S. Sincere thanks to the Department of Biotechnology of Bangabandhu Sheikh Mujibur Rahman Agricultural University (BSMRAU), Bangladesh for generously providing the chitosan sample and other materials used in this experiment. Special thanks are due to Mr. Mosaddiqur Rahman of the Department of Biotechnology of BSMRAU for helping prepare working solutions of chitosan. Ms Jean Gurr, Acting Principal of CISB deserves special thanks for valuable comments and encouragement for completing this work.

REFERENCES

1. Anonymous (1998) Directive 98/8/EC of the European Parliament and of the Council of 16 February 1998 Concerning the Placing of Biocidal Products on the Market. OJEU L123, 1-63. http://eur-lex.europa.eu/LexUriServ /LexUriServ.do?uri=CELEX:31998L0008:EN:HTML
2. Costa, S., Coelho, P. and Costa, C. (2008) Genotoxic Damage in Pathology Anatomy Laboratory Workers Exposed to Formaldehyde. Toxicology, 252, 40-48. http://dx.doi.org/10.1016/j.tox.2008.07.056
3. Ashfaq, S. (2013) Formalin in Fruits Can Be Fatal. Dhaka Tribune. June 15. http://www. dhakatribune.com/food/2013/jun/15/formalin-fruits-can-be-fatal
4. Haque, M. (2014) Safe Alternative to Formalin. The Daily New Age. November 28. http://newagebd.net/71190/safe-alternative-to-formalin/
5. Aider, M. (2010) Chitosan Application for Active Bio-Based Films Production and Potential in the Food Industry: Review. Food Science and Technology-LEB, 43, 837-842. http://dx.doi.org/10.1016/j.lwt.2010.01.021
6. Romanazzi, G., Feliziani, E., Banos, S.B. and Sivakumar, D. (2015) Shelf Life Extension of Fresh Fruit and Vegetables by Chitosan Treatment. Critical Reviews in Food Science and Nutrition, in press. http://dx.doi.org/10.1080/10408398.2014.900474
7. Jianglin, D. and Shaoying, Z. (2013) Application of Chitosan-Based Coating in Fruit and Vegetables Preservation: A Review. Journal of Food Process and Technology, 4, 227. http://dx.doi.org/10.4172/2157-7110.1000227

8. Alexander, L. and Grierson, D. (2002) Ethylene Biosynthesis and Action in Tomato: A Model for Climacteric Fruit Ripening. Journal of Experimental Botany, 53, 2039-2055. http://dx.doi.org/10.1093/jxb/erf072
9. El Ghaouth, A., Arul, J., Ponnampalam, R. and Boulet, M. (1991) Chitosan Coating Effect on Storability and Quality of Fresh Strawberries. Journal of Food Science, 56, 1618-1620. http://dx.doi.org/10.1111/j.1365-2621.1991.tb08655.x
10. Baez-Sanudo, M., Siller-Cepeda, J., Muy-Rangel, D. and Heredia, J.B. (2009) Extending the Shelf-Life of Bananas with 1-Methylcyclopropene and a Chitosan-Based Edible Coating. Journal of Science, Food and Agriculture, 89, 2343- 2349. http://dx.doi.org/10.1002/jsfa.3715
11. Landi, L., Feliziani, E. and Romanazzi, G. (2014) Expression of Defense Genes in Strawberry Fruit Treated with Different Resistance Inducers. Journal of Agricultural and Food Chemistry, 62, 3047-3056. http://dx.doi.org/10.1021/jf404423x
12. Zhang, D. and Quantick, P.C. (1998) Antifungal Effects of Chitosan Coating on Fresh Strawberries and Raspberries during Storage. Journal of Horticultural Science and Biotechnology, 73, 763-1748. http://dx.doi.org/10.1080/14620316.1998.11511045
13. Rapussi, C., Pascholati, S.F., Aparecida Benato, E. and Cia, P. (2009) Chitosan Reduces Infection by Guignardia citricarpa in Postharvest "Valencia" Oranges. Brazilian Archives of Biological Technology, 52, 513-521. http://dx.doi.org/10.1590/S1516-89132009000300001
14. Maqbool, M., Ali, A., Ramachandran, S., Smith, D.R. and Alderson, P.G. (2010) Control of Postharvest Antrachnose of Banana Using a New Edible Composite Film. Crop Protection, 29, 1136-1141. http://dx.doi.org/10.1016/j.cropro.2010.06.005
15. Maqbool, M., Ali, A. and Alderson, P.G. (2010b) A Combination of Gum Arabic and Chitosan Can Control Antrachnose Caused by Colletotrichum musae and Enhance 1456 the Shelf-Life of Banana Fruit. Journal of Horticultural Science and Biotechnology, 85, 432-436. http://dx.doi.org/10.1080/14620316.2010.11512693
16. Kittur, F.S., Saroja, N., Habibunnisa, Ms. and Tharanathan, R.N. (2001) Polysaccharide-Based Composite Formulations for Shelf-Life Extension of Fresh Banana and Mango. European Food Research and Technology, 213, 306-311. http://dx.doi.org/10.1007/s002170100363
17. Win, N.K.K., Jitareerat, P., Kanlayanarat, S. and Sangchote, S. (2007) Effects of Cinnamon Extract, Chitosan Coating, Hot Water Treatment and Their Combinations on Crown Rot Disease and Quality of Banana Fruit. Postharvest Biology and Technology, 45, 333-340. http://dx.doi.org/10.1016/j.postharvbio.2007.01.020
18. Terry, L.A. and Joyce, D.C. (2004) Elicitors of Induced Disease Resistance in Postharvest Horticultural Crops: A Brief Review. Postharvest Biology and Technology, 32, 1-13. http://dx.doi.org/10.1016/j.postharvbio.2003.09.016
19. Perdones, A., Sánchez-González, L., Chiralt, A. and Vargas, M. (2012) Effect of Chitosan-Lemon Essential Oil Coatings on Storage-Keeping Quality of Strawberry. Postharvest Biology and Technology, 70, 32-41. http://dx.doi.org/ 10.1016/j.postharvbio.2012.04.002
20. Yu, Y.W., Zhang, S.Y., Ren, Y.Z., Li, H., Zhang, X.N., et al. (2012) Jujube Preservation Using Chitosan Film with Nano-Silicon Dioxide. Journal of Food Engineering, 113, 408-414. http://dx.doi.org/10.1016/j.jfoodeng.2012.06.021
21. Pichyangkura, R. and Chadchawan, S. (2015) Biostimulant Activity of Chitosan in Horticulture. Scientia Horticulturae, 196, 49-65. http://dx.doi.org/ 10.1016 /j.scienta.2015.09.031
22. Zhang, D., Wang, H., Hu, Y. and Liu, Y. (2015) Chitosan Controls Postharvest Decay on Cherry Tomato Fruit Possible via the Mitogen-Activated Protein Kinase Signaling Pathway. Journal of Agricultural Food Chemistry, 63, 7399-7404. http://dx.doi.org/10.1021/acs.jafc.5b01566

CHAPTER 14

Active Packaging Coatings

Luis J. Bastarrachea, Dana E. Wong, Maxine J. Roman, Zhuangsheng Lin and Julie M. Goddard

Department of Food Science, University of Massachusetts Amherst, 102 Holdsworth Way, Amherst, MA 01003, USA

ABSTRACT

Active food packaging involves the packaging of foods with materials that provide an enhanced functionality, such as antimicrobial, antioxidant or biocatalytic functions. This can be achieved through the incorporation of active compounds into the matrix of the commonly used packaging materials, or by the application of coatings with the corresponding functionality through surface modification. The latter option offers the advantage of preserving the packaging materials' bulk properties nearly intact. Herein, different coating technologies like embedding for controlled release, immobilization, layer-by-layer deposition, and photografting are explained and their potential application for active food packaging is explored and discussed.

Keywords: active food packaging; antimicrobial; antioxidant; biocatalytic; surface modification

1. INTRODUCTION

Active packaging, in which the packaging material performs an additional function beyond containment and basic protection, remains an area of active research with great potential for commercial applications. Active packaging has application in packaging of food, pharmaceutical, and consumer goods products, with a common goal of improving shelf life, safety, or quality of packaged goods. A number of excellent reviews have been written on targeted applications of active packaging materials with less focus on material synthesis techniques [1,2,3,4,5]. Synthesis can be achieved by incorporating an active agent (e.g., antioxidant, enzyme, antimicrobial, oxygen scavenger) within or at the product contact surface of a packaging material. Positioning the active agent at the product contact side (*versus* bulk incorporation) has several benefits, including retention of bulk material properties and minimizing the amount (and therefore cost) of active agent required to impart efficacy. Understanding the technologies and challenges associated with various coating methods for preparation of active packaging materials will support effective technology transfer to commercial applications. The goal of this review is to describe key technologies for preparing active packaging coatings, including embedding, layer-by-layer deposition, and photografting, with a discussion of the difference between covalent (non-migratory) and non-covalent (migratory) immobilization chemistries. We then survey the current literature for active packaging technologies using these coating methods to impart antimicrobial, antioxidant, and biocatalytic activity. Smart/intelligent packaging [6,7], in which indicating devices are incorporated into the packaging structure, and sachet-based technologies [8] are well covered in the current literature and are outside the scope of this review. We conclude with a discussion of the challenges that remain in achieving commercial translation for active packaging coatings.

2. OVERVIEW OF COATING TECHNOLOGIES

In the following subsections, the main approaches employed to prepare coatings for active packaging that have been reported in the literature are explained. These are the most frequent techniques found in the field of coatings for active packaging. Figure 1 shows a graphic summary of them.

2.1. Coatings with Embedded Agents for Controlled Release

As illustrated in Figure 1, active agents can be incorporated into coatings for active packaging by a number of technologies, including those in which the active agent is intended to migrate to the packaged good (embedding, non-covalent immobilization, some layer-by-layer deposition techniques) and non-migratory technologies in which the

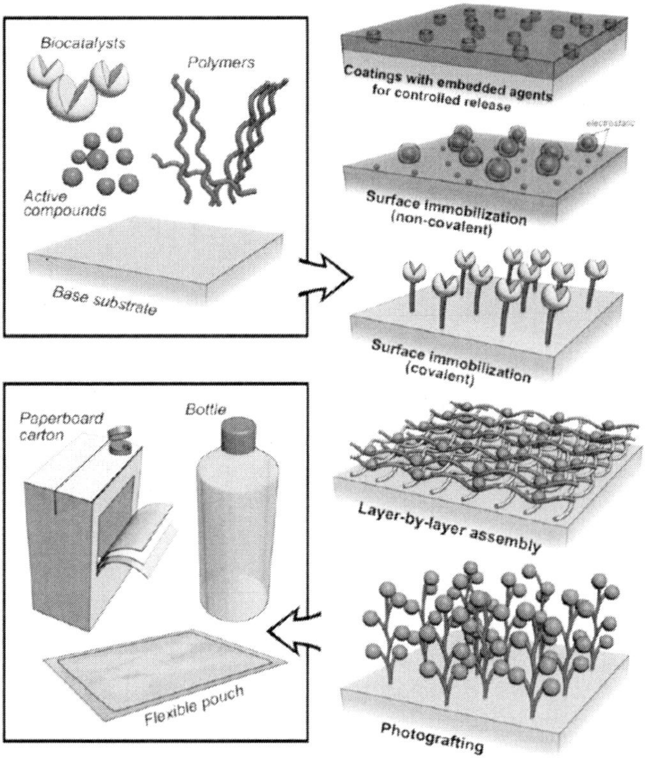

Figure 1. Summary of coating technologies.

active agent is intended to remain stable in the packaging matrix (covalent immobilization, some layer-by-layer deposition techniques, photografting). Controlled release coatings are those in which active agents have been incorporated into the matrices of polymeric materials, from which they are expected to migrate and exert their specific function within a packaged good, whether antimicrobial, antioxidant, biocatalytic, or neutraceutical [6]. The main reported mechanisms through which active compounds have been incorporated into polymeric materials have been extrusion/blending and solution casting [9,10]. Solution casting consists of dissolving the polymer intended for packaging in a suitable solvent and simultaneously incorporating the active compound of interest, followed by pouring the solubilized polymer and active agent onto a surface for the solvent to evaporate, resulting in formation of plastic film with targeted functionality (antimicrobial, pharmaceutical, biocatalytic, *etc.*). In extrusion, the active compound is incorporated with polymeric material melted by heat transfer forming a blend from which films can be formed. The first method, although extensively used at a laboratory scale, exhibits some limitations for its practical and commercial application as most of the polymers intended for food packaging can only be dissolved at high temperatures with organic solvents [11], which would compromise the stability and effectiveness of many active compounds of interest. This is probably the main reason why incorporation of active compounds through solution casting has been studied more frequently with biodegradable polymers that can be dissolved at milder temperatures, for example poly(lactic acid), poly(butylene adipate-*co*-terephtalate), and cellulose derivatives [9]. Extrusion and blending, although industrially scalable, can present serious limitations due to the intrinsic lack of thermal stability of many active compounds, which can be lost through degradation and evaporation during the heat transfer these unit operations involve [3,9]. Ensuring uniform distribution of the active compound and retaining thermomechanical film properties are additional challenges to preparation of extruded active packaging materials. Controlled release can rely on physical and chemical phenomena, like the degree of affinity between the active compound and the matrix of the packaging material, as well as its morphology and porosity, which if low could allow a slower release [3,9,12]. In addition, controlled release can be provided by a multilaminated system in which the layer that harbors the active compound is covered by an adjacent layer that could serve as a barrier that controls the rate of release of such compound [9,13]. Temperature also plays a relevant role in the release of active compounds embedded in polymer packaging materials. It has been observed extensively that diffusion increases with temperature, which could represent an advantage specifically for the case of antimicrobial packaging as microorganisms also reproduce faster with an increase in temperature [9,13]. However, even if after incorporation of the compound of interest its activity and effectiveness is demonstrated at a laboratory scale, it is well known that the incorporation of a foreign compound into the matrix of a plastic material (even at small concentrations, at which no activity may be observed) can substantially affect its properties relevant to processing, production and machinability. These include the tensile and mechanical properties (tensile strength, elastic modulus, elongation at break, *etc.*), the thermal properties (melting point, glass transition temperature, crystallization temperature, heat capacity, *etc.*), and the gas barrier properties (water vapor permeability, O_2 and CO_2 permeability, *etc.*) [9,14]. As the goal of active packaging is to enhance material performance, losing desirable material properties as a result of active agent incorporation limits commercial applicability. There is therefore an interest in the application of thin coatings rather than bulk modifications, as thin coatings are not expected to affect relevant physical and chemical properties [15]. Nevertheless, a major benefit of embedding active agents into a packaging film by coating or coextrusion is their use of currently available converting equipment.

2.2. Surface Immobilization

In addition to embedding, active packaging materials can be prepared by immobilization of an active agent on the surface of a bulk packaging material. Although less studied as a material for active packaging, paperboard can be functionalized to immobilize active compounds by modifying the chemistry of its main component cellulose, with the formation of reactive groups like aldehydes, epoxy, carboxylic acids, *etc.* [16]. More commonly reported is the surface modification of polymer materials for use in active packaging. Polymer packaging materials are typically inert, and require an initial functionalization to enable immobilization of an active agent. Surface activation techniques of plastic polymer substrates can be divided into physical and wet methods. Physical methods include flame, corona discharge, UV radiation, and plasma [5,15,17]. Wet methods involve the use of corrosive liquids to which the polymer substrates are directly exposed, like piranha solution (dissolved hydrogen peroxide and sulfuric acid), combined sodium hydroxide and sulfuric acid, chromic acid, potassium permanganate, and nitric acid [18,19,20,21,22]. The main effect of these surface activation methods is the formation of reactive oxygenated species on the polymer surface like carbonyl, hydroxyl and carboxylic acid groups [21,22]. Some methods of industrial relevance to activate the surface of polymeric packaging materials are flame, corona, and atmospheric or vacuum plasma [23], which remove contaminants from the surface of polymer films, increase their surface energy and wettability (in order to apply coatings of different types, like inks for printing or metallic coatings), and increase the level of oxygen on their surface. In contrast to the wet methods, the physical methods are more widely used due to their industrial scalability, given the fact that no liquid reagents of any kind are needed in their application, avoiding accumulation and handling of harmful waste [22]. Once a polymeric surface has been functionalized and the mentioned oxygen-containing moieties haven been incorporated on it, compounds of different functionality can be applied through either covalent or non-covalent immobilization. Non-covalent binding relies mainly on electrostatic interactions and affinity [22]. In the first case, the polymer surface possesses a certain net charge which can attract molecules with an opposite net charge. The second case refers to specific ligand–receptor interactions, like in the case of biotin–avidin [24]. Covalent immobilization as its name implies involves the covalent coupling of the active compound of interest onto the polymer surface. A potential benefit of covalent immobilization is that the active compound is not expected to migrate to the packaged product matrix, which could otherwise compromise its commercial application [22]. Active agents can be covalently linked directly to the polymer surface or by use of a crosslinker, which may either share a permanent covalent bond with both the polymeric substrate and the bioactive compound, or may promote covalent bond formation between the activated substrate and the bioactive compound without forming part of that link ("zero-length" crosslinkers) [22,25]. The covalent attachment between a polymeric surface and an active compound relies mainly on the formation of amide, ether, ester, and thioether bonds, created between the hydroxyl, amine, imine, carboxylic acid, and thiol groups the active compounds of interest may possess intrinsically (or are incorporated in their structure) and the functional groups created on the substrate [22,25,26]. A benefit to immobilization technologies is the potential versatility: once functional groups are introduced to the polymer surface, a range of bioactive agents (e.g., enzymes, peptides) can be immobilized through standard bioconjugation techniques.

2.3. Layer-by-Layer Assembly

Layer-by-layer assembly is a versatile method of surface modification that relies on the deposition and mutual attraction of alternating polyelectrolytes with opposite net charges onto a solid support [27,28]. A polymeric substrate can be made reactive through the previously explained methods of surface activation, imparting a certain charge on it, or a polyelectrolyte can be covalently attached onto its surface to further apply the alternate polyelectrolyte layers. The polyelectrolytes can be polymers of any kind (proteins, polysaccharides, synthetic polymers, *etc.*) that harbor a net charge, and they can be modified by the incorporation of functional groups to facilitate their deposition [27,28,29]. Deposition can be accomplished either by submersion of the substrate into polyelectrolyte solutions or by spraying of solutions onto the substrate; both techniques have potential for scalability as high throughput coating technologies. The deposition of the polyelectrolytes can be optimized by adjusting the pH of their solutions to a point of full protonation or deprotonation, to maximize the presence of charge [30]. The deposition can rely exclusively on electrostatic interactions [29,31], or covalent bonds can be built between the alternating layers through the use of crosslinkers [25,29,32,33]. Although theoretically an indefinite number of polyelectrolyte layers can be applied on a substrate [15], a state of saturation can be reached in which no more polyelectrolytes will deposit [32]. There seems to be a positive association between molecular weight of the polyelectrolytes and the thickness and stability of the system [32,34], although the type of bonds that hold together the layers (when covalently bound) and their likelihood to undergo hydrolysis under different environmental conditions also dictates stability [32]. Layer-by-layer deposition can be used to prepare active packaging coatings by the incorporation of active agents either between layers or within the structure of an individual polyelectrolyte. Pilot scale layer-by-layer deposition tools have been developed, suggesting the potential for scalability to high through put production of active packaging coatings via layer-by-layer deposition [35].

2.4. Photografting

In photografting, polymer chains are grafted from a surface by exposure to UV light in the 315–400 nm range in the presence of photoinitiators and monomers [36]. UV irradiation generates free radicals on the polymeric surface after which photoinitiators, reactive compounds that generate unpaired electrons upon UV light exposure, initiate polymerization of monomers [28]. Photografting may take place through three types of mechanisms: hydrogen abstraction, electron transfer, and cleavage [37]. In hydrogen abstraction, the energy provided by the UV light removes a hydrogen atom from a substrate (either from a hydrocarbon, an alcohol, or an ether), which leaves unpaired electrons on its surface. At the same time, the ketone group of a photoinitiator undergoes reduction by coupling with the removed hydrogen from the substrate and coupling its unpaired electron to the unpaired electrons on the substrate, which is followed by polymerization of any monomer of interest. In electron transfer, the carbonyl group of the photoinitiator is also reduced in an electron acceptor-donor system, in which an aminated compound (normally a tertiary amine, but secondary and primary amines can also take place in the reaction) provides the hydrogen from the α carbon of its structure. In the cleavage mechanism, the photoinitiator undergoes homolytic scission upon UV light exposure, which generates free radical species able to initiate polymerization between vinyl monomers. Some common photoinitiators are benzophenone [38], anthraquinone [37], thioxantone [37], and phenyl azide [39]. Polymeric photoinitiators have also been studied [37], as well as naturally derived photoinitiators like curcumin [40,41]. Photografting can be used to prepare active packaging coatings either by direct incorporation of the active agent during photografting, or by subsequent immobilization of the active compound after grafting of a polymer chain with reactive functional groups (e.g., acrylic acid).

3. APPLICATIONS

There are a number of active agents that can be incorporated into or onto coatings for active packaging using the technologies described previously, as summarized in Table 1. Major classes of active agents include antimicrobials, antioxidants, and enzymes, with applications in active packaging coatings ranging from controlling microbial growth, inhibiting oxidative degradation reactions, and targeted biocatalysis. The nature of the bulk packaging material onto which the active coating is deposited varies, including materials such as polyethylene (PE), low density polyethylene (LDPE), polypropylene (PP), polyethylene terephthalate (PET), ethylene vinyl alcohol (EVOH), poly(lactic acid) (PLA), polyamide (PA), polystyrene (PS), polyvinyl alcohol (PVOH), *etc.* Much of the published and commercialized research on active packaging coatings has targeted food packaging applications, largely for maintaining safety and quality and controlling spoilage. While the focus of the applications surveyed below is largely on food applications, these technologies have relevance to packaging of consumer products (e.g., health and beauty) as well as pharmaceuticals and biomedical devices. Smart packaging intended for monitoring of temperature, oxygen, pH, moisture, *etc.* [42,43] of packaged products represents a promising area for active packaging coatings, but falls outside the scope of this review.

3.1. Antimicrobial

Probably the most studied application of active packaging technologies has been antimicrobial packaging to control growth of pathogenic and/or spoilage microorganisms in packaged products. Antimicrobial agents such as essential oils, organic acids, peptides, enzymes, and biopolymers have been introduced into antimicrobial active packaging by a range of coating technologies. Regarding controlled release, a European patent describes a liquid formulation composed of essential oils, adhesion promotors (e.g., acrylic or vinyl resins or nitrocellulose), and fixatives to modulate the release of the essential oils for application on common packaging materials [44]. A recent study by Manso *et al.* demonstrated the anti-fungal character of the patented coating [45], in which a 30 μm thick coating containing cinnamon essential oil (coating grammage of 2.5 $g \cdot m^{-2}$, with essential oil concentrations of 2%, 4% and 6%) was applied to PP and evaluated against different species of *Aspergillus* and *Penicillium*. Seeded agar was exposed to coated films at a distance normally given between the agar surfaces and Petri dish lids, and it was observed that total inhibition was obtained at 4% and 6% concentrations for all the fungi evaluated regardless of temperature and pH. In another work also involving controlled release of thyme and oregano essential oils [46], Valderrama Solano *et al.* reported coating onto corona treated LDPE films at concentrations of 1% and 4% (based on the weight of LDPE). Antimicrobial activity was demonstrated by zone of inhibition assays against *Escherichia coli*0157:H7, *Salmonella* Typhimurium and *Listeria monocytogen*es, showing substantial inhibition at a concentration of 4% for all the pathogens, and no inhibition at 1% concentration. The authors further demonstrated that the reported coating method had not significant changes in the film mechanical properties (tensile strength and elongation at break) and although the changes in barrier properties (oxygen and water vapor transmission rates) were significant, they were probably not relevant enough to compromise their usage since a reduction in both parameters was observed. Another approach for controlled release involves the application of the sol-gel technique, which consists of the condensation of hydroxylated monomers and polymers into a network that can harbor active substances [47]. Lantano *et al.* recently reported on the preparation of a sol-gel coating in which the antifungal agent natamycin was embedded in a tetraethyl orthosilicate/EVOH gel and applied to plasma treated PLA films [48]. The natamycin loaded PLA films demonstrated release rates that were in accordance to European standards and were able to inhibit mold growth on cheese stored for 30 days at 4 °C. In another study, Zhu *et al.*showed the potential to apply antimicrobial photocatalysts (e.g., TiO_2) using the sol-gel technique [49].

Table 1. Summary of technologies and applications in active packaging coatings.

Coating Technology	Application		
	Antimicrobial	**Antioxidant**	**Biocatalytic**
Controlled release	Essential oils [45,46]	Citrus oil [50–52]	Lactase [53]
	Natamycin [48]	Rosemary extract [54]	Laccase [55]
	TiO_2 [49]	α-tocopherol [56]	Oxalate oxidase [57]
	Cinnamaldehyde [58]	–	–
	Sorbic acid and lauric arginate ester [59]	–	–
	Lauric arginate ester [60]	–	–
	Nisin [61]	–	–
Immobilization	3-aminopropyltrimethoxysilane [62]	Gallic acid [63]	Lactase [64,65]
	Lysozyme [66]	Aluminum oxide [67]	Catalase [68]
	(3-bromopropyl)triphenylphosphonium [69]	–	Naringinase [70–72]
	SO_2 [73]	–	–
Layer-by-layer assembly	Chitosan [74,75]	Tannic acid [76]	Lactase [77]
	Lysozyme [78]	–	–
Photografting	–	Caffeic acid [79]	Trypsin [80–82]
	–	Acrylic acid [83–85]	Urease [86]
	–	Hydroxamic acid [87,88]	–

There have been several reports in which controlled release coating technologies have been combined with surface immobilization to prepare antimicrobial coatings. In a recent study [58], Makwana *et al.*, covalently attached polydiacetylene liposomes containing cinnamaldehyde onto an amine-functionalized silane monolayer on piranha treated glass. The authors also explored immobilizing the liposomes on amine-functionalized PLA films. Approximately three logarithmic reductions were obtained against *Escherichia coli* after 50 min of exposure to the liposome-encapsulated cinnamaldehyde (for both glass and poly(lactic acid)). For the case of *Bacillus cereus*, PLA was more effective (~4 logarithmic reductions) than glass (~3 logarithmic reductions). What may be remarkable about this study is that it suggests the possibility of applying an approach for controlled release of antimicrobial substances on both organic and inorganic substrates. In another study, Guo *et al.* reported the direct coating of PLA with a mixture of chitosan (a widely studied polycation with known antimicrobial activity), sorbic acid, sodium lactate, and lauric arginate ester [59]. Antimicrobial effectiveness was confirmed both in bacterial suspensions (to an undetectable level for all microorganisms) and on ready to eat meat against *Listeria innocua* (2–3 logarithmic reductions), *Listeria monocytogenes* (2–3 logarithmic reductions) and *Salmonella* Tiphymurium (1–1.5 logarithmic reductions) under storage at 10 °C. In other work [60], Theinsathid *et al.* spray coated corona treated PLA with solutions having varying concentrations of lauric arginate ester (0%–10%). Between two and three logarithmic reductions were reached against *Listeria monocytogenes* and *Salmonella* Tiphymurium in wrapped ham samples after seven days of storage at 4 °C under the lowest antimicrobial concentration tested (0.07% based on the weight of PLA). The authors demonstrated that the coating treatment did not affect the mechanical properties of poly(lactic acid); however, the gas barrier properties measured as CO_2 permeation were adversely affected. In another work, the antimicrobial peptide nisin was entrapped in polyethylene oxide brushes grown on silane modified silicon wafers [61]. It was hypothesized that the polyethylene oxide brushes would protect nisin from being eluted by foreign proteins and subsequently losing its activity. Antimicrobial activity of the modified silicon wafers was confirmed through zone of inhibition assay against the Gram positive bacterium *Pediococcus pentosaceous* over a period of seven days. Retention of antimicrobial activity even after introduction of bovine serum albumin to the agar media supported their hypothesis that the polyethylene oxide brushes protected nisin. Although this coating was tested on an inorganic substrate, the authors proposed its potential for application on polymer packaging films.

Preparation of antimicrobial active coatings via covalent immobilization of antimicrobial agents is probably the least studied approach explored for antimicrobial packaging applications, likely because the most commonly studied antimicrobial agents (e.g., nisin, essential oils) must migrate from the packaging material to be effective. Cationic polymers and some antimicrobial enzymes, however, may retain efficacy after covalent immobilization onto a solid support. The cationic property of amines was tested in a study by Fernandes *et al.*, in which pullalan powder was rendered cationic by reaction with an amine terminated silane, after which films were formed by solution casting [62]. The cationic pullalan films demonstrated more than three logarithmic reductions *Staphylococcus aureus* and *Escherichia coli*. In addition, the cationic pullulan films showed improved mechanical and thermal properties as compared to unmodified pullulan. In recent work by Muriel-Galet *et al.*, the antimicrobial enzyme lysozyme was covalently attached onto UV-ozone treated EVOH films via carbodiimide chemistry and reported more than one logarithmic reduction of *Listeria monocytogenes* in bacterial suspension [66]. In another work, Anthierens *et al.* tested the possibility of functionalizing the flexible polymer poly(butylene adipate-*co*-terephtalate) with a quaternary phosphonium compound, (3-bromopropyl)triphenylphosphonium [69]. For that purpose, the biodegradable polymer was modified to introduce alkyne bonds to enable an azide-alkyne "click" reaction with the quaternary compound. The modified polyester was challenged against *Escherichia coli* both in bacterial suspension and under direct contact, reaching ~4 logarithmic reductions after 1 h in bacterial suspension and after 24 h under direct contact. In another recent study by Mackiw *et al.* [73], a multilayered film made of PA and PE was subjected to atmospheric plasma treatment (Ar, Na_2O and SO_2) on the PA side of the films. Na_2O was applied either alone with Ar or combined with Ar and SO_2. The temperature for Na_2O sublimation in the reactor was varied from 300–640 °C. It was observed that the highest inhibitions were obtained from films possessing the highest concentrations of SO_2 (the antimicrobial agent) against *Escherichia coli* (82%), *Staphylococcus aureus* (86%), *Listeria monocytogenes* (63%), *Bacillus subtilis* (79%), and *Candida albicans* (75%). Even though less than one logarithmic reduction was reached for every microorganism, this type of approach seems promising due to its potential industrial scalability, as plasma treatment is a widely used method of surface activation for food packaging films [23].

The possibility of using the layer-by-layer approach has been explored through several studies. In one recent work published by Pinheiro *et al.*, κ-carrageenan (polyanion) and chitosan (polycation) were applied onto aminated PET resulting in improved gas barrier properties (reduced oxygen and water vapor transmission rates) [74]. In other work presented by Carneiro-da-Cunha *et al.* [75], alginate was employed as the polyanion, suggesting it as an attractive alternative as there was not significant difference in water vapor permeability between the aminated PET film used and its coated counterpart. Finally, Medeiros *et al.*, have reported on layer-by-layer deposition of three layers of κ-carrageenan alternated with two layers of the antimicrobial enzyme lysozyme on amine-functionalized PET films [78]. The modified PET exhibited improved values of oxygen and water vapor permeability as compared to aminated PET. To date, layer-by-layer deposition technologies have largely focused on improving barrier properties, but those described above incorporate components with antimicrobial character, suggesting their potential application in antimicrobial active packaging coatings. While antimicrobial active packaging technologies have focused largely on inhibiting microbial growth or inactivating viable organisms, there may be opportunity in adapting food processing equipment coatings technologies which seek to inhibit microbial adhesion.

3.2. Antioxidant

Oxidative degradation is a key limiting factor in overall shelf life of many food and consumer products, as well as nutritional supplements. Many packaged products undergo quality deterioration during transport and storage due to oxidative reactions that cause lipid rancidity (e.g., mono and polyunsaturated fatty acids), color loss (e.g., carotenoids, chlorophyll, anthocyanins), and vitamin degradation (e.g., Vitamin A and its precursors, Vitamin C, Vitamin D, Vitamin E). There are several preservation strategies for packaged goods that contain oxidation-sensitive ingredients based on direct addition of antioxidants (e.g., free radical scavengers, metal chelators, singlet oxygen quenchers, oxygen scavengers) and design of appropriate packaging systems. One such preservation strategy is antioxidant active packaging, wherein an antioxidant has been incorporated into a packaging material to enhance food preservation [14]. The most widely used commercially available antioxidant active packaging technologies include oxygen scavengers, manufactured as sachets or labels [89]. However, there is emerging research in the development of migratory and non-migratory antioxidant coatings for active packaging, as surveyed in this section.

Migratory antioxidant active coatings are designed for controlled release of an antioxidant over the course of product shelf life. The majority of research for antioxidant active packaging that is designed for controlled release has been focused on blending of antioxidants with polymers followed by extrusion [4,90]. However, inclusion of antioxidants in packaging may have a negative impact on bulk material properties that may limit applications. Research in the design of active coatings for controlled release of antioxidant is promising for overcoming this challenge. Corlini *et al.* prepared antioxidant packaging materials by spray deposition of citrus oil in methanol onto the surface of PET trays and observed that plasma pretreatment enhanced the adhesion of citrus oil to the surface of PET trays, which demonstrated higher overall antioxidant activity [50]. This coating exhibited antioxidant activity with cooked turkey meat and retained activity after six months of storage [51,52]. Bolumar *et al.* coated rosemary extract directly onto LDPE plastic wrap (0.45 mg rosemary cm^{-2}) and demonstrated significant delay in the onset of surface lipid oxidation of wrapped chicken patties. Controlled released of migratory antioxidant coatings may be enhanced by using a polymeric carrier for antioxidants rather than direct surface application [54]. For example, Lee *et al.* applied α-tocopherol to paperboard using a vinyl acetate-ethylene copolymer as a carrier for controlled release ($D = 2.91 \times 10^{-11}\ m^2 \cdot s^{-1}$) and demonstrated its ability to retard oxidation in an oil-in-water emulsion [56].

Non-migratory antioxidant coatings are designed to scavenge prooxidants, such as free radicals, oxygen and transition metals, to extend the shelf life of oxidation-sensitive foods. Garces *et al.* patented an antioxidant coating composed of a polymer blend that contains plant extracts that scavenge free radicals [91]. Tovar *et al.* demonstrated that this antioxidant coating may be classified as non-migratory as it demonstrated migration that was 20 times less than the legal limits for the European Union [92]. Furthermore, this antioxidant active packaging coating derives activity from free radical scavenging in the packaging headspace, thus not requiring direct contact with the food [93,94,95]. Shutava *et al.* imparted free radical scavenging properties on glass slides via layer-by-layer deposition of tannic acid and poly(allylamine hydrochloride) and found that 2,2′-azino-bis(3-ethylbenzothiazoline-6-sulphonic acid) (ABTS) radical scavenging activity per mole immobilized tannic acid decreased with increasing number of bilayers, most likely due to inability of ABTS to diffuse through bilayers [76]. However, increasing the number of bilayers did result in a linear increase in overall scavenging activity (mol ABTS cm^{-2}). There has also been significant research on metal oxide coatings that have been applied to packaging materials to reduce oxygen diffusion into packaging to extend shelf life [67,96].

Non-migratory antioxidant coatings may also be applied by covalent immobilization of functional groups to the surface of packaging materials. Gallic acid was immobilized to chitosan by carbodiimide assisted reaction that resulted in an active packaging material that demonstrated significant reduction in oxidation of ground peanuts [63]. In other work, Arrua *et al.*utilized photoinitiated graft polymerization of a polymer functionalized with caffeic acid to coat polypropylene packaging materials and demonstrated its ability to prevent oxidative degradation of ascorbic acid in orange juice [79]. Photoinitiated graft polymerization has also been used to fabricate metal chelating active packaging coatings that extended the lag phase of lipid oxidation in oil-in-water emulsions by chelating transition metals, which are the most influential prooxidants in food emulsions [83,87]. Further research in this area demonstrated that active packaging coatings composed of photografted polyhydroxamate chelators retain activity in a wide range of pH values (3–5), viscosity (~1–10^4 CP) and competing ion (Na, Mg, Ca) conditions typically found in food and consumer products [97]. In addition to potential "clean label" regulatory benefits of non-migratory antioxidant coatings, such technologies have product quality advantages over migratory antioxidant packaging coatings, as migratory antioxidant coating technologies typically use active agents which must be used at a concentration which alters sensory perception (flavor, color, viscosity) of the packaged product. Nevertheless, technical hurdles remain in adapting non-migratory active coatings for commercial application.

3.3. Biocatalytic

The partnership of enzymes and solid support materials provides many opportunities for biocatalytic coatings for active packaging. Enzymes are proteins with enhanced specificity to a substrate, which catalyze reactions by lowering their activation energy to create products. Enzymes are employed in numerous processes used in the food, pharmaceutical, and diagnostic industries. In food processing, biocatalysts are used in production of ingredients, enhancement of product quality, and breakdown of undesirable components that may be harmful or may decrease product quality [98]. As the working conditions for biocatalytic materials can be very specific and variable with the immobilization method and material, there is much research on the immobilization of enzymes onto and into solid supports in order extend their thermostability, pH optima, and solvent stability. Integrating biocatalysts into packaging materials facilitates their use as active packaging coatings [99] which may enable so-called "in-package processing" where food constituents undergo changes to improve quality or shelf-life while in transport and storage.

The goal of many biocatalytic active packaging technologies is to control the growth of spoilage and/or pathogenic microorganisms. Lysozyme is a commonly utilized enzyme for such antimicrobial active packaging. Such antimicrobial enzymes have been incorporated into active packaging coatings via blending, non-covalent binding for controlled release, and covalent immobilization [100,101,102]. For additional information, refer to the comprehensive section on antimicrobial coatings of this review. Herein outlines methods for the incorporation of enzymes into biocatalytic coatings for active packaging, and a few works which have the potential for application in active packaging coatings.

Incorporation of enzymes into active packaging coatings by embedding and blending requires compatibility between the enzyme and the coating matrix. Compatibility can be achieved by surface functionalization, altering bulk material composition, and enzyme modification. Embedding and blending may simplify commercialization and scale-up as one-pot preparation methods become more available [16]; however, these techniques typically result in the enzyme becoming a part of the packaged product, reducing catalytic activity of the coating over time [103]. A targeted application for several biocatalytic active packaging coatings technologies involves the removal of oxygen for increased food shelf stability. Efforts towards incorporation of glucose oxidase in a low-density polyethylene and paper board multilaminate have included details for scaled-up production. Various compositions of LDPE, glucose oxidase, and catalase were produced in Tetra Pak's pilot plant, and showed up to 97% activity could initially be achieved even after exposure to 325 °C during production. Control of production parameters was key to maintaining the package's oxygen scavenging capability [104]. Johansson *et al.* worked to improve embedding the glucose oxidase and catalase oxygen scavenging pair by varying combinations of LDPE, polypropylene, and polylactic acid [55]. Variations of these embedding methods for oxygen scavenging by laccase and oxalate oxidase have shown similar results [55,57]. Recent work by Talbert *et al.* demonstrates that enzymes may be modified by hydrophobic ion pairing to be soluble and retain activity in solvents used in ink formulations, enabling the preparation of biocatalytic active packaging coatings using existing printing technology [105].

Much of the work on biocatalytic coatings for food packaging involves covalent immobilization. Covalent immobilization allows for biocatalytic coatings to interact with packaged products without being incorporated into the food matrix, enabling their regulation as contact materials rather than direct additives. Non-migratory biocatalytic coatings therefore offer a potential regulatory benefit, as there is growing demand for "clean label" products. For example, Soares and Hotchkiss developed non-migratory packaging films to de-bitter fruit juices by covalently immobilizing fungal naringinase [70,71]. Naringinase activity was maintained for 15 days storage after preparation of the biocatalytic active packaging coating, and k_M values were lower when compared to free enzyme. Nunes *et al.*, achieved similar success by cross-linking naringinase to polyvinyl alcohol and alginate using boric acid for debittering [72]. In other work, β-galactosidase was bound to amine-functionalized polyethylene films by a dialdehyde tether to reduce milk lactose in package [64]. Further research demonstrated that tether length and chemistry can influence retained activity of immobilized lactase [65], but more significant enhancements in retained activity can be achieved by immobilization onto nanostructures [106].

Layer-by-layer deposition allows for more enzyme to be incorporated in a biocatalytic coating compared to traditional immobilization techniques. Indeed, increasing the number of functional layers increases total protein content. Biocatalytic coatings prepared by layer-by-layer deposition of polyethylenamine, glutaraldehyde, and lactase exhibited increased protein loading with each layer [77]. However, overall activity did not increase with layers, likely a result of reduced substrate accessibility by enzyme entrapped in sub-interfacial layers. In other work, Shutava *et al.* layered hemoglobin, PS, and catalase in order to create a physical and chemical protective barrier [68]. Although the layer-by-layer technique has been shown to increase protein loading, diffusion of substrate to enzyme can become difficult. Increasing substrate diffusion through formed layers would improve activity retention, and therefore commercial potential, of biocatalytic coatings prepared by layer-by-layer deposition.

Photografting is often used in combination with other surface modification methods for preparation of biocatalytic coatings. Garnett *et al.* published an early report on immobilization of the protease trypsin to PP, PVOH, and PS by enhancing photografting with the use of metal salts [80]. Carboiimide chemistry can be used to covalently couple enzymes (via amine groups) to carboxylic acid groups introduced to a material surface via photografted polymerization [81]. Krenkova *et al.*, functionalized poly(ethylene) glycol methacrylate to immobilize trypsin and endoproteinases for antibody analysis in enzyme reactors. In that work, 4-vinyl-2,2-dimethylazolactone was photografted to introduce a porous surface morphology with the goal of optimizing enzyme orientation to improve performance. As a result, non-specific protein binding was reduced and the protein substrate was successfully digested [82]. In other work, the hydrolase urease was immobilized to photgrafted polytetrafluoroethylene to remove urea from beverages and foods by Yamada *et al.*, After immobilization, urease exhibited lowered activity at higher protein contents due to high protein densities as a result of grafted layers [86]. Indeed, a common challenge in enzyme immobilization is tailoring immobilization density and material chemistry to reduce adverse protein–protein and protein–surface interactions [107].

Emerging technologies in biocatalytic materials such as electrospun nanofibers and biopatterning offer new opportunities in preparing biocatalytic active packaging coatings. Electrospinning is a method by which enzymes or other active agents can be incorporated into polymer nanofibers. In one report, lactase blended into polyethylene oxide nanofibers retained up to 93% of free enzyme activity with significant retention of activity after dry storage [53]. In other work by Ge *et al.*, glucose oxidase was immobilized onto electrospun polyvinyl acetate/chitosan/tea extract fibers to reduce oxygen in packaged foods to extend shelf life [108]. The immobilized glucose oxidase in this system retained over 68% of free enzyme activity. Drug delivery with controlled release has sparked new interest in various fiber morphologies, as well as micro and nanoparticles and hydrogels. Embedding and blending is often paired with cross-linking techniques to improve retention of the support material's physical properties [109]. Biopatterning, in which specific patterns of immobilized biomolecules are defined with micron or submicron resolution, also has potential application in active packaging coatings. For example, patterning cells to the interior of microfluidic channels by photografting poly(ethylene glycol) reduced non-specific protein binding [110,111]. Creating specific patterns for enzyme immobilization can focus activity on targeted regions in a packaging matrix, thereby reducing material waste during functionalization and controlling coating costs. Because of the interdependence of the enzyme structure on biocatalytic activity, preparation of biocatalytic coatings demands a whole systems approach, with consideration given to the enzyme and the coating matrix as well as the packaged product. Consideration must be given to reducing diffusional limitations of substrate and product, as well as activity retention for increased thermostability and pH stability. The overall stability of the bound enzyme determines the success of the coating method.

4. CHALLENGES AND PERSPECTIVES

Exciting opportunities exist in the development of active packaging coatings for improving the safety, quality, and shelf life of packaged goods, which is brought about by an increasing preference and demand from consumers for additive-free foods and a continuously expanding transport chain and standards of quality driven by global commerce [112]. The global trade of active food packaging was estimated to be close to US $9,000,000,000.00 in 2011 and it is expected to be around US $12,000,000,000.00 in 2017 [113]. Nevertheless, the active food packaging market is currently widely dominated by oxygen scavenging and moisture absorption applications [112,114]. A number of challenges remain prior to the commercial adoption of the types of applications studied in this review. A key consideration in developing coatings for active packaging applications is considering toxicity and potential for regulatory approval of not just the active agent but any tether molecules or cross-linkers that may be employed [115]. Consideration must also be given to requirements of different regulatory agencies, whether under the jurisdiction of the Food and Drug Administration (in the United States), the European Food Safety Authority (in the European Union), or elsewhere. Non-migratory technologies produced by either covalent immobilization, cross-linked layer-by-layer deposition, or some photografted coatings offer a potential regulatory benefit in US regulated food packaging applications, as they would require approval as food contact materials rather than direct additives [116]. Migration testing using standardized simulants (water, 3% acetic acid, 15% ethanol, olive oil, iso-octane, and 95% ethanol) must be performed to quantify levels of migrants in packaged product systems [117]. It is also critical to consider the influence of the active coating on other performance characteristics of the packaging material. A benefit to coatings over bulk material modification is that bulk material properties should remain intact. However, the influence of the coating on processability, thermomechanical properties, barrier properties, and seal strength must be characterized [118]. Rigorous application tests must also be performed to ensure that neither material conversion steps nor end use result in delamination of active coatings. Many of the coatings technologies surveyed in this report have potential for scalability to roll-to-roll, high throughput coating operations. Finally, while incorporation of active agents and specialized packaging processes will indeed increase material cost, the opportunities for new products, enhanced safety, and reduced waste of packaged goods highlight the potential for increasing product value through smart integration of active packaging coatings.

ACKNOWLEDGMENTS

This material is based upon work supported by the National Institute of Food and Agriculture, U.S. Department of Agriculture, under project numbers 2012-67017-30157, 2011-65210-20059, 2014-67021-21584, and 2015-67017-23119. M.J. Roman would like to acknowledge the Peter Salmon Graduate Fellowship (Department of Food Science, UMass Amherst) and Northeast Alliance for Graduate Education and the Professoriate Fellowship for their support. D.E. Wong acknowledges Agriculture and Food Research Initiative Grant No. 2015-67011-22820 for NIFA Predoctoral Fellowship support.

REFERENCES

1. Appendini, P.; Hotchkiss, J.H. Review of antimicrobial food packaging. *Innov. Food Sci. Emerg. Technol.* 2002, *3*, 113–126
2. Vermeiren, L.; Devlieghere, F.; van Beest, M.; de Kruijf, N.; Debevere, J. Developments in the active packaging of foods. *Trends Food Sci. Technol.* 1999, *10*, 77–86
3. Suppakul, P.; Miltz, J.; Sonneveld, K.; Bigger, S.W. Active packaging technologies with an emphasis on antimicrobial packaging and its applications. *J. Food Sci.* 2003, *68*, 408–420
4. Tian, F.; Decker, E.A.; Goddard, J.M. Controlling lipid oxidation of food by active packaging technologies. *Food Funct.* 2013, *4*, 669–680
5. Ozdemir, M.; Floros, J. Active food packaging technologies. *Crit. Rev. Food Sci. Nutr.* 2004, *44*, 185–193
6. Yam, K.; Takhistov, P.; Miltz, J. Intelligent packaging: Concepts and applications. *J. Food Sci.* 2005, *70*, R1–R10
7. Kuswandi, B.; Wicaksono, Y.; Jayus; Aminah, A.; Heng, L.Y.; Ahmad, M. Smart packaging: Sensors for monitoring of food quality and safety. *Sens. Instrum. Food Qual. Saf.* 2011, *5*, 137–146
8. Smith, J.P.; Hoshino, J.; Abe, Y. Interactive packaging involving sachet technology. In *Active Food Packaging*, 1st ed.; Rooney, M.L., Ed.; Springer: Berlin/Heidelberg, Germany, 1995; pp. 143–173
9. Bastarrachea, L.; Dhawan, S.; Sablani, S.S. Engineering properties of polymeric-based antimicrobial films for food packaging. *Food Eng. Rev.* 2011, *3*, 79–93
10. Han, J.H.; Floros, J.D. Casting antimicrobial packaging films and measuring their physical properties and antimicrobial activity. *J. Plast. Film Sheeting* 1997, *13*, 287–298
11. Vasile, C. General survey of the properties of polyolefins. In *Handbook of Polyolefins*, 2nd ed.; Vasile, C., Ed.; Marcel Dekker: New York, NY, USA, 2000; pp. 401–416
12. Min, S.; Krochta, J.M. Edible coatings containing bioactive antimicrobial agents. In *Packaging for Nonthermal Processing of Food*; Han, J.H., Ed.; Blackwell Publishing: Hoboken, NJ, USA; IFT Press: Ames, IA, USA, 2007; pp. 29–52
13. Han, J.H. Antimicrobial food packaging. In *Novel Food Packaging Techniques*; Ahvenainen, R., Ed.; CRC Press: Boca Raton, FL, USA; Woodhead Pub. Ltd.: Cambridge, UK, 2003; pp. 50–65
14. Robertson, G.L. *Food Packaging: Principles and Practice*, 2nd ed.; Taylor & Francis/CRC Press: Boca Raton, FL, USA, 2006; p. 550
15. Ratner, B.D. Surface modification of polymers: Chemical, biological and surface analytical challenges. *Biosens. Bioelectron.* 1995, *10*, 797–804
16. Kong, F.; Hu, Y.F. Biomolecule immobilization techniques for bioactive paper fabrication. *Anal. Bioanal. Chem.* 2012, *403*, 7–13
17. Chan, C.M.; Ko, T.M.; Hiraoka, H. Polymer surface modification by plasmas and photons. *Surf. Sci. Rep.* 1996, *24*, 1–54
18. Chatelier, R.C.; Xie, X.; Gengenbach, T.R.; Griesser, H.J. Quantitative analysis of polymer surface restructuring. *Langmuir* 1995, *11*, 2576–2584
19. Jansen, B.; Kohnen, W. Prevention of Biofilm Formation by Polymer Modification. *J. Ind. Microbiol.* 1995, *15*, 391–396
20. Sato, T.; Hiruma, K.; Shirai, M.; Tominanga, K.; Haraguchi, K.; Katsuyama, T.; Shimada, T. Site-controlled growth of nanowhiskers. *Appl. Phys. Lett.* 1995, *66*, 159–161
21. Barish, J.A.; Goddard, J.M. Topographical and chemical characterization of polymer surfaces modified by physical and chemical processes. *J. Appl. Polym. Sci.* 2011, *120*, 2863–2871

22. Goddard, J.M.; Hotchkiss, J.H. Polymer surface modification for the attachment of bioactive compounds. *Prog. Polym. Sci.* 2007, *32*, 698–725
23. Wagner, J.R., Jr. *Multilayer Flexible Packaging Technology and Applications for the Food, Personal Care and Over-the-Counter Pharmaceutical Industries*, 1st ed.; Elsevier Science: Oxford, NY, USA, 2010; p. 258
24. Moy, V.T.; Florin, E.L.; Gaub, H.E. Intermolecular forces and energies between ligands and receptors. *Science* 1994, *266*, 257–259
25. Hermanson, G. *Bioconjugate Techniques*, 2nd ed.; Academic Press: Boston, MA, USA, 2008; p. 1202
26. Farkaš, P.; Bystrický, S. Chemical conjugation of biomacromolecules: A mini-review. *Chem. Pap.* 2010, *64*, 683–695
27. Decher, G. Fuzzy nanoassemblies: Toward layered polymeric multicomposites. *Science* 1997, *277*, 1232–1237
28. Ratner, B.D.; Hoffman, A.S. Physichochemical surface modification of materials used in medicine. In *Biomaterials Science: An Introduction to Materials in Medicine*, 3rd ed.; Ratner, B.D., Hoffman, A.S., Schoen, F.J., Lemons, J.E., Eds.; Academic Press: Oxford, UK; Waltham, MA, USA, 2013; pp. 259–276
29. Bastarrachea, L.J.; Denis-Rohr, A.; Goddard, J.M. Antimicrobial food equipment coatings: Applications and challenges. *Annu. Rev. Food Sci. Technol.* 2015, *6*, 97–118
30. Yang, Y.; Haile, M.; Park, Y.T.; Malek, F.A.; Grunlan, J.C. Super gas barrier of all-polymer multilayer thin films. *Macromolecules* 2011, *44*, 1450–1459
31. Cerkez, I.; Kocer, H.B.; Worley, S.D.; Broughton, R.M.; Huang, T.S. N-Halamine biocidal coatings via a layer-by-layer assembly technique. *Langmuir* 2011, *27*, 4091–4097
32. Bastarrachea, L.J.; McLandsborough, L.A.; Peleg, M.; Goddard, J.M. Antimicrobial N-Halamine modified polyethylene: Characterization, biocidal efficacy, regeneration, and stability. *J. Food Sci.* 2014, *79*, E887–E897
33. Bastarrachea, L.J.; Peleg, M.; McLandsborough, L.A.; Goddard, J.M. Inactivation of *Listeria Monocytogenes* on a polyethylene surface modified by layer-by-layer deposition of the antimicrobial N-Halamine. *J. Food Eng.* 2013, *117*, 52–58
34. Haynie, D.; Zhang, L.; Rudra, J.; Zhao, W.; Zhong, Y.; Palath, N. Polypeptide multilayer films. *Biomacromolecules* 2005, *6*, 2895–2913
35. SPALASTM (Spray Assisted Layer-by-Layer Assembly) Coating System. Available online: http://www.agiltron.com(accessed on 20 September 2015)
36. Odian, G.G. *Principles of Polymerization*, 4th ed.; Wiley InterScience: Hoboker, NJ, USA, 2004; p. 812
37. Carlini, C.; Angiolini, L. Polymers as free radical photoinitiators. In *Synthesis and Photosynthesis*; Springer-Verlag: Heidelberg, Germany, 1995; pp. 127–214
38. Dunkirk, S.G.; Gregg, S.L.; Duran, L.W.; Monfils, J.D.; Haapala, J.E.; Marcy, J.A.; Clapper, D.L.; Amos, R.A.; Guire, P.E. Photochemical coatings for the prevention of bacterial colonization. *J. Biomater. Appl.* 1991, *6*, 131–156
39. Matsuda, T.; Inoue, K. Novel photoreactive surface modification technology for fabricated devices. *ASAIO Trans.*1990, *36*, 161–164
40. Mishra, A.; Daswal, S. Curcumin, A Novel natural photoinitiator for the copolymerization of styrene and methylmethacrylate. *J. Macromol. Sci. A* 2005, *42*, 1667–1678
41. Zhao, J.; Lalevée, J.; Lu, H.; MacQueen, R.; Kable, S.H.; Schmidt, T.W.; Stenzel, M.H.; Xiao, P. A new role of curcumin: As a multicolor photoinitiator for polymer fabrication under household UV to red LED bulbs. *Polym. Chem.* 2015, *6*, 5053–5061
42. Brody, A.L.; Strupinsky, E.R.; Kline, L.R. *Active Packaging for Food Applications*; Technomic Pub. Co.: Lancaster, PA, USA, 2001; p. 218
43. Kerry, J.; Butler, P. *Smart Packaging Technologies for Fast Moving Consumer Goods*; John Wiley: Chichester, UK; Hoboken, NJ, USA, 2008; p. 340

44. Garces, L.O.; de la Puerta, C.N. Antimicrobial packaging based on the use of natural extracts and the process to obtain this packaging. European Patent EP1657181-B1, 13 January 2010
45. Manso, S.; Becerril, R.; Nerin, C.; Gomez-Lus, R. Influence of pH and temperature variations on vapor phase action of an antifungal food packaging against five mold strains. *Food Control* 2015, *47*, 20–26
46. Valderrama Solano, A.C.; Rojas de Gante, C. Two different processes to obtain antimicrobial packaging containing natural oils. *Food Bioprocess Technol.* 2012, *5*, 2522–2528
47. Minelli, M.; de Angelis, M.G.; Doghieri, F.; Rocchetti, M.; Montenero, A. Barrier properties of organic-inorganic hybrid coatings based on polyvinyl alcohol with improved water resistance. *Polym. Eng. Sci.* 2010, *50*, 144–153
48. Lantano, C.; Alfieri, I.; Cavazza, A.; Corradini, C.; Lorenzi, A.; Zucchetto, N.; Montenero, A. Natamycin based sol-gel antimicrobial coatings on polylactic acid films for food packaging. *Food Chem.* 2014, *165*, 342–347
49. Zhu, Y.; Buonocore, G.G.; Lavorgna, M. Photocatalytic activity of PLA/TiO_2 nanocomposites and TiO_2-active multilayered hybrid coatings. *Ital. J. Food Sci.* 2012, *24*, 102–106
50. Contini, C.; Katsikogianni, M.G.; O'Neill, F.T.; O'Sullivan, M.; Boland, F.; Dowling, D.P.; Monahan, F.J. Storage stability of an antioxidant active packaging coated with citrus extract following a plasma jet pretreatment. *Food Bioprocess Technol.* 2014, *7*, 2228–2240
51. Contini, C.; Álvarez, R.; O'Sullivan, M.; Dowling, D.P.; Gargan, S.O.; Monahan, F.J. Effect of an active packaging with citrus extract on lipid oxidation and sensory quality of cooked turkey meat. *Meat Sci.* 2014, *96*, 1171–1176
52. Contini, C.; Katsikogianni, M.G.; O'Neill, F.T.; O'Sullivan, M.; Dowling, D.P.; Monahan, F.J. PET trays coated with citrus extract exhibit antioxidant activity with cooked turkey meat. *LWT Food Sci. Technol.* 2012, *47*, 471–477
53. Wong, D.E.; Dai, M.; Talbert, J.N.; Nugen, S.R.; Goddard, J.M. Biocatalytic polymer nanofibers for stabilization and delivery of enzymes. *J. Mol. Catal. B Enzym.* 2014, *110*, 16–22
54. Bolumar, T.; Andersen, M.L.; Orlien, V. Antioxidant active packaging for chicken meat processed by high pressure treatment. *Food Chem.* 2011, *129*, 1406–1412
55. Johansson, K.; Winestrand, S.; Johansson, C.; Jarnstrom, L.; Jonsson, L.J. Oxygen-scavenging coatings and films based on lignosulfonates and laccase. *J. Biotechnol.* 2012, *161*, 14–18
56. Lee, C.H.; An, D.S.; Lee, S.C.; Park, H.J.; Lee, D.S. A Coating for use as an antimicrobial and antioxidative packaging material incorporating nisin and α-tocopherol. *J. Food Eng.* 2004, *62*, 323–329
57. Winestrand, S.; Johansson, K.; Järnström, L.; Jönsson, L.J. Co-immobilization of oxalate oxidase and catalase in films for scavenging of oxygen or oxalic acid. *Biochem. Eng. J.* 2013, *72*, 96–101
58. Makwana, S.; Choudhary, R.; Dogra, N.; Kohli, P.; Haddock, J. Nanoencapsulation and immobilization of cinnamaldehyde for developing antimicrobial food packaging material. *LWT Food Sci. Technol.* 2014, *57*, 470–476
59. Guo, M.; Jin, T.Z.; Yang, R. Antimicrobial polylactic acid packaging films against listeria and salmonella in culture medium and on ready-to-eat meat. *Food Bioprocess Technol.* 2014, *7*, 3293–3307
60. Theinsathid, P.; Visessanguan, W.; Kruenate, J.; Kingcha, Y.; Keeratipibul, S. Antimicrobial activity of lauric arginate-coated polylactic acid films against listeria monocytogenes and salmonella typhimurium on cooked sliced ham. *J. Food Sci.* 2012, *77*, M142–M149
61. Auxier, J.A.; Schilke, K.F.; McGuire, J. Activity retention after nisin entrapment in a polyethylene oxide brush layer. *J. Food Protect.* 2014, *77*, 1624–1629
62. Fernandes, S.C.M.; Sadocco, P.; Causio, J.; Silvestre, A.J.D.; Mondragon, I.; Freire, C.S.R. Antimicrobial pullulan derivative prepared by grafting with 3-

aminopropyltrimethoxysilane: characterization and ability to form transparent films. *Food Hydrocoll.* 2014, *35*, 247–252

63. Schreiber, S.B.; Bozell, J.J.; Hayes, D.G.; Zivanovic, S. Introduction of primary antioxidant activity to chitosan for application as a multifunctional food packaging material. *Food Hydrocoll.* 2013, *33*, 207–214
64. Goddard, J.M.; Talbert, J.N.; Hotchkiss, J.H. Covalent attachment of lactase to low-density polyethylene films. *J. Food Sci.* 2007, *72*, E36–E41
65. Mahoney, K.W.; Talbert, J.N.; Goddard, J.M. Effect of polyethylene glycol tether size and chemistry on the attachment of lactase to polyethylene films. *J. Appl. Polym. Sci.* 2013, *127*, 1203–1210
66. Muriel-Galet, V.; Talbert, J.N.; Hernandez-Munoz, P.; Gavara, R.; Goddard, J.M. Covalent immobilization of lysozyme on ethylene vinyl alcohol films for nonmigrating antimicrobial packaging applications. *J. Agric. Food Chem.*2013, *61*, 6720–6727
67. Struller, C.F.; Kelly, P.J.; Copeland, N.J. Aluminum oxide barrier coatings on polymer films for food packaging applications. *Surf. Coat. Technol.* 2014, *241*, 130–137
68. Shutava, T.G.; Kommireddy, D.S.; Lvov, Y.M. Layer-by-layer enzyme/polyelectrolyte films as a functional protective barrier in oxidizing media. *J. Am. Chem. Soc.* 2006, *128*, 9926–9934
69. Anthierens, T.; Billiet, L.; Devlieghere, F.; du Prez, F. Poly(butylene adipate) functionalized with quaternary phosphonium groups as potential antimicrobial packaging material. *Innov. Food Sci. Emerg. Technol.* 2012, *15*, 81–85
70. Soares, N.F.F.; Hotchkiss, J.H. Bitterness reduction in grapefruit juice through active packaging. *Packag. Technol. Sci.*1998, *11*, 9–18
71. Soares, N.; Hotchkiss, J. Naringinase immobilization in packaging films for reducing naringin concentration in grapefruit juice. *J. Food Sci.* 1998, *63*, 61–65
72. Nunes, M.A.P.; Vila-Real, H.; Fernandes, P.C.B.; Ribeiro, M.H.L. Immobilization of naringinase in PVA-alginate matrix using an innovative technique. *Appl. Biochem. Biotechnol.* 2010, *160*, 2129–2147
73. Mackiw, E.; Maka, L.; Sciezynska, H.; Pawlicka, M.; Dziadczyk, P.; Rzanek-Boroch, Z. The impact of plasma-modified films with sulfur dioxide, sodium oxide on food pathogenic microorganisms. *Package Technol. Sci.* 2015, *28*, 285–292
74. Pinheiro, A.C.; Bourbon, A.I.; Medeiros, B.G.D.S.; da Silva, L.H.M.; da Silva, M.C.H.; Carneiro-da-Cunha, M.G.; Coimbra, M.A.; Vicente, A.A. Interactions between κ-carrageenan and chitosan in nanolayered coatings-structural and transport properties. *Carbohydr. Polym.* 2012, *87*, 1081–1090
75. Carneiro-da-Cunha, M.G.; Cerqueira, M.A.; Souza, B.W.S.; Carvalhoc, S.; Quintas, M.A.C.; Teixeira, J.A.; Vicente, A.A. Physical and thermal properties of a chitosan/alginate nanolayered PET film. *Carbohydr. Polym.* 2010, *82*, 153–159
76. Shutava, T.G.; Prouty, M.D.; Agabekov, V.E.; Lvov, Y.M. Antioxidant properties of layer-by-layer films on the basis of tannic acid. *Chem. Lett.* 2006, *35*, 1144–1145
77. Wong, D.E.; Talbert, J.N.; Goddard, J.M. Layer by layer assembly of a biocatalytic packaging film: Lactase covalently bound to low-density polyethylene. *J. Food Sci.* 2013, *78*, E853–E860
78. Medeiros, B.G.D.S.; Pinheiro, A.C.; Teixeira, J.A.; Vicente, A.A.; Carneiro-da-Cunha, M.G. Polysaccharide/protein nanomultilayer coatings: Construction, characterization and evaluation of their effect on "Rocha" pear (*Pyrus communis* L.) shelf-life. *Food Bioprocess Technol.* 2012, *5*, 2435–2445
79. Arrua, D.; Strumia, M.C.; Nazareno, M.A. Immobilization of Caffeic acid on a polypropylene film: Synthesis and antioxidant properties. *J. Agric. Food Chem.* 2010, *58*, 9228–9234
80. Garnett, J.L.; Jankiewicz, S.V.; Long, M.A.; Sangster, D.F. Radiation and photografting as complementary techniques for immobilizing bioactive materials. *Int. J. Radiat. Appl. Instrum. C Radiat. Phys. Chem.* 1986, *27*, 301–309

81. Yamada, K.; Nakasone, T.; Nagano, R.; Hirata, M. Retention and reusability of trypsin activity by covalent immobilization onto grafted polyethylene plates. *J. Appl. Polym. Sci.* 2003, *89*, 3574–3581
82. Krenkova, J.; Lacher, N.A.; Svec, F. Highly efficient enzyme reactors containing trypsin and endoproteinase LysC immobilized on porous polymer monolith coupled to MS suitable for analysis of antibodies. *Anal. Chem.* 2009, *81*, 2004–2012
83. Tian, F.; Decker, E.A.; Goddard, J.M. Control of lipid oxidation by nonmigratory active packaging films prepared by photoinitiated graft polymerization. *J. Agric. Food Chem.* 2012, *60*, 7710–7718
84. Roman, M.J.; Tian, F.; Decker, E.A.; Goddard, J.M. Iron chelating polypropylene films: Manipulating photoinitiated graft polymerization to tailor chelating activity. *J. Appl. Polym. Sci.* 2014, *131*
85. Tian, F.; Decker, E.A.; McClements, D.J.; Goddard, J.M. Influence of non-migratory metal-chelating active packaging film on food quality: Impact on physical and chemical stability of emulsions. *Food Chem.* 2014, *151*, 257–265
86. Yamada, K.; Iizawa, Y.; Yamada, J.; Hirata, M. Retention of activity of urease immobilized on grafted polymer films. *J. Appl. Polym. Sci.* 2006, *102*, 4886–4896
87. Tian, F.; Decker, E.A.; Goddard, J.M. Controlling lipid oxidation via a biomimetic iron chelating active packaging material. *J. Agric. Food Chem.* 2013, *61*, 12397–12404
88. Tian, F.; Roman, M.J.; Decker, E.A.; Goddard, J.M. Biomimetic design of chelating interfaces. *J. Appl. Polym. Sci.* 2015, *132*, 1–8
89. López-Rubio, A.; Lagarón, J.M.; Ocio, M.J. *Active Polymer Packaging of Non-Meat Food Products*; John Wiley & Sons, Ltd.: West Sussex, UK, 2008; pp. 19–32
90. Gomez-Estaca, J.; Lopez-de-Dicastillo, C.; Hernandez-Munoz, P.; Catala, R.; Gavara, R. Advances in antioxidant active food packaging. *Trends Food Sci. Technol.* 2014, *35*, 42–51
91. Garces, O.; Nerin, C.; Beltran, J.; Roncales, P. Antioxidant active varnish. European Patent EP1477159-A1, 22 December 2003
92. Tovar, L.; Salafranca, J.; Sánchez, C.; Nerín, C. Migration studies to assess the safety in use of a new antioxidant active packaging. *J. Agric. Food Chem.* 2005, *53*, 5270–5275
93. Nerin, C.; Tovar, L.; Salafranca, J. Behaviour of a new antioxidant active film *versus* oxidizable model compounds. *J. Food Eng.* 2008, *84*, 313–320
94. Pezo, D.; Salafranca, J.; Nerín, C. Design of a method for generation of gas-phase hydroxyl radicals, and use of HPLC with fluorescence detection to assess the antioxidant capacity of natural essential oils. *Anal. Bioanal. Chem.* 2006, *385*, 1241–1246
95. Pezo, D.; Salafranca, J.; Nerín, C. Determination of the antioxidant capacity of active food packagings by in situ gas-phase hydroxyl radical generation and high-performance liquid chromatography-fluorescence detection. *J. Chromatogr. A* 2008, *1178*, 126–133
96. Chatham, H. Oxygen diffusion barrier properties of transparent oxide coatings on polymeric substrates. *Surf. Coat. Technol.* 1996, *78*, 1–9
97. Roman, M.J.; Decker, E.A.; Goddard, J.M. Performance of nonmigratory iron chelating active packaging materials in viscous model food systems. *J. Food Sci.* 2015, *80*, 1965–1973
98. Fernández, A.; Cava, D.; Ocio, M.J.; Lagarón, J.M. Perspectives for biocatalysts in food packaging. *Trends Food Sci. Technol.* 2008, *19*, 198–206
99. Brody, A.L.; Budny, J.A. Enzymes as active packaging agents. In *Active Food Packaging*; Rooney, M.L., Ed.; Blackie Academic and Professional: New York, NY, USA, 1995; pp. 174–192
100. Gemili, S.; Yemenicioğlu, A.; Altınkaya, S.A. Development of cellulose acetate based antimicrobial food packaging materials for controlled release of lysozyme. *J. Food Eng.* 2009, *90*, 453–462

101. Barbiroli, A.; Bonomi, F.; Capretti, G.; Iametti, S.; Manzoni, M.; Piergiovanni, L.; Rollini, M. Antimicrobial activity of lysozyme and lactoferrin incorporated in cellulose-based food packaging. *Food Control* 2012, *26*, 387–392
102. Mendes de Souza, P.; Fernandez, A.; Lopez-Carballo, G.; Gavara, R.; Hernandez-Munoz, P. Modified sodium caseinate films as releasing carriers of lysozyme. *Food Hydrocoll.* 2010, *24*, 300–306
103. Del Nobile, M.A.; Conte, A. *Packaging for Food Preservation*; Springer: New York, NY, USA, 2013; p. 270
104. Andersson, M.; Andersson, T.; Adlercreutz, P.; Nielsen, T.; Hornsten, E. Toward an enzyme-based oxygen scavenging laminate. Influence of industrial lamination conditions on the performance of glucose oxidase. *Biotechnol. Bioeng.*2002, *79*, 37–42
105. Talbert, J.N.; He, F.; Seto, K.; Nugen, S.R.; Goddard, J.M. Modification of glucose oxidase for the development of biocatalytic solvent inks. *Enzyme Microb. Technol.* 2014, *55*, 21–25
106. Talbert, J.N.; Goddard, J.M. Influence of nanoparticle diameter on conjugated enzyme activity. *Food Bioprod. Process.*2013, *91*, 693–699
107. Talbert, J.N.; Goddard, J.M. Enzymes on material surfaces. *Colloid Surf. B Biointerfaces* 2012, *93*, 8–19
108. Ge, L.; Zhao, Y.; Mo, T.; Li, J.; Li, P. Immobilization of glucose oxidase in electrospun nanofibrous membranes for food preservation. *Food Control* 2012, *26*, 188–193
109. Moskovitz, Y.; Srebnik, S. Mean-field model of immobilized enzymes embedded in a grafted polymer layer. *Biophys. J.* 2005, *89*, 22–31
110. Larsen, E.K.U.; Mikkelsen, M.B.L.; Larsen, N.B. Protein and cell patterning in closed polymer channels by photoimmobilizing proteins on photografted poly(ethylene glycol) diacrylate. *Biomicrofluidics* 2014, *8*
111. Larsen, E.K.U.; Mikkelsen, M.B.L.; Larsen, N.B. Facile photoimmobilization of proteins onto low-binding PEG-coated polymer surfaces. *Biomacromolecules* 2014, *15*, 894–899
112. Realini, C.E.; Marcos, B. Active and intelligent packaging systems for a modern society. *Meat Sci.* 2014, *98*, 404–419
113. BCC Research. The Advanced Packaging Solutions Market Value for 2017 is Projected to be nearly $44.3 Billion. 2015. Available online: http://www.bccresearch.com (accessed on 25 October 2015)
114. Day, B.P.F. Active packaging of foods. In *Smart Packaging Technologies for Fast Moving Consumer Goods*; Kerry, J.P., Butler, P., Eds.; Wiley & Sons, Ltd.: West Sussex, UK, 2008; pp. 1–18
115. Lopez-Rubio, A.; Gavara, R.; Lagaron, J.A. Bioactive packaging: Turning foods into healthier foods through biomaterials. *Trends Food Sci. Technol.* 2006, *17*, 567–575
116. Koontz, J. Active Packaging Materials to Inhibit Lipid Oxidation: US Regulatory Framework. *Int. News Fats Oils Relat. Mater.* 2012, *23*, 598–600
117. Barnes, K.A.; Sinclair, C.R.; Watson, D.H. *Chemical Migration and Food Contact Materials*, 1st ed.; CRC: Boca Raton, FL, USA, 2007; p. 464
118. Farris, S.; Introzzi, L.; Piergiovanni, L.; Cozzolino, C.A. Effects of different sealing conditions on the seal strength of polypropylene films coated with a bio-based thin layer. *Ital. J. Food Sci.* 2011, *23*, 111–114

CHAPTER 15

Evaluation of Some Postharvest Storage Approaches on Essential Oil Characteristics of Fresh Organic Damask Rose (*Rosa damascena* Mill.) Flowers

Maryam Mirzaei[1], Nima Ahmadi[1,*], Fatemeh Sefidkon[2], Abdolali Shojaeiyan[1] and Alireza Mazaheri[3]

[1]*Department of Horticultural Sciences, Tarbiat Modares University, Tehran 336-14115, Iran*

[2]*Research Institute of Forests and Rangelands, Tehran 13185-116, Iran*

[3]*R & D Department, Golkaran Essential Oils and Herbal Extracts Company, Kashan 1178, Iran*

ABSTRACT

Damask rose (*Rosa damascena* Mill.) is an economically important species in the Rosaceae family for rose oil and rose water production, obtained exclusively from freshly-gathered Damask rose flowers. Iran is famous for producing organic rose essential oil, mainly used in the perfume industry and for folk medicines due to its healing properties. Due to their high perishability, postharvest handling of the flowers prior to oil extraction are a key point in rose essential oil production. In this study, postharvest handling conditions of Damask rose flowers were evaluated for the effect on yield and quality of the extracted essential oils. Organically-grown flowers were stored under different conditions included packaging in polyethylene bags and immersing into water containers, and then held at 25 ± 2 °C or 4 ± 1 °C for 1, 2, or 3 days. Identification of the chemical composition of the essential oils was performed by GC and GC/MS. The results indicated that petal storage in water resulted in the highest essential oil content and quality, and storage in polyethylene bags resulted in the greatest loss, compared to those from unstored petals. The results provide a reference for interested groups such as producers and consumers who are concerned about Damask rose flower storage and subsequent oil extraction.

Keywords: essential oil characteristics; organic damask rose; post-harvest handling

1. INTRODUCTION

Rosa damascena Mill., commonly known as the oil-bearing rose, is one of the most important Rosa species cultivated for production of rose oil and rose water [1]. Rose oil, one of the most expensive base materials in the flavor and perfume industries, is characterized by a high percentage of monoterpene alcohols including citronellol, nerol, geraniol, linalool, and phenyl ethyl alcohol. These components are the main contributors to the perfume value of rose oil [2]. According to the international standard of rose essential oil [3], rose oil requires 20%–34% citronellol, 15%–22% geraniol, and 8%–15% nonadecane as the main compounds [3]. As harvested Damask rose flowers are highly perishable, postharvest handling of the flowers prior to oil extraction is a vital step in rose essential oil production. Due to the huge volume of Damask rose flowers delivered to the factories during the harvest period, it often takes a long time before flowers are eventually extracted and distilled. Excessive temperatures inside the stacks of harvested flowers caused by their high respiration rate results in losses of essential oil quantity and quality [2]. Providing optimal storage conditions throughout postharvest handling could reduce losses of yield and quality of essential oils [4,5]. The high value of the essential oil of Damask rose is an important reason for finding proper storage techniques for the flower crop after harvest. Therefore, the objectives of this study were to determine the effects of ambient and reduced temperatures as well as different storage durations and surrounding mediums on the yield and chemical components of organic Damask rose flowers throughout a postharvest handling period of 3 days.

2. EXPERIMENTAL SECTION

2.1. Plant Material

Organically-grown fresh Damask roses petals were handpicked from a Rosarium located in Mashhad-e Ardehal, Kashan, Iran, and transported immediately to the laboratory. The essential oils obtained from unstored petals (0 day storage duration), which were hydrodistilled shortly after harvest, were considered as the reference point in order to determine the effect of storage conditions on essential oil content and oil composition.

2.2. Storage Conditions

The petals were stored under the following conditions: packaged using sealed polyethylene (PE) bags (60 μm thick), or immersed in water inside sealed plastic containers (20 cm × 25 cm × 30 cm), and then held at 25 °C (25 ± 2 °C, 45% ± 5% RH) or 4 °C (4 ± 1 °C, 90% ± 5% RH) for 1, 2 and 3 days.

2.3. Essential Oil Extraction

The essential oil was extracted from 250 g petals by hydrodistillation with 750 mL of distilled water using a 4 L Clevenger-type apparatus for each experimental replication. The distillation time was 1.5 h and the oil content was measured as percentage (*v*/*w*%). In the case of the water storage method, the water used for storage was also hydrodistilled.

2.4. Characterization of Essential Oil Components

2.4.1. Gas Chromatography (GC)

The essential oils were analyzed on a Thermo-UFM (Ultra-fast model, Milan, Italy) gas chromatograph using a 10 m × 0.1 mm capillary column PH-5 (nonpolar), with an inner surface with a stationary phase of 5% dimethyl siloxane phenyl. The oven temperature was held at 60 °C for 3 min, programmed to 285 °C at the rate of 80 °C/min, and detector (FID) and injection chamber temperatures were 280 °C. Helium was used as the carrier gas with inlet pressure of 0.5 mL/min.

2.4.2. Gas Chromatography-Mass Spectrometry (GC-MS)

The GC-MS analysis was performed using a Varian 3400 gas chromatograph connected to a mass spectrometer model Saturn II, using an ion trap system having ionization energy of 70 eV, with a DB-5 column (semi-polar), with dimensions of 30 m × 0.25 mm, and a film thickness 0.25 mm. Gas pressure was 2.46 kg/cm^2, column temperature was 40 to 250 °C with a rate of increase of 3 °C/min, and the injection chamber temperature and the transfer line were set at 260 °C and 270 °C, respectively. The carrier gas was helium with a linear velocity of 31.5 cm/s, and a split ratio of 1:60. A scan time of 1 s with a mass range 40–300 were used. Spectra identification were carried out using retention indices and by injection of normal hydrocarbons (C7-C25) under the same conditions and confirmed with the Wiley 275-L library and the data reported in the literature [6,7,8].

2.5. Data Analysis

The experiment was carried out as a factorial with the factors of storage duration, petal storage medium, and storage temperature in a randomized complete block design (RCBD). There were three replications of essential oil extraction from the petals for every treatment combination. Analysis of data from the essential oil content was performed using the general linear model (GLM) of SAS 9.2 [9]. Mean values were compared at the 5% ($p \leq 0.05$) and 1% ($p \leq 0.01$) levels of significance using Fisher's Least Significant Difference (LSD) test. There were three replications of every treatment combination. Component analyses of the oils were performed by injecting a mix of the three replications of each treatment combination on both GC and GC/MS.

3. RESULTS AND DISCUSSION

3.1. Essential Oil Contents

The petal essential oil content was significantly affected by storage condition and duration, but not temperature alone. There was a significantly greater content of essential oil stored in water at 4 °C and 25 °C at 1, 2 and 3 day compared to petals stored in PE bags (Figure 1). Storage of flowers in water recorded a significant increase ($p \leq 0.01$) of 16%, 19% and 23% in essential oil content after 1, 2, and 3 day, respectively, over the unstored petals with an average value of 0.069 (v/w%). This increase could be explained by considering the limitation of

O_2 inside the containers along with the presence of water that may have caused fermentation, as it has been indicated that rose petals produce more essential oil with a fermentation period after harvest [10]. The mean essential oil content obtained from petals stored in PE for 1 day were close to the value at harvest (unstored petals) but were significantly lower after 2 or 3 day of storage. Cold-stored flowers in water contained higher oil content than those in PE at the same storage temperature, which might be explained by the ability of water to maintain volatile oil content of the petals during the storage period, while the permeability of the PE bags might result in losses from volatilization of essential oils, particularly with the extension of the storage period.

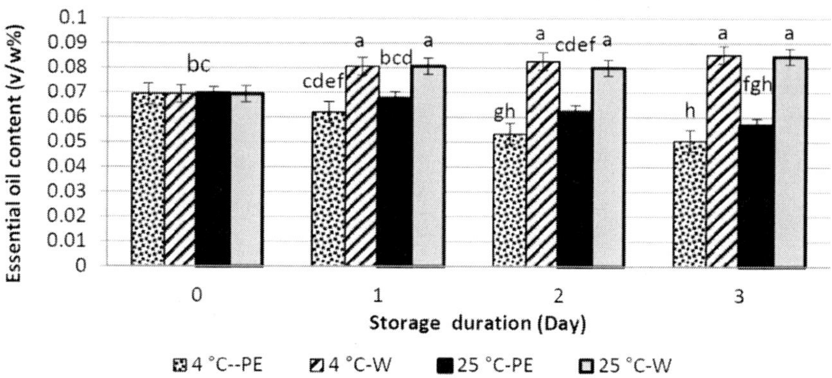

Figure 1. Mean essential oil content (*v/w*%, fresh weight basis) of organic Damask rose under different storage conditions. PE: (Polyethylene bag); W: (Water).

3.2. The Essential Oil Components

The changing patterns of the essential oil components under the storage conditions indicated that the chemical profile of rose oil throughout storage of petals in PE bags was affected more than storage in water under either 25 °C or 4 °C storage conditions, with a higher degradation of geraniol in favor of citronellol at 25 °C (Table 1). Storage of petals in water impacted the maintenance of the major essential oil compounds. The component analyses of oils from 2 and 3 day samples indicated that storage of the petals in the cold water (4 °C-W) maintained the rose oil characteristics better than the other handling method, with approximately 38% lower levels of long-chain aliphatic hydrocarbons after 1 day of storage, which resulted in improvement of the oil quality. A possible reason for the rise of monoterpene alcohols might be ascribed to the proper conditions for the action of enzymes, such as β-glucosidase involved in hydrolysis and liberation of monoterpene alcohols from their glycosidic forms [11].

Table 1. Damask rose essential oil components under different storage conditions.

Factors of Variance		Compound	Phenyl Ethyl Alcohol	*cis*-Rose Oxide	Dihydrolinalool	Citronellol	Neral	Geraniol	Geranial	Citronellyl Acetate	*E*-*β*-Damascone	*n*-Heptadecane	*E,E*-Farnesyl Acetate	Nonadecane	*n*-Eicosane	*n*-Heneicosane	*n*-Tricosane
Time	Storage conditions	RI	1109	1113	1135	1228	1240	1255	1269	1355	1416	1693	1728	1902	2003	2102	2298
0 day	Unstored petals		1.3	0.3	0.8	35.5	13.3	16.8	3.1	0.4	0.2	2.3	1.1	15.6	0.7	2.6	0.8 *
1 day	4 °C-PE		1.0	0.3	0.8	42.5	3.9	13.0	0.3	-	-	2.3	0.2	17.9	1.4	8.1	1.2
	4 °C-W		1.3	0.3	0.9	38.2	21.5	13.4	2.9	0.2	0.3	2.5	0.9	9.1	0.5	1.0	0.3
	25 °C-PE		1.6	0.4	1.3	47.4	16.5	9.8	5.8	-	0.3	2.6	0.4	8.8	0.3	4.1	0.8
	25 °C-W		1.1	0.7	0.9	42.3	11.4	12.6	2.0	0.1	0.4	2.9	0.7	20.5	1.0	6.3	0.9
2 days	4 °C-PE		0.9	0.4	0.8	45.6	12.5	10.8	1.6	-	0.4	2.6	0.5	18.9	2.1	12.3	2.2
	4 °C-W		1.7	0.5	1.4	48.2	15.6	8.1	6.2	-	0.3	2.8	0.7	7.3	0.4	0.6	0.8
	25 °C-PE		1.2	0.3	0.5	36.5	18.7	10.0	2.0	0.2	0.1	2.6	0.5	12.8	0.4	1.7	0.5
	25 °C-W		1.1	1.1	1.1	43.4	10.2	11.8	1.2	-	0.6	3.3	0.4	22.2	1.1	5.0	1.1
3 days	4 °C-PE		1.0	0.3	0.7	47.4	7.4	9.0	0.6	-	-	2.9	-	20.7	2.7	14.3	4.1
	4 °C-W		0.7	0.4	1.0	35.6	16.4	15.4	3.4	-	0.4	3.1	-	7.6	0.6	2.2	0.4
	25 °C-PE		1.9	0.6	0.8	36.5	12.8	5.2	9.9	-	0.2	3.4	0.8	12.6	0.4	3.6	0.7
	25 °C-W		1.0	1.5	1.4	46.7	9.2	10.0	0.9	-	0.8	3.9	0.1	23.6	1.2	4.6	1.3

RI: Retention Index; * Relative percentage of each component in relation to total essential oil. W: water; PE (polyethylene).

4. CONCLUSIONS

As the water used for the storage of the flowers was then used in the hydrodistillation, water, particularly at low temperatures, might be an effective storage method for oil-bearing rose petals after harvest to maintain essential oil content and component quantities close to the content of freshly-harvested flowers. These results are useful for choosing the best storage techniques for Damask rose flowers prior to processing.

ACKNOWLEDGMENTS

The authors would like to express their appreciation to Barij Essence Company for their support and for providing facilities for this study.

REFERENCES

1. Widrlechner, M.P. History and utilization of *Rosa damascena*. *Econ. Bot.* **1981**, *35*, 42–58.

2. Baydar, H.; Baydar, N. The effects of harvest date, fermentation duration and Tween 20 treatment on essential oil content and composition of industrial oil rose (*Rosa damascena* Mill.). *Ind. Crops Prod.* **2005**, *21*, 251–255.
3. Oil of Rose (*Rosa × damascena* Mill.). International Standards for Business, Government and Society. Available online:http://www.iso.org (accessed on 10 August 2015).
4. Najafian, S. Storage conditions affect the essential oil composition of cultivated Balm Mint Herb (Lamiaceae) in Iran. *Ind. Crops Prod.* **2014**, *52*, 575–581.
5. Kazaz, S.; Sabri, E.; Baydar, H.; Dilmacunal, T.; Koyuncu, M.A. Cold storage of oil rose (*Rosa damascena* Mill.) flowers. *Sci. Hortic.* **2010**, *126*, 284–290.
6. Adams, R.P. *Identification of Essential Oils by Ion Trap Mass Spectroscopy*; Academic Press: New York, NY, USA, 1989; p. 302.
7. Shibamoto, T. *Retention Indices in Essential Oil Analysis in Capillary Gas Chromatography in Essential Oil Analysis*; Sandra, P., Bicci, C., Eds.; Verlag: New York, NY, USA, 1987; pp. 259–274.
8. McLafferty, F.W.; Stauffer, D.B. *The Wiley/NBS Registry of Mass Spectral Data*; Wiley and Sons: New York, NY, USA, 1989.
9. SAS Institute Inc. *SAS Online Doc® 9.2*; SAS Institute Inc.: Cary, NC, USA, 2008.
10. Menary, R.; Mactavish, H.S. Rural industries research and development corporation. In *Boronia Extraction: Increasing Yield and Quality: A Report for the Rural Industries*; Research and Development Corporation: Hobart, Australia, 2000.
11. Baldermann, S.; Yang, Z.; Sakai, M.; Fleischmann, P.; Watanabe, N. Volatile constituents in the scent of roses. *Floric. Ornam. Biotechnol.* **2009**, *3*, 89–97.

CHAPTER 16

Fruit Setting Behaviour of Passion Fruit

Mira Rani Das[1*], Tofazzal Hossain[2], M. A. Baset Mia[2], J. U. Ahmed[2], A. J. M. Sirajul Karim[3], M. Mofazzal Hossain[4]

[1]*Agriculture Training Institute, Gazipur, Bangladesh;*

[2]*Department of Crop Botany, Bangabandhu Sheikh Mujibur Rahman Agricultural University, Gazipur, Bangladesh;*

[3]*Department of Soil Science, Bangabandhu Sheikh Mujibur Rahman Agricultural University, Gazipur, Bangladesh;*

[4]*Department of Horticulture, Bangabandhu Sheikh Mujibur Rahman Agricultural University, Gazipur, Bangladesh.*

ABSTRACT

Passion fruit has great prospective to fascinate fruit consumer for its taste and delicious fruit juice and improvement of the economic condition of the fruit grower in the developing countries. The self-incompatibility in the passion fruits is an imperative reason to be considered regarding fruit production. Pollination is an essential for self-sterile crops as passion fruit (Passiflora edulis Sims.). The experiment was conducted in the field and laboratories of the Bangabandhu Sheikh Mujibur Rahman Agricultural University, Salna, Gazipur, Bangladesh to investigate the fruit setting behavior of passion fruits at during five flashes. We estimated to study fruit setting behavior of passion fruit at different flashes and determine effective pollination method and suitable flashes among five flashes. Result revealed that percent of fruit set of passion fruit was recorded highest; Seeds per plant were recorded highest in third flash when flowers were pollinated by hand compared with self-, and natural pollinations. Length-breadth ratios of fruits in third flashes were recorded higher when pollinates by hand. Individual fruit weight was also recorded higher at third flash. Plants required minimum days from flower anthesis to full maturity during third flash. On the other hand, fruit growth behavior of hand pollinated flowers was recorded higher during third flash of passion fruit. Results indicated that all studied characteristics of fruit and seed of passion fruit of third flash performed best.

Keywords: Passion Fruit; Fruit Setting; Flash; Pollination

1. INTRODUCTION

The passion fruit belongs to family passionflowers is an allogamous species [1]. In Passiflora genus contain more than 450 species but 12 species are cultivated. Only one species is Passiflora edulis Sims that grown as perennial, possess large flowers, and are cross-pollinated and vastly commercialized fruits [2]. It largely distributed around the tropics and warm humid subtropics [3-5]. It is an agronomically important delicious fruits [6] and cultivated for its ornamental, medicinal and nutritive characteristics [7]. Purple and yellow passion fruits are commonly cultivated in northern region of India [8].

Passion fruit has hermaphrodite, solitary flowers, located in the leaf axils. There are usually five stamens and ovary is borne over the androgynophore. There are three styles united at base, and at the top of style there are three bifurcated stigmas [9]. Passion fruits are protandrous as anther dehiscence before stigma becomes receptive and stigma remains receptive from time of flower opening to closing. Rhythmic movement of the style of passion flower has been shown self-incompatibility because of the style is in upright position and it starts curving in due course of time. Due to it floral morphology yellow passion fruit is an allogamous plant and self-incompatibility of sporophytic type [10] and cultivated for its edible fruits [11]. Self-incompatibility is an important factor for passion fruit production [12].

Pollination is important for fruit production on passion vines and extent of fruit set is dependent on effective pollination. Many yellow passion fruits do not set fruit unless their flowers are dusted with pollen from a different vine that is genetically compatible. Hand pollination is the easiest way to ensure fruit production. Hand pollination increases fruit yield in passion fruit. Artificial pollination is necessary in passion fruit because of its floral morphology where the anthers placed bellow the stigma. Anther of passion fruit is versatile in nature and turns upside down at the time of anthesis. Pollen grains are large, highly sticky and self-incompatible. The self-incompatibility in the passion fruits is an important factor to be considered regarding fruit production and studies on the heredity [10,12].

The flowers of passion fruit are large, attractive, colorful and fragrant. The flowers produce a plenty of pollen and nectar that facilitate insect pollination. The principal insects visiting passion flower include Apis mellifera (honey bee) and Xylocopa vanpuncta (carpenter bees). Carpenter bee is most effective pollinator as it has large body and its body brushes along the anther and stigma while collecting nectar. On the other hand honey bees are not effective pollinator because of their foraging habit [13]. Though the passion flowers are hermaphrodite, they are self-sterile and self-incompatible which lead to poor fruit set [10]. The anther at the top of filament is versatile in nature. The anther just at the time anthesis turns upset down. The pollens of passion flower are very sticky in nature which may be another cause of poor fruit set naturally. Wind is ineffective for pollination because of heaviness and stickiness of the pollen though flowers have fertile pollen.

The passion flower usually opens around mid-day that is generally the warmest time of day, until the end of the afternoon. During this period pollinators collect nectar, transfer pollen from one flower to another. The effective pollination occurs in the period after the style curves completely [14].

Fruit setting behavior is an important criterion for plant breeder in the process of development of a variety. The development of variety has been associated with yield and fruit quality [15]. The agro-ecological conditions mainly of hill tract regions of Bangladesh are amiable for passion fruit cultivations [16]. Passion fruit in Bangladesh is grown popularly in home gardens for its lucrative color of flower and flavor and tasty yellow juice. Though the fruit setting behavior of passion fruit has been studied in details in other countries, very little work has been done in Bangladesh [16,17]. Due to above mentioned natural constraints pollination and fruit set of passion fruit is hampered. Considering the above facts the present study was undertaken to find out suitable pollination method for successful fruit setting.

2. MATERIALS AND METHODS

The study was conducted at the experimental farm of Bangabandhu Sheikh Mujibur Rahman Agricultural University, Gazipur, situated at 24.09° North latitude and 90.26° East longitudes with an elevation of 8.4 meter from the sea level. The climate of experimental site is subtropical characterized by heavy rainfall during April to September and scanty during the rest of the year. Annual rainfall is favorable for passion fruit growing. The soil of the experimental field is clay loam in texture and acidic in nature with pH of around 5.8. The yellow passion fruit (Passiflora edulis Sims.) was used for the present experiment. Vine cuttings of two years old plant earlier collected from Bangladesh Agricultural Research Institute were used for the this study. Passion fruit were planted in trellised in rows 4.5 m apart and spaced 4.5 m apart within rows. The experimental plot was welldrained high land and where pit were prepared on the raised beds. Recommended fertilizers were applied in the pits (50 × 50 × 50 cm^3) 10 days before transplanting of cuttings. Flowers were pollinated by hand compared with self-, and natural pollinations.

Natural pollination: For natural pollination 100 well developed flower buds were tagged for each flash from healthy plants. At 15th day after tagging, fruits were counted. Natural pollination is usually caused by wind, water and insects.

Self-pollination: For self-pollination another 100 well developed flower buds were tagged and bagged from each flash on previous day of blooming from healthy plants. After 5 days the tags were removed. At 15th day after tagging, fruits were counted.

Hand Pollination: For hand pollination another 100 well developed flower buds were tagged from each flash on previous day of blooming from healthy plants. On next day the tagged well bloomed flowers were pollinated by hand. At 15th day after hand pollination fruits were counted.

Data on the Fruit weight, Fruit length, Fruit breadth, Fruit and seed setting and Number of seeds per fruits were recorded.

Fruit and seed setting after open-, self-, and hand pollination: Flowers tagged for natural, selfand hand pollination were done at different times of the day. Twenty flowers of each type were considered for each treatment. The flowers were bagged 24 hrs before anthesis and they were rebagged for another 5 days after pollination. Open pollinated flowers were only tagged. Fruit set was observed by counting the fruits harvested at maturity.

Number of seeds per fruits: To observe the seed setting ability of both types 40 - 45 days old fruits were harvested and numbers of seeds per fruit were counted. Total numbers of seeds per fruit were recorded.

The collected data were analyzed statistically using MSTAT-C computer package (Michigan State University, East Lansing, MI, USA) following the methods ofGomez and Gomez (1984) [18]. The analysis of variance procedure (ANOVA), differences among treatment means were determined using the Least Significant Difference (LSD) at 5% level of significance.

3. RESULTS

3.1. Fruit Set

Fruit set using three pollination methods at different flashes of passion fruit was described here.

3.1.1. Self-Pollination

Fruit set percent at different flashes using self-pollination ranged from 1.71% to 4.51% (**Table 1**). Plants produced maximum (4.51%) fruit during third flash using selfpollination. Before and after third flash using self-pollination, fruit set percent was noticed to be declined gradually. Plants produced significantly higher fruits during third flash compared with other flashes using self-pollination. No significant change of fruit set was observed between third and fifth flashes using self-pollination.

3.1.2. Natural Pollination

At different flashes imposing natural pollination fruit set percent of passion fruit varied from 12.60 to 25.67. As self-pollination plants produced maximum fruit (25.67%) during third flash imposing natural pollination (**Table 1**). Fruit set percent during third flash was recorded highest (25.67) compared with other flashes by natural pollination. Higher rate of fruit set of passion fruit was recorded in all flashes during natural pollination compared with self-pollination.

3.1.3. Hand Pollination

Among the studied methods, hand pollination showed to set highest percent of fruit at all flashes. Imposing hand pollination fruit set percent ranged from 31.61% to 46.71% (**Table 1**). Among the three pollination methods, hand pollination was noticed to be best in respect of fruit set. As other two methods (self-pollination and natural pollination), hand pollination produced maximum fruit during third flash. Among the three pollination methods and five flashes, plant produced highest fruit imposing hand-pollination during third flash. Among the flashes, flowers pollinated by hand produced maximum fruits during third flash followed by fourth, second, fifth and first flashes.

3.2. Seeds Per Fruit

Effect of the pollination methods on seeds per fruit at different flashes was presented in **Table 2**.

3.2.1. Self-Pollination

Seeds per fruit in multiple seeded fruits depend upon number of pollens availability on the stigma of flower. Seeds per fruit of self-pollinated flower of passion fruit at different flashes were recorded variable. Seeds per fruit at different flashes ranged from 3.59 to 10.51 (**Table 2**). Seeds per fruit were recorded maximum (10.51) at third flashes followed by fourth, second, fifth and first flashes. From the finding it was noticed that imposing self-pollination early and late flowers produced minimum seeds per fruit.

3.2.2. Natural Pollination

From the findings it was observed that seeds per fruit imposing natural pollination at different flashes were recorded higher than that of self-pollination (**Table 2**). As self-pollination seeds per fruit in open pollinated flower at third flash were recorded maximum (16.51 seeds/fruit). The trend of seeds per fruit in open pollinated flowers at different flashes was noticed more or less as self-pollinated flowers. Seeds per fruit were recorded minimum (8.56) at fifth flash compared with self-pollination.

3.2.3. Hand Pollination

Among the three pollination methods, hand pollination produced maximum seeds per plant at all flashes (**Table 2**). It was recorded that among the three pollination methods and five flashes plant produced maximum seeds per fruit imposing hand-pollination at third flashes. Stigma of passion flower received maximum number of pollen which is the cause of maximum seeds per fruit. On the other hand, hot-humid weather was noticed favorable for better growth and development of passion fruit which prevailed during third flash. On the contrary average seeds per fruit imposing hand pollination were recorded maximum (**Figure 1**).

3.3. Fruit Weight

Effect of pollination methods on individual fruit weight (g) of passion fruit at different flashes at maturity on fresh weight basis was reported in **Table 3**.

3.3.1. Self-Pollination

Individual fruit weight of passion fruit at different flashes at maturity using self-pollination was noticed variable (**Table 3**). Individual fruit weight at different flashes ranged from 25.28 to 35.81 g. Individual fruit weight was noticed maximum at third flash compared with other flashes. At first and fifth flashes individual fruit weight was recorded minimum as compared with other flashes imposing self-pollination.

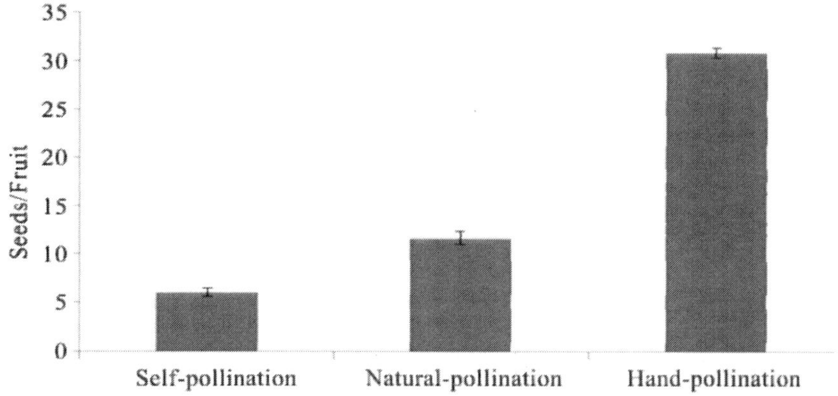

Figure 1. Effect of pollination methods on seeds per fruit (mean value).

Table 1. Effect of pollination methods on fruit sets of passion fruit at different flashes (%).

Flashes	Self-pollination	Natural-pollination	Hand-pollination	Mean
First	1.71d	15.50 c	31.61d	16.27
Second	2.16c	20.45b	35.45c	19.35
Third	4.51a	25.67a	46.71a	25.63
Fourth	3.67b	15.18c	40.12b	19.66
Fifth	1.90d	12.60d	32.15d	15.55
Mean	2.79	17.88	37.21	19.29

Mean followed by same letter in the column are statistically similar

Table 2. Effect of pollination methods on seeds per fruit of passion fruit at different flashes (seeds/fruit).

Flashes	Self-pollination	Natural-pollination	Hand-pollination	Mean
First	3.59 ± 0.76	10.18 ± 0.55	30.64 ± 0.47	14.80 ± 59
Second	5.84 ± 0.52	12.25 ± 0.63	30.90 ± 0.66	16.33 ± 0.60
Third	10.51 ± 0.67	16.51 ± 0.64	35.18 ± 0.52	20.73 ± 0.61
Fourth	6.38 ± 0.67	10.91 ± 0.76	31.65 ± 0.75	16.31 ± 0.73
Fifth	4.12 ± 0.66	8.56 ± 0.46	25.68 ± 0.66	12.79 ± 0.59
Mean	6.09 ± 0.66	11.68 ± 0.61	30.81 ± 0.63	16.19 ± 0.63

Table 3. Effect of pollination methods on individual fruit weight of passion fruit at different flashes at maturity on fresh weight basis (grams).

Flashes	Self-pollination	Natural-pollination	Hand-pollination	Mean
First	26.42 ± 0.57	34.18 ± 0.62	30.41 ± 0.65	30.34
Second	32.91 ± 0.51	36.76 ± 0.49	33.56 ± 0.62	34.41
Third	35.81 ± 0.54	42.67 ± 0.53	39.40 ± 0.66	39.29
Fourth	33.95 ± 0.56	40.01 ± 0.60	35.63 ± 0.40	36.53
Fifth	25.28 ± 0.44	34.91 ± 0.53	31.18 ± 0.62	30.46
Mean	30.87	37.71	34.04	34.21

3.3.2. Natural Pollination

Imposing natural pollination individual fruit weight of passion fruit at different flashes varied from 34.18 to 42.67 grams at maturity (**Table 3**). Results revealed that plant produced largest fruit (42.67 grams) using natural pollination followed by fourth, second, fifth and first flashes. As self-pollination, plant produced largest fruit during third flash imposing natural pollination. Comparing with self-pollination, individual fruit weight at different flashes imposing natural pollination was recorded higher. Individual fruit weight is associated with the number of seeds per fruit. Fruit set of passion fruit was mainly caused by Apis mellifera (data not shown).

3.3.3. Hand Pollination

Individual fruit weight imposing hand pollination at different flashes ranged from 30.41 to 39.40 grams (**Table 3**). As self and natural pollinations, individual fruit weight at third flash was also recorded higher. Individual fruit weight of passion fruit at all flashes was noticed comparatively lower than natural pollination. Individual fruit weight depends upon number of fruits per plant. As fruit set it was higher imposing hand pollination. So the individual fruit weight was recorded comparatively lower than natural pollination.

Among the pollination methods at different flashes plant produced biggest fruit (42.67 g/fruit) during third flash pollinated with natural pollination. During third flash, plant showed vigorous growth which may be the cause of formation of longer fruit during third flash.

3.4. Days Required from Anthesis to Maturity

Fruit maturity of a crop is influenced by genetic make-up, physiological condition of the specific crop as well as environmental factors such as rainfall, humidity, temperature, day length etc. Fruit maturity of passion fruit at different flashes was noticed variable (**Figure 2**). Results revealed that plants required minimum days (42 days) from flower anthesis to full maturity during third flash. Days to maturity from anthesis was observed to be enhanced. On the other hand maturity of fruit was noticed to be delayed during first and fifth flashes of passion fruit. During third flash hot and humid weather as well as longest day-length prevailed that might be the cause of enhancement of fruit maturity.

3.5. Fruit Length and Diameter

Effect of pollination methods on fruit length and diameter of passion fruit at different flashes was presented in **Table 4**.

3.5.1. Self-Pollination on Fruit Length

Fruit length of passion fruit imposing self-pollination at different flashes was recorded not uniform (**Table 4**). Fruit length at different flashes ranged from 3.19 to 4.16 cm. Fruit length was recorded maximum (4.16 cm) at third flash followed by fourth, second, fifth and first flashes. No significant change of fruit length between third and fourth flashes imposes self-pollination.

3.5.2. Natural Pollination on Fruit Length

Similar trend of fruit length at different flashes imposing natural pollination was noticed. But fruit length at all flashes was recorded little bigger than that of self-pollinated fruits. As self -pollination, plant produced longest fruits (5.91 cm) imposing natural pollination at third flash. Among the flashes, plant produced smallest fruit (3.36 cm) at first flash. Fruit length of passion fruit between third and fourth flashes did not show any signifycant change.

3.5.3. Hand Pollination on Fruit Length

Plants produced longest fruit at all flashes imposing handpollination compared with self-and open pollinations. Plant produced longest fruit imposing hand pollination as self and open pollination at third flash (**Table 4**). Among these methods of pollination at different flashes plant showed longest fruit imposing hand pollination at third flash. Fruit length of multiple seeded fruit depends upon number seeds set in the fruit.

3.6. Fruit Diameter

Fruit diameter of passion fruit imposing self-, natural and hand pollinations at different flashes was mentioned in **Table 4**. No noticeable change of fruit diameter of passion fruit was observed imposing the above mentioned pollination methods during five flashes. Fruit diameter was affected by pollination methods and flashes.

3.7. Length-Breadth Ratio

Length-breadth ratio of passion fruit imposing different methods of pollination at different flashes was shown in **Figure 3**. It was noticed that length-breadth ratios of fruits of all flashes were recorded higher when passion flowers were pollinated by hand pollination followed by natural and self-pollinations.

3.8. Fruit Yield

Fruits per plant ranged from 2 to 28 by natural pollination. Plant produced maximal number of fruit by natural pollination during third flash. Minimal number of fruits per plant was noticed in fifth flash by natural pollination (**Table 5**). Fruit set percent at different flashes varied from 7.41 to 20.00 by natural pollination. As fruits per plant, fruit set percent was recorded highest at third flash. Fresh weight of individual fruit (g) was recorded maximum during third flash though it ranged from 21 to 34 by natural pollination.

Table 4. Effect of pollination methods on fruits length and diameter of passion fruit at different flashes on maturity (cm).

Flashes	Self-pollination		Open-pollination		Hand-pollination		Mean
	L	B	L	B	L	B	L
First	3.19 ± 0.22	3.10 ± 0.29	4.25 ± 0.37	3.20 ± 0.43	3.36 ± 0.27	3.15 ± 0.26	3.60
Second	3.85 ± 0.26	3.12 ± 0.26	5.25 ± 0.34	3.18 ± 0.40	3.95 ± 0.20	3.17 ± 0.21	4.35
Third	4.16 ± 0.26	3.10 ± 0.16	5.91 ± 0.24	3.25 ± 0.46	4.72 ± 0.28	3.16 ± 0.25	4.93
Fourth	4.03 ± 0.35	3.13 ± 0.29	5.37 ± 0.30	3.16 ± 0.35	4.39 ± 0.30	3.20 ± 0.25	4.60
Fifth	3.79 ± 0.33	3.11 ± 0.31	4.76 ± 0.40	3.16 ± .46	3.80 ± 0.30	3.15 ± 0.30	4.12
Mean	3.80	3.11	5.11	3.19	4.04	3.17	4.32

Table 5. Yield of passion fruit per plant by natural pollination.

Flashes	Fruit set %	Fruit per plant	Fresh weight of individual fruit (g)
First flash	15.38 b	8b	21b
Second flash	18.66 ab	25a	31a
Third flash	20.00 a	28a	34a
Fourth flash	17.24 ab	25a	32a
Fifth flash	7.41 c	2c	23b

Mean followed by same letter in the column are statistically similar

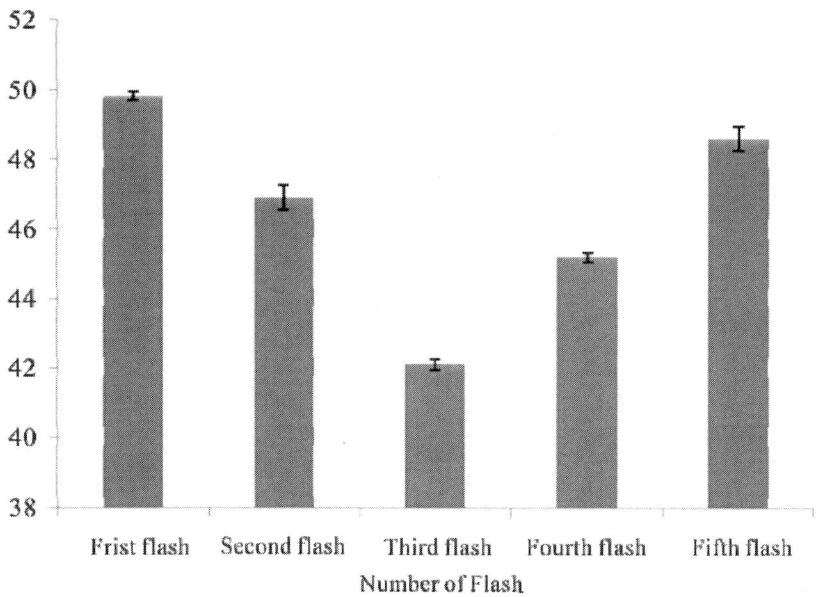

Figure 2. Effect of different flashes on days to fruit maturity from anthesis (days) of passion fruit.

4. DISCUSSION

Pollination is an important criterion for fruit set in passion fruit. Our main aim was to determine the extent to which pollination methods are effective for successful fruit production. Kishore et al. (2010) [19] reported that maximum number of bees was observed between 07:00 - 08:00 h in purple and giant passion fruit, but between 13:00 - 14:00 h in yellow passion fruit. In the present study the maximum number of bees was noticed in the field during 13:00 - 14:00 h. For open pollination the most common pollinating bee for purple, giant and yellow passion fruit was A. mellifera, while A.

cerena was for P. foetida [19]. In this study Apis mellifera was noted the most common pollination bee.

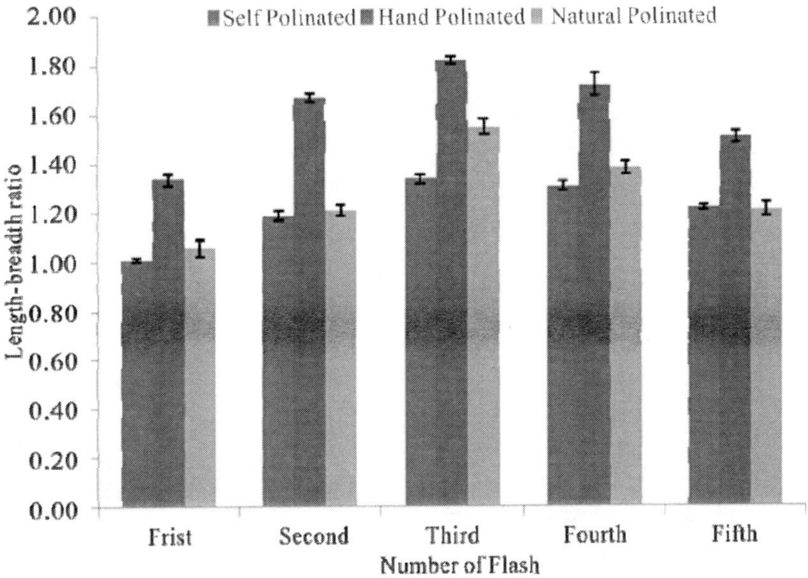

Figure 3. Length-breadth ratios at different flashes by applying different pollination.

Floral morphology of passion flower is the barrier for self-pollination due to pollen not reaching the stigma. Except self-pollination (0% fruit set), manual pollination gave greater fruit set than natural pollination (about 55% vs. 44.6%), resulting in 90-g fruits (20% - 25% heavier than those of the control). Juice percentage and seed number were increased by hand pollination by about 40 and 10%. In general, cross pollinating at around 16.00 h gave the best results (Duarte and Sierra, 1997) [20] which was little contradictory with the present experiment. A high correlation between fruit weight and seed number, fruit length and diameter were noticed [21].

Though passion fruit (P. edulis f. flavicarpa) is capable of a very small degree of self-compatibility, cross-pollination results in increased fruit set, fruit weight, seed set, juice volume and juice sugar concentration. Although large bees such as Xylocopa spp. were expected as pollinators, honeybees (A. mellifera) were the only pollinators observed. The foraging habits of honeybees, not their size, may cause less than expected percentage fruit set [13]. Fruit set of passion fruit in the present study in open pollinated flowers was caused mainly by honey bees (data were not shown) which was supported by Hammer (1987) [13].

Hand pollination alone or in combination with GA3 increased fruit set, fruiting and fruit quality (weight, juice weight and TSS) compared with open pollination. Hand pollinated fruits contained more seeds than open pollinated fruits. Hand pollination decreased rind weight and total acidity content [14]. Seeds per fruit by hand pollination in the present study were observed highest (35.18) compared with selfand natural pollinations which were supported by Saleh and Zarad (1996) [14].

Nazrul et al. (2003) [16] in yellow passion fruit reported 30% fruit set by hand pollination. In the present investigation, fruit set percent was recorded maximum (20) during third flash by natural pollination mainly caused by Apis mellifira. In another study, Kishore et al. (2010) [19] mentioned 42.2% fruit set of yellow passion by using A. mellifera as a natural pollinator. Photoperiod [22], air temperature [23], pesticides [24] and soil moisture are factors that determine the yield of yellow passion fruit. Due to high humidity in anther opening time, cell content of pollen grain contain high osmotic pressure as well as low resistance of their walls, reduced pollen viability as a result affect the frutification and fertilization percentage in pollination [12]. Biotic factors such as the physiological stages of the plant interfere with the frutification and seed production [25]. The biochemical pathways that control fruit setting and maturity in yellow passionfruit [15]. The low yield before and after third flash is due to the lack of flowering and fructification of the plant because of climatic conditions. Flowering of third flash during June-July, that time cause of rainfall and high humidity temperature was relatively low in summer time. Relatively low temperatures are suitable for flowering and fruit set, and moderately high temperatures are favorable for fruit growth and quality in purple passion fruit [26].

5. CONCLUSION

M. Mofazzal Hossain (self-, natural and hand pollination), passion fruit produced maximum fruits pollinating flowers by hand. Hand pollination showed maximum fruits per plant at all flashes. Seeds per plant were recorded highest when flowers are pollinated by hand. Individual fruit weight was recorded maximum with naturally pollinated flower. Plants required minmum period day from anthesis to full maturity at third flash compared with other flashes. Plants produced longest fruit by natural pollination followed by hand pollination during third flash.

REFERENCES

1. W. B. Storey, "Chromosome Numbers of Some Species of Passiflora Occurring in Hawaii," Pacific Science, Vol. 4, No. 1, 1950, pp. 37-42.
2. R. Yockteng, G. C. Eeckenbrugge and T. T. Souza-Chies, "Genomic and Breeding Resources," Wild Crop Relatives, 2011, pp. 129-171. doi:10.1007/978-3-642-20447-0_7
3. J. Vanderplank, "Passion Flowers," MIT Press, Cambridge, 2000.
4. T. Ulmer and J. MacDougal, "Passiflora: Passionflowers of the World," Timber Press, Portland, USA, 2004.
5. J. Morton, "Passion Fruit," In: C. F. Dowling, Eds., Fruits of Warm Climates, Julia F. Morton, Miami, 1987, pp. 320-328.
6. D. W. Turner, C. M. Menzel and D. R. Simpson, "Short Term Drying of Half the Root System Reduces Growth but Not Water Status or Photosynthesis in Leaves of Passion Fruit (Passiflora sp.)," Scientia Horticulturae, Vol. 65, No. 1, 1996, pp. 25-36.
7. K. R. Shivanna, "Reproductive Assurance through Unusual Autogamy in the Absence of Pollinators in Passiflora edulis (Passion Fruit)," Current Science, Vol. 103, No. 9, 2012, pp. 1091-1096
8. K. Kishore, R. Bharali, K. A. Pathak and D. S. Yadav, "Studies on Ripening Changes in Purple Passion Fruit (Passiflora edulis Sims)," Journal of Food Science and Technology-Mysore, Vol. 43, No. 6, 2006, pp. 599-602.
9. J. Vanderplank, "Passion Flowers," MIT Press, UK, 1996.
10. C. H. Bruckner, V. W. D. Casali Moaraes, A. J. Ragazzi and E. M. A. Silva, "Selfincompatibility in Passion Fruit (Passiflora edulis Sims)," Acta Horticulturae, Vol. 370, 1995, pp. 45-57. http://www.actahort.org
11. M. M. Souza, T. N. S. Pereira, A. J. B. Dias, B. F. Ribeiro and A. P. Viana, "Structural, Hystochemical and Cytochemical Characteristics of the Stigma and Style in Passiflora edulis f. flavicarpa (Passifloraceae)," Braz. arch. biol.technol. Vol. 49, No. 1, 2006. doi:10.1590/S1516-89132006000100011
12. M. M. Souza, A. P. Viana and T. N. S. Pereira, "A Putative Mutant of a Self-Compatible Yellow Passion Fruit with the Corona Color as a Phenotypic Marker," Bragantia, Vol. 69, No. 1, 2010. doi:10.1590/S0006-87052010000100003
13. H. Hammer, "The Pollinators of the Yellow Passion Fruit Do They Limit the Success of Passiflora edulis f. flavicarpa as a Tropical Crop," Proceedings of the Florida State Horticultural Society, Vol. 100, 1987, pp. 283-287.
14. M. Saleh and S. S. Zarad, "Effect of Hand Pollination and GA3 Application on Fruit Set, Fruiting and Fruit Quality of Passionfruit (Passiflora edulis)," Annals of Agricultural Science, Cairo, Vol. 41, 1996, pp. 303-312.
15. M. F. Zerbini, W. C. Otoni and L. C. Vieira, "Compendium of Transgenic Crop Plants," Passion Fruit, 2009. http://onlinelibrary.wiley.com
16. M. I. Nazrul, M. M. Zaman, H. P. Mondal and M. Dhar, "Comparative Reproductive Biology of Passion Flower (Passiflora edulis Sims.)," Bangladesh Journal of Botany, Vol. 32, No. 2, 2003, pp. 95-99.
17. M. B. Banu, M. Q. I. Matin, T. H. Hossain and M. M. Hoassain, "Flowering Behaviour, Flower Morphology of Passion Fruit (Passiflora edulis Sims.)," International Journal of Sustainable Crop Production, Vol. 4, No. 4, 2009, pp. 5-7.
18. M. A. Gomez and A. A. Gomez, "Statistical Procedures for Agriculture Research," 2nd Edition, John Wiley and Sons, New York, 1984, p. 640.
19. K. Kishore, A. Pathak, R. Shukla and R. Bharali, "Study on Floral Biology of Passion Fruit (Passiflora spp.)," Pakistan Journal of Botany, Vol. 42, No. 1, 2010, pp. 21-29. http://www.parc.gov.pk/NARC/narc.html
20. O. Duarte and O. Sierra, "Effect of Hand Pollination on Yellow Passion Fruit (Passiflora edulis f. flavicarpa) in El Zamorano," Proceedings of the Inter-American Society for Tropical Horticulture 43rd Annual Meeting, Guatemala City, 1997.

21. M. M. Souza, T. N. S. Pereira, A. P. Viana, M. G. Pereira, A. T. do Amaral Jr. and H. C. Madureira, "Flower Receptivity and Fruit Characteristics Associated to Time of Pollination in the Yellow Passion Fruit Passiflora edulis Sims f. flavicarpa Degener (Passifloraceae)," Scientia Horticulturae, Vol. 101, No. 4, 2004, pp. 373-385. doi:10.1016/j.scienta.2003.11.020
22. J. C. Cavichioli, C. Ruggiero, C. A. Volpe, E. M. Paulo, J. L. Fagundes and F. S. Kasai, "Flowering and Fructification of Yellow Passion Fruit Submitted to Artificial Light, Irrigation and Shade," Revista Brasileira de Fruticultura, Vol. 28, No. 1, 2006, pp. 92-96. doi:10.1590/S0100-29452006000100026
23. G. F. Iapichino and J. B. Loy, "High Temperature Stress Affects Pollen Viability in Bottle Gourd," Journal of the American Society for Horticultural Science, Vol. 112, 1987, pp. 372-374. http://www.nal.usda.gov/
24. M. M. Silva, C. H. Bruckner, M. Picanco and A. J. M. Rugama, "Number of Flowers, Climate, Population Density of Xylocopa spp. (Hymenoptera: Anthophoridae) and Pollination of Passionflower," Revista de Biologia Tropical, Vol. 47, No. 4, 1999, pp. 711-718.
25. E. Lederman, "The Involvement of Ethylene in Fruit Development, Maturation and Ripening of the Passion Fruit, Passiflora edulis Sims," The Hebrew University of Jerusalem, Jerusalem, 1987, p. 280.
26. N. Utsunomiya, "Effect of Temperature on Shoot Growth, Flowering and Fruit Growth of Purple Passionfruit (Passiflora edulis Sims var. edulis)," Scientia Horticulturae, Vol. 52, No. 1-2, 1992, pp. 63-68. doi:10.1016/0304-4238(92)90008-Z

CHAPTER 17

Improvement of Postharvest Quality of Asian Pear Fruits by Foliar Application of Boron and Calcium

Kobra Khalaj, Nima Ahmadi* and Mohammad Kazem Souri

Department of Horticultural Sciences, Tarbiat Modares University, Tehran 336-14115, Iran;

ABSTRACT

The aim of this study was to evaluate the effects of foliar application of boron and calcium on postharvest storage characteristics of Asian pear (*Pyrus pyrifolia*) fruit. This experiment was carried out in the experimental orchard of Asian pear at Tarbiat Modares University, Tehran, Iran. The treatments were two concentrations of boric acid (0% and 0.5%) and three concentrations of calcium chloride (0%, 0.5%, and 0.7%) which were sprayed on trees during the growing season. The data showed that foliar application of these two elements resulted in improvement in physicochemical characteristics of the pear fruit, particularly fruit firmness and polyphenol oxidase activity. At harvest, fruit firmness and total soluble solids (TSS) were significantly improved by B and Ca application, compared to control fruit. Although fruit firmness decreased during 3 months of cold storage, fruit which received higher rates of Ca and B retained more firmness. During the storage period, total phenolic content was significantly reduced in control fruit compared to fruit which were sprayed with Ca and/or B. The highest level of polyphenol oxidase activity was in B and Ca sprayed fruit and the lowest was in control fruit. In general, foliar application of Ca and B resulted in improvement of postharvest characteristics of Asian pear fruit and reduced internal browning symptoms. Thus, application of Ca and B could be recommended as treatments for increasing postharvest life of Asian pear in similar semi-arid climates.

Keywords: boric acid; calcium chloride; foliar spray; internal browning; total soluble solids

1. INTRODUCTION

Asian pear (*Pyrus pyrifolia*), originating from the eastern part of Asia, has been cultivated for centuries. Due to the high quality of the fruit, cultivation and consumption of Asian pear fruit is increasing around the world [1]. The nutrient balance is the main factor that plays a role in postharvest quality of Asian pear fruit.

However, nutrients deficiency disorders are very common in many horticultural crops, particularly under arid and calcareous soil conditions. These disorders play important roles in quality and therefore in marketability of the crops. Calcium and boron are key elements in maintaining the quality and shelf life of fruits, particularly apple and pear. Both elements generally have similar uptake, translocation and correlated physiological roles. Cell strength, cell wall thickness and tissue firmness are highly dependent upon adequate amounts of calcium, mainly forming pectin in the middle lamella [2]. Optimum levels of calcium application have led to better stability of cell membranes and cell walls, and consequently resulted in increasing fruit quality [3]. Boron is another important element in quality parameters of fruits, in particular in pome fruit such as pear. Boron participates in many physiological reactions, particularly cell wall formation and metabolism of sugars and phenolic compounds. It helps maintain cell membrane structures integrity, and is also considered an antioxidant that prevents fruit internal browning [4].

So, the aim of this study was to evaluate the foliar application of calcium and boron on quality characteristics of Asian pear in fruit under the semi-arid climatic conditions of Tehran, Iran.

2. EXPERIMENTAL SECTION

2.1. Plant Materials

This study was conducted during 2010 and 2011 in an Asian pear orchard located at the Faculty of Agriculture, Tarbiat Modares University, in Tehran, Iran. The early Asian pear cultivar "KS_{10}" was used. The effect of foliar applications of boric acid (0%, 0.5%) and calcium chloride (0%, 0.5%, and 0.7%) were evaluated on harvest and postharvest storage quality. Trees were sprayed starting 20 days after full bloom three times with boric acid and/or six times with calcium chloride solutions at 2-week intervals (Table 1) in a factorial design within randomized complete blocks, three replications of each treatment combination, one replication (one tree) per block. Fruit were harvested approximately 90 days after full bloom, and held in cold storage at 2 °C and 85% relative humidity for 3 months. Fruit from each treatment were immediately sliced at harvest or after cold storage and frozen in liquid nitrogen, and then stored at −80 °C for further use.

Table 1. Different treatments and their abbreviations used in this experiment.

Treatment	Calcium Chloride (%)	Boric Acid (%)
c0b0	0	0
c0b1	0	0.5
c1b0	0.5	0
c1b1	0.5	0.5
c2b0	0.7	0
c2b1	0.7	0.5

2.2. Physicochemical Quality Characteristics

Physicochemical quality characteristics of fruit at harvest and after three months of cold storage were characterized. The fruit length and diameter were determined using digital calipers, and weight loss during cold storage was measured by weighing each fruit individually with a precision digital balance at the beginning and end of storage. After cutting the fruit into small pieces, they were dried at 70 °C in an oven to measure fruit dry weight. Fruit firmness and total soluble solids were evaluated using a portable penetrometer and refractometer, respectively.

Titratable acidity was determined by titration of fruit juice by adding 0.1 N NaOH solution to reach pH of 8.3.

2.3. The Total Phenol Content

The total phenol content of fruit at harvest and three months after storage was determined by the Folin-Ciocalteu method [5]. To prepare the sample for total phenol quantification, frozen pieces of fruit were ground to a fine powder using a mortar and pestle cooled by liquid nitrogen during grinding. For quantification of total phenol content, 6 mL of 80% methanol was added to 1 g of powdered sample. Then samples were centrifuged at 14,000 rpm for 20 min at 4 °C. Extract (20 μL) was added to 1.58 mL of distilled water containing 100 μL Folin-Ciocalteu reagent. After 8 min, 300 μL of 20% (*w/v*) sodium carbonate was added to the sample and placed 2 h at room temperature in darkness. The absorption of the sample was measured with a spectrophotometer at 765 nm. The results were calculated based on using garlic acid as a standard and are expressed as mg garlic acid per g fresh weight of fruit.

2.4. Measuring the Activity of Polyphenol Oxidase Enzyme

For measuring the activity of polyphenol oxidase enzyme activity after 3 months of storage, 200 mg of fruit tissue powder was mixed with 3 mL sodium phosphate buffer (25 mM, pH 6.8) at 4 °C, and the homogenate was centrifuged for 15 min at 4 °C. The supernatant was used for measurement of enzyme activity. Activity of polyphenol oxidase in the supernatant was measured according to Ghanati et al. [6] with minor changes. Sodium phosphate buffer (50 mM, pH 6.8) was used. Methyl catechol (0.15 M) was used as substrate. The reaction mixture contained 2 mL sodium phosphate buffer (50 mM, pH 6.8), 600 μL methyl catechol (0.15 M), and 400 μL enzyme extract. Changes in the oxidation of catechol were monitored at 410 nm for 1 min. Protein content of the supernatant was measured using the Bradford assay [7]. Polyphenol oxidase enzyme activity was determined as change in absorption at 410 nm for 1 min per mg protein.

2.5. Statistical Analysis

Statistical analysis was performed using SAS software (SAS Institute Inc., Cary, NC, USA) and diagrams were drawn using Excel software. Comparison of means was conducted using Duncan's New Multiple Range Test at $p = 0.05$.

3. RESULTS

Spray applications of boric acid and calcium chloride had no significant effect on fruit length and diameter at harvest and three months after storage (Table 2 and Table 3). Fruit fresh weight was affected only by Ca spray. The highest fruit fresh weight (54.72 g) was with the 0.7% calcium treatment, and the lowest (40.77 g) was the control (Figure 1). The fruit dry weight at harvest was only affected by the interaction of Ca and B. The highest fruit dry weight (17.48 g per 100 g FW) was with 7% Ca by 0.5% B. After 3 months of storage, significant differences in fruit fresh weight were observed for the effect of Ca, and for the interaction between Ca and B (Table 3). The highest fruit fresh weight (56.63 g) was at 0.7% Ca (Figure 2).

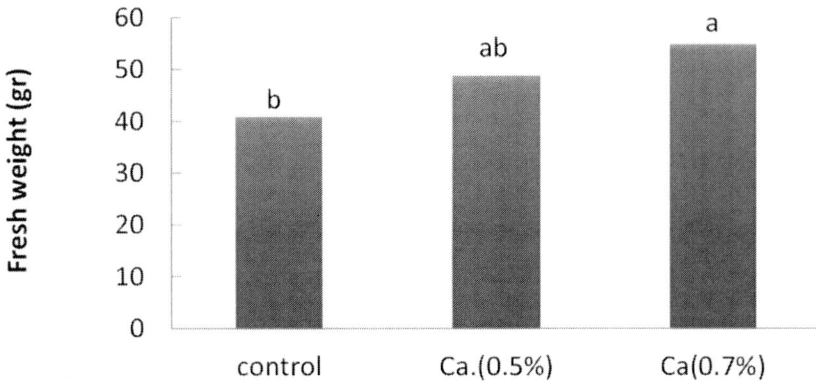

Figure 1. The effect of calcium (%) on fruit fresh weight at harvest. Means with different letters were significantly different by Duncan's New Multiple Range Test at $p = 0.05$.

Fruit firmness at harvest was affected by both Ca and B, but not by their interaction (Table 2). The highest fruit firmness (5.83 kg/cm^2) was with 0.7% Ca, and the lowest firmness (4.1 kg/cm^2) was the control (Figure 3). Similarly, fruit firmness was significantly higher at 0.5% B (5.17 kg/cm^2) compared to the control (4.73 kg/cm^2) (data not shown). Analysis of variance showed that after three months of cold storage, firmness was only affected by Ca (Table 3). The highest fruit firmness (3.23 kg/cm^2) was at 0.7% Ca, and the lowest (2.3 kg/cm^2) was the control (Figure 3).

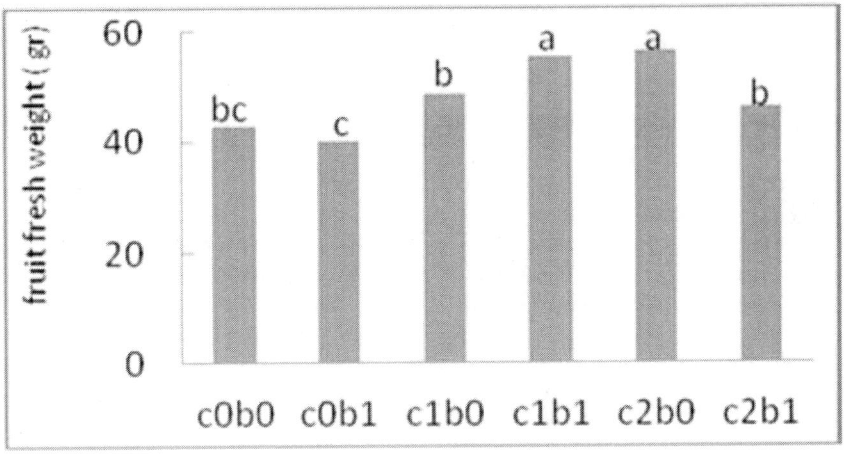

Figure 2. Interaction between calcium and boron on fruit fresh weight after three months of cold storage. See Table 1 for definitions of treatment abbreviations. Means with different letters were significantly different by Duncan's New Multiple Range Test at $p = 0.05$.

Table 2. Sums of squares in the analysis of variance for the effects of calcium and boron on physicochemical properties of pear fruit at harvest.

Source of Variation	df	Fruit Length	Fruit Diameter	Fruit Fresh Weight	Fruit Dry Weight	Fruit Firmness	Total Soluble Solids	Titratable Acidity
Block	2	9.25 ns [z]	7.68 ns	49.14 ns	0.06 ns	0.06 ns	0.09 ns	729.5 ns
Calcium	2	22.31 ns	30.09 ns	293.9 *	1.47 ns	4.5 **	7.72 **	909.6 ns
Boron	1	1.91 ns	9.7 ns	54.21 ns	0.02 ns	0.88 **	2.72 **	122.3 ns
Calcium × Boron	2	1.48 ns	2.8 ns	22.43 ns	2.99 *	0.1 ns	4.05 **	1029.1 ns
Error	10	9.02	11.55	67.9	0.4	0.06	0.14	399.6
CV(%) [z]	-	7.97	7.56	17.14	3.84	5.08	2.83	20.72

*, ** and ns indicate significant differences at $p \leq 0.05$, 0.01, and no significant difference, respectively;
[z] Coefficient of variation.

Table 3. Sums of squares in the analysis of variance for the effects of calcium and boron on physicochemical properties of pear fruit after three months of cold storage.

Source of Variation	df	Fruit Length	Fruit Diameter	Fruit Fresh Weight	Fruit Firmness	TSS	TA
Block	2	0.66 ns	0.99 ns	5.07 ns	0.07 ns	1.39 *	1.93 ns
Calcium	2	16.61 ns	7.52 ns	216.64 **	1.39 **	31.35 **	630.396 ns
Boron	1	3.78 ns	0.01 ns	21.15 ns	0.02 ns	6.12 **	360.08 ns
Calcium × Boron	2	3.59 ns	0.65 ns	112.80 **	0.02 ns	12.12 **	1109.26 *
Error	10	5.15	7.98	10.04	0.03	0.29	164.43
CV(%) [z]	-	6.80	6.99	6.57	5.73	3.36	13.99

*, ** and ns indicate significant differences at $p \leq 0.05$, 0.01, and no significant difference, respectively;
[z] Coefficient of variation.

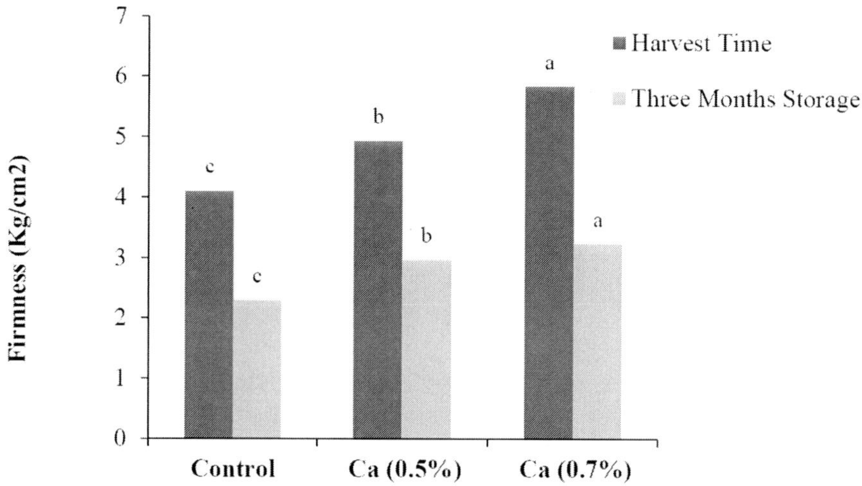

Figure 3. The effect of calcium (%) on fruit firmness at harvest and after three months of cold storage. Means with different letters within harvest or storage time were significantly different by Duncan's New Multiple Range Test at $p = 0.05$.

Total soluble solids at harvest showed significant effects of Ca, B and their interaction (Table 2). The highest TSS (16 °Brix) was the control (Figure 4). After 3 months of cold storage, TSS showed significant effects of Ca, B, and their interaction (Table 3). The highest TSS (20 °Brix) was in control fruit (Figure 4). At harvest, there were no significant differences among treatments in TA values. However, after 3 months of storage, TA was significantly affected by the interaction of B and Ca (Table 3), as the highest was measured in the 0.7% Ca treatment.

At harvest, control fruit had the highest total phenol content, while application of Ca and B reduced total phenol content. Fruit treated with 0.7% Ca and 0.5% B (c_2b_1) showed the lowest phenol content (Figure 5). In addition, total phenol content showed a significant reduction during cold storage, and the reduction in the control samples was more than in B and Ca treated fruit. Control fruit showed the lowest total phenol after 3 months of cold storage. For fruit receiving 0.7% Ca and 0.5% B (c_2b_1) treatment, the total phenol was maintained (Figure 5). Ca application reduced the activity of polyphenol oxidase at 3 months (Figure 6). Thus, after storage, PPO activity was highest in control fruit, which was accompanied by a reduction in phenolic content. Thus, treating fruit with B and Ca during the growing season reduced the activity of PPO at the end of the storage period (Figure 6).

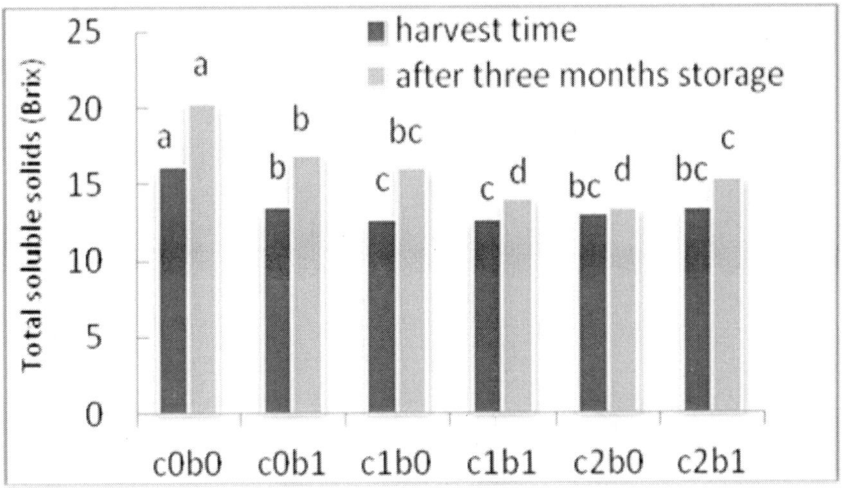

Figure 4. Interaction between calcium and boron on total soluble solids at harvest and after three months of cold storage. See Table 1 for definitions of treatment abbreviations. Means with different letters within harvest or storage time were significantly different by Duncan's New Multiple Range Test at $p = 0.05$.

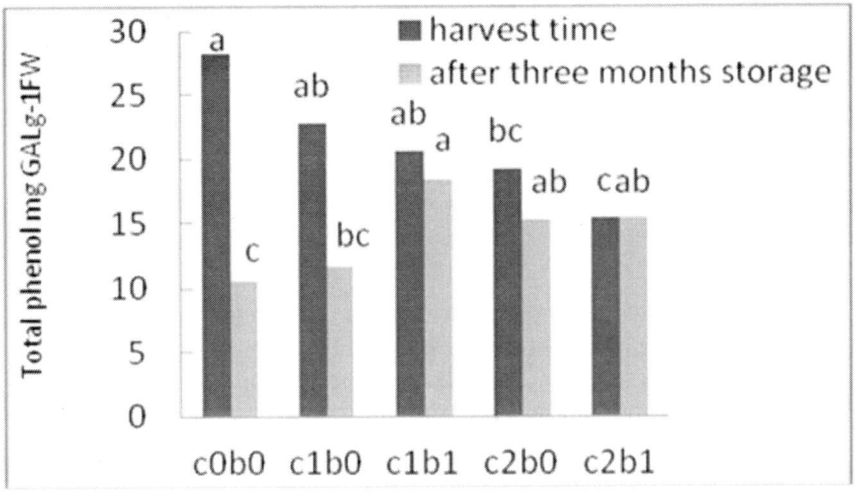

Figure 5. Total phenol at harvest and after three months of cold storage. See Table 1for definitions of treatment abbreviations. Means with different letters within harvest or storage time were significantly different by Duncan's New Multiple Range Test at $p= 0.05$.

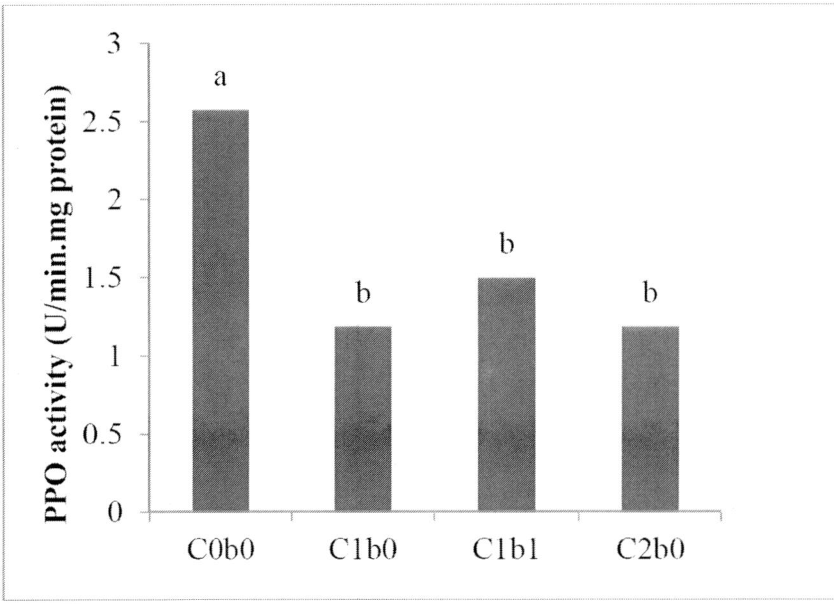

Figure 6. Activity of polyphenol oxidase after three months of cold storage. See Table 1 for definitions of treatment abbreviations. Means with different letters were significantly different by Duncan's New Multiple Range Test at $p = 0.05$.

4. DISCUSSION

In this study, fresh weight was increased by Ca, and dry weight of fruit was increased by Ca × B applications. Increasing cell wall and pectin amounts probably were involved in these responses. Other studies also reported improvement of fruit weight with foliar sprays of Ca and B [8,9]. Fruit firmness was also significantly improved by Ca and B sprays. Increasing pectin levels in the cell wall also probably played a role in this regard [10]. Fruit firmness declined during storage. However, Ca application resulted in a higher fruit firmness compared to the control both at harvest as well as after 3 months of cold storage. Treatment of apple fruit two weeks before maturity with 8% calcium chloride has significantly increased fruit firmness [11]. Also, by application of 4% calcium chloride, Asian pear fruit firmness increased compared to a control [12]. In this study, Ca reduced fruit TSS, similar to results obtained by some others [13], although other research has had different results [8,9,14]. Betts and Bramlage [14] reported that in apple fruit, Ca did not affect TSS as application of Ca reduced respiratory rate and consequently degradation of starch to simple sugars, as well as total soluble solids. Total soluble solids increased during storage, probably due to conversion of starch to sugar. Titratable acidity of fruit decreased during storage. Previous reports indicated that acidity in pear and other fruits gradually increased until final fruit size, while it decreased during maturation, ripening and storage of apple [15]. Similar results were obtained in the present study. Other studies also reported reduction of organic acids during storage of Yali pear [16], although there are reports showing that acidity was not affected by Ca [11]. The role of calcium in cell wall strength is probably responsible for most quality-related properties of this element. Calcium was shown to reduce the production of phenolic compounds in fruit [17], and also increase the polysaccharides and non-alcohol-soluble solids in the fruit cell walls. It also maintained cell membrane stability [3,18,19]. The reduction in total phenolic levels in control fruit after three months of cold storage was likely due to the higher activity of polyphenol oxidase enzyme (Figure 6), resulting in oxidation of phenolic compounds and internal browning in fruit tissue (Figure 7). Generally, prevention of phenol oxidation by application of Ca and B has mainly been related to their effects on cell membrane integrity and membrane permeability [20,21]. Boron, on the other hand, facilitates the role of calcium in cell membrane and cell wall interactions. Nevertheless, both Ca and B have restricted transport within plants, in particular under semi-arid and calcareous soil conditions. Therefore, foliar sprays of these two nutrient elements several times during a growing season can improve Ca and B concentration in fruit tissues [21].

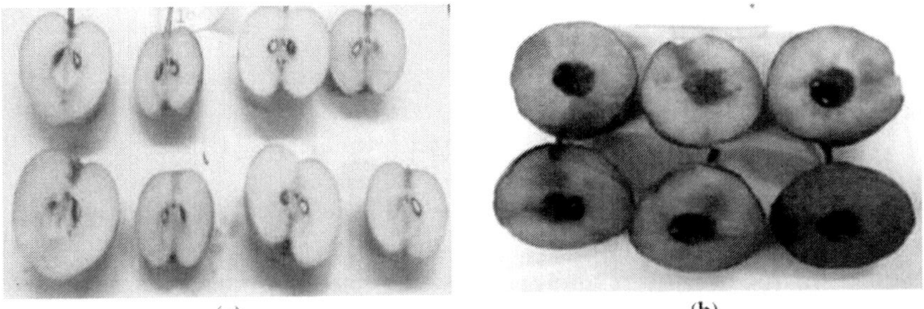

(a) (b)

Figure 7. Internal browning after three months of cold storage (2 °C, 85% relative humidity). (**a**) Ca at 0.7%; (**b**) Control.

5. CONCLUSIONS

In the present study, application of both Ca and B improved the physical and chemical characteristics of fruit compared to a control. Increasing fruit firmness and TSS due to Ca and B applications could have important implications, particularly in terms of fruit quality and storage life. Foliar sprays of both Ca and B reduced the activity of polyphenol oxidase enzyme, which is mainly responsible for internal browning of fruit. The role of calcium in cell wall strength was probably responsible for most quality-related properties of this element. So, using foliar applications of Ca and B during the growing season of Asian pear trees is recommended for improving postharvest quality parameters of fruit in semi-arid regions.

ACKNOWLEDGMENTS

We would like to thank Kazem Arzani, Department of Horticultural Sciences, Tarbiat Modares University, for providing the plant materials and samples used in this research.

REFERENCES

1. Chen, J.; Wang, Z.; Wu, J.; Wang, Q.; Hu, X. Chemical compositional characterization of eight pear cultivars. *J. Food Chem.* **2007**, *104*, 268–275.
2. Subburamu, K.; Singaravelu, M.; Nazar, A.; Irulappan, I. Pre-harvest spray of calcium in grapes (Vitis vinifera). *South Indian Hort.* **1990**, *5*, 268–269.
3. Sharpels, R.O.; Johnson, D.S. The influence of calcium on senescence changes in apples. *Ann. Appl. Biol.* **1977**, *85*, 150–453.
4. Dzondo-Gadet, M.; Mayap-Nzietchueng, R.; Hess, K.; Nabet, P.; Belleville, F.; Dousset, B. Action of boron at the molecular level: Effects on transcription and translation in an acellular system. *Biol. Trace Elem. Res.* **2002**, *85*, 23–33.
5. Singh, R.P.; Murthy, K.N.C.; Jayaprakasha, G.K. Studies on the antioxidant activity of pomegranate (*Punica granatum*) peel and seed extracts using in vitro models. *J. Agric. Food Chem.* **2002**, *50*, 81–86.
6. Ghanati, F.; Morita, A.; Yokota, H. Induction of suberin and increase of lignin content by excess boron in tobacco cell. *Soil Sci. Plant Nutr.* **2002**, *48*, 357–364.
7. Bradford, M. A rapid and sensitive method for the quantitation of protein utilizing the principle of protein-dye binding. *Anal. Biochem.* **1976**, *72*, 248–254.
8. Khoshgalb, H. The Effects of Calcium, Zinc and Boron on the Chemical Composition of Fruits, Postharvest Shelf Life and Reduce Symptoms of Internal Browning of Fruits of Two Varieties of Asian Pear (*Pyrus serotina* Rehd.) in Tehran Climate. Ph.D. Thesis, Faculty of Agriculture, Tarbiat Modarres University, Tehran, Iran. (In Persian)

9. Akl, A.M.; Eid, A.F.M.; Hegab, M.Y. Effect of urea, some micronutrients and growth-regulators foliar spray on the yield, fruit quality, and some vegetative characteristics of "Washington Navel" orange trees. *Hort. Sci.* **1995**, *30*, 774–780.
10. Veltman, R.H.; Sanders, M.G.; Persijn, S.T.; Peppelenbos, H.W.; Oosterhaven, J. Decreased ascorbic acid levels and brown core development in pears (*Pyrus communis* cv. "Conference"). *Physiol. Plant.* **1999**, *107*, 39–45.
11. Conway, W.S.; Sams, C.E. Effect of postharvest calcium treatment on decay of Delicious apples. *Plant Dis.* **1982**, *66*, 402–403.
12. Dhatt, A.S.; Mahajan, B.V.C.; Bhatt, A.R. Effect of pre and post-harvest calcium treatments on the storage life of Asian pear. *ISHS Acta Hortic.* **2005**, *696*, 497–501.
13. Dolati Baneh, H.; Hacani, A.; Majidi, A.; Zomorrodi, S.; Hacani, G.; Malakoti, M.J. Effect of calcium chloride on foliar concentration and frequency characteristics of stiffness and a Lebanese Red apples stored in Urmia region. *J. Agric. Sci.* **2002**, *12*, 47–54. (In Persian)
14. Betts, H.A.; Bramlag, W.J. Uptake of calcium by apples from postharvest dips in calcium chloride solutions. *J. Am. Soc. Hortic. Sci.* **1979**, *102*, 785–788.
15. Ackerman, J.; Fischer, M.; Amado, R. Changes in sugars, acids, and amino acids during ripening and storage of apples (cv. Glockenapfel). *J. Agric. Food Chem.* **1992**, *40*, 1131–1134.
16. Chen, J.L.; Yan, S.J.; Feng, Z.; Xiao, L.; Hu, X.S. Changes in the volatile compounds and chemical and physical properties of "YaLi" pear (*Pyrus bertschneideri* Rehd) during storage. *Food Chem.* **2006**, *97*, 248–255.
17. Coseteng, M.Y.; Lee, C.Y. Changes in apple polyphenoloxidase and polyphenol concentrations in relation to degree of browning. *J. Food Sci.* **1987**, *52*, 985–989.
18. Malakoti, M.J.; Tehrani, M.M. *The Role of Micronutrients in Increasing Yield and Improving the Quality of Agricultural Products*, 3rd ed.; Tarbiat Modares University: Tehran, Iran, 2001.
19. Dunn, J.L.; Able, A.J. Pre-harvest calcium effects on sensory quality and calcium mobility in strawberry fruit. *Acta Hortic.* **2006**, *708*, 307–312.
20. Tomas-Barberan, F.A.; Gil, M.I.; Castaner, M.; Artes, F.; Saltveit, M.E. Effect of selected browning inhibitors on phenolic metabolism in stem tissue of harvested lettuce. *J. Agric. Food Chem.* **1997**, *45*, 583–589.
21. Tohidloo, G.; Souri, M.K. Uptake and translocation of boron in two different tomato genotypes. *Hortic. Environ. Biotechnol.* **2009**, *50*, 487–491.

CHAPTER 18

Osmotic Dehydration of Pineapple Stems in Hypertonic Sucrose Solutions

Ronaldo dos Santos Falcão Filho[1], Rennan Pereira de Gusmão[2], Wilton Pereira da Silva[2], Josivanda Palmeira Gomes[2], Edvaldo Vasconcelos Carvalho Filho[3], Anoar Abbas El-Aouar[4]

[1]Federal Institute of Rio Grande do Norte, Currais Novos, Brazil

[2]Federal University of Campina Grande, Campina Grande, Brazil

[3]Federal University of Rio Grande do Norte, Santa Cruz, Brazil

[4]Federal University of Paraíba, João Pessoa, Brazil

ABSTRACT

The present study provides the evaluation in the influence of the variables: temperature (T), concentration of the osmotic solution (C) and ratio of fruit and osmotic solution (F:OS) during the osmotic dehydration of slices of pineapple stem variety Perola, on the responses: water loss (WL), mass loss (ML), gain of solids (GS) and ratio of gain of solids and water loss (GS/WL). The centesimal composition was determined both in the raw material and in the dehydrated product. To optmize the process, the studied factors were: temperature (T), with factorial points −1 equal to 30°C and +1 equal to 50°C; concentration of sucrose solution (C), with factorial points −1 equal to 40 $g \cdot 100\ g^{-1}$ and +1 equal to 60 $g \cdot 100\ g^{-1}$ and ratio of fruit and osmotic dehydration solution (F:OS), with factorial points −1 and +1 equal to 1:20 and 3:20, respectively. In all essays, the dehydration time was 4 hours. The essays showed that F:OS was not significant in any responses; the models adopted were predictive and fitted for WL and ML, and reasonable for GS and GS/WL. The temperature was the most significant variable in the process; the optimized values were: T= 50°C, C = 40 $g\ 100\ g^{-1}$ e F:OS = 3:20. The product needs a complementary drying to adapt itself to the legislation demands.

Keywords: Pineapple, Dried Fruit, Experimental Design

1. INTRODUCTION

Brazil is the largest tropical fruit producer in the world, and due to its diversified conditions of soil and climate, it can also cultivate fruits typical of temperate and subtropical climates, which have a high potential in the international market [1].

In 2008, Brazil was the largest pineapple producer in the world, with a total produce of 2.5 million tons, corresponding to approximately 482 million dollars, overcoming countries in fast development, such as China and India, 5th and 6th largest producers, respectively [2].

In 2009, Brazil's Northeast Region was the second ranked among the country regions in fruit production, the Southeast being the first, but when it comes to pineapple the Northeast leads with 46% of national producing and Paraíba, the state with the largest produce, accounting for about 20% of the Brazilian production [3].

The pineapple has long been the favorite non-citrus fruit in the tropical and subtropical countries, especially because of the attractiveness of its flavor and aroma, and it contains a wide variety of vitamins and mineral nutrients, particularly vitamins A and C, and potassium [4].

Pineapple (Ananas comosus L., Merrill) belongs to the monocotyledon family bromeliaceae and is the third most abundant tropical fruit cultivated in Brazil. The Pérola variety has high sugar content (14 - 16 Brix), and low acidity and it is mainly consumed in natura or used for juice production (single strength or concentrate) [5].

It has been shown that 30% of the foods are wasted, like: peels, stems, leaves and seeds. That happens due to a lack of information about the dietary value of these products and their correct processing techniques [6]. Thus, it is important to make rational use of all the fractions of the foods.

It is an established fact that a large portion of the food production in Brazil is wasted, especially fruits. Therefore, the process of osmotic dehydration is a good technique for pineapple conservation through moisture reduction. This technique is widely used for a pre-drying followed by a conventional convective drying, or solar drying, so that moisture content below 25 $g \cdot 100\ g^{-1}$ is reached. Using this technology, the pineapples can be preserved longer than the in natura fruit, making it possible to sell them in non-producing regions, aggregating value for the producer and avoiding part of that unwanted wasting.

Solar-osmotic dehydration has attracted interest, particularly in countries with a wide diversity of fruits, ample availability of sugar and high incidence of solar rays throughout the year [7].

Using the response-surface methodology (RSM) as an optimization technique, as described by [8] , it is possible to obtain the best relation and learn which of the variables in the process have a significant influence upon osmotic dehydration, as showed in the studies of [9] [10].

The aim of this work was to assess the influence of the variables: temperature, concentration of the sucrose solution and ratio of fruit and solution during osmotic dehydration of slices of pineapple stem from the variety "Pérola" [Ananas comosus (L.) Merrill cv. "Pérola"], as well as the following responses: water loss, mass loss, gain of solid material and the optimization parameter (ratio of gain of solids and water loss) and chemical composition of the raw material and the dehydrated product.

2. MATERIALS AND METHODS

The study was carried out at the Fruit and Vegetable Processing Laboratory of the IFRN at city of Currais Novos. The pineapples of the variety Pérola, cultivated around the municipality of Santa Rita, Paraíba, Brazil, were purchased at the market in the city of Currais Novos, Rio Grande do Norte, Brazil.

The selection of pineapples was done according to color intensity (green, slightly yellowish), content of soluble solid material (12 - 16 $g \cdot 100\ g^{-1}$), same size and shape, in order to obtain homogeneous samples [4].

2.1. Chemical Characterization

The raw material and the product obtained through osmotic dehydration were analyzed as to the relative contents of moisture, ashes, crude fiber, proteins, lipids and total sugars, as stipulated by Brazil [11].

2.2. Factorial Design for Optimizing Osmotic Dehydration

After preliminary tests it was verified that the studied variables (temperature, concentration of sucrose solution and ratio of fruit and osmotic dehydration solution) had a linear behavior, then, the factorial design 2^3 with four repetitions at the central point was utilized [8]. The studied factors were: temperature (T), with factorial points −1 equal to 30°C and +1 equal to 50°C; concentration of sucrose solution (C), with factorial points −1 equal to 40 g·100 g^{-1} and +1 equal to 60 g·100 g^{-1} and ratio of fruit and osmotic dehydration solution (F:OS) with factorial points −1 and +1 equal to 1:20 and 3:20, respectively.

Twelve essays were carried out in triplicate, eight factorial trials and four repetitions at the central point; the design is reported in Table 1.

Table 2 shows the sequence of the 12 essays of factorial design for optimizing osmotic dehydration of slices of pineapple stem from the variety "Pérola".

2.3. Procedure of Osmotic Dehydration

After removing the crowns, the fruits were washed in flowing water and immersed in a solution of commercial neutral detergent at 1% for 15 minutes, for a first cleaning, and then, submerged in chlorinated water (8 to 10 ppm of active chlorine) for 10 minutes, to disinfection. That procedure was followed by rinsing with distilled water [12].

The pineapples were cut to isolate the stems, and these were sliced in the shape of infinite plane plates, with approximately 50 mm length and 5 mm thick.

The dehydration solutions were produced by diluting commercial sucrose in distilled water, using a mechanical agitator MARCONI®; solutions were calculated according to the concentrations in the design, and controlled with a handheld refractometer, model R-5000 ATAGO®.

The samples of pineapple stem slices underwent osmotic dehydration in closed glass pots containing dehydration solution and fruit in the ratios, concentrations and temperatures described in the experimental design. Osmotic dehydration was conducted without agitation in a drying oven for dehydration for four hours.

After dehydration time, the samples were rinsed with distilled water for about 5 seconds in order to remove the excess of solute retained on the product surface, and then they were drained with a paper towel to remove the excess of surface water [12].

Table 1. Real and coded levels of independent variables in the factorial design.

Independent variables	Levels		
	−1	0	+1
T (°C)	30	40	50
C (g·100 g^{-1})	40	50	60
F:OS	1:20	1:10	3:20

Table 2. Factorial design 2^3 with 4 repetitions at the central point with coded and decoded factors.

Essay	T (°C)	C (g·100 g^{-1})	F:OS
1	30 (−1)	40 (−1)	1:20 (−1)
2	30 (−1)	40 (−1)	3:20 (+1)
3	30 (−1)	60 (+1)	3:20 (+1)
4	30 (−1)	60 (+1)	1:20 (−1)
5	50 (+1)	40 (−1)	1:20 (−1)
6	50 (+1)	40 (−1)	3:20 (+1)
7	50 (+1)	60 (+1)	3:20 (+1)
8	50 (+1)	60 (+1)	1:20 (−1)
9	40 (0)	50 (0)	1:10 (0)
10	40 (0)	50 (0)	1:10 (0)
11	40 (0)	50 (0)	1:10 (0)
12	40 (0)	50 (0)	1:10 (0)

Right after the osmotic dehydration, the samples were analyzed for moisture content, according to [11].

2.4. Mathematical-Statistical Treatment of Data

The calculations of water loss (WL), mass loss (ML), solids gain (SG) and ratio of solids gain and water loss (SG/WL) in the process of osmotic dehydration was done by using the Equations (1), (2) and (3) prescribed by [10].

$$\mathrm{ML}(\%) = \frac{\left(w_i - w_f\right)\cdot 100}{w_i} \quad (1)$$

$$\mathrm{WL}(\%) = \frac{\left(w_i X_i - w_f X_f\right)\cdot 100}{w_i} \quad (2)$$

$$\mathrm{SG}(\%) = \frac{\left[w_f \cdot \left(1 - \frac{X_f}{100}\right) - w_i \cdot \left(1 - \frac{X_i}{100}\right)\right]\cdot 100}{w_i} \quad (3)$$

The linear models described by Equation (4) were filted to the experimental data by means of linear regression, using the software Statistica 7.0®.

$$y = \beta_0 + \beta_1 \cdot x_1 + \beta_2 \cdot x_2 + \beta_1 \cdot x_1 + \beta_3 \cdot x_3 + \beta_{1,2} \cdot x_1 \cdot x_2 + \beta_{1,3} \cdot x_1 \cdot x_3 + \beta_{1,2} \cdot x_1 \cdot x_2 + \beta_{2,3} \cdot x_2 \cdot x_3 + \beta_{1,2,3} \cdot x_1 \cdot x_2 \cdot x_3 + \varepsilon \quad (4)$$

The models were tested for the prediction of regression and the lack of fit from the variance analysis (ANOVA) through the Fischer test (F), and the determination coefficients (R^2) were calculated to check the significance of the linear models used in the factorial design.

The effects and interactions with $p \geq 0.05$ were discarded from the models, since they are statistically insignificant. The models considered predictive were those with an F_{calc}/F_{tab} ratio of at least 4 for the regression, and only those with $R^2 \geq 0.7$ were considered significant. Lack of fit was also used as a criterion to assess the prediction capability of the statistic model, when it showed a value of F_{calc} less than the value of F_{tab} for a significance level of $p \geq 0.05$ [8].

3. RESULTS AND DISCUSSION

Table 3 shows the experimental data of WL, ML, GS and GS/WL regarding the independent variables of the process. The table reports that highest temperatures, as well as largest concentrations of dehydration solutions and highest ratios of fruit and dehydration solution led to the largest contents of water loss (WL)-which reached 2.63 times higher than the minimal loss condition—whereas the mass loss (ML) was 2.92 times higher than the minimal loss condition. But these conditions had the largest gain of solids (GS), which is not interesting, since in this case, the gain of solids contents represents just a gain of sugar, and that is not necessary because the in natura pineapples already contains a high concentration of sugars. Therefore, the optimization of the dehydration process is best attained through the variable GS/WL. According to [9] , the optimization is reached in the region where you have the highest WL and lowest GS, so that the ratio GS/WL is minimal.

The water transfer is one of the most important aspects of osmotic dehydration. The main mechanism of mass transfer is diffusion, due to the concentration gradient between the osmotic solution and the food [13]. Figure 1 (A) shows that the significant effects for WL, at 5% of significance level, were temperature and concentration of dehydration solution: both effects have a strong influence on water loss, because they are positive and proximal. [9] [14] , studying osmotic dehydration of tomatoes and mangoes (of the variety Tommy), respectively, also found out that the process temperature and the concentration of the dehydrating solution were among the significant and positive factors in water loss. However, [15] treated pineapples with osmotic solutions in various concentrations and obtained a constant reduction of the water content, regardless of the osmotic treatment used. [12] evaluated water loss in pineapple slices through osmotic dehydration and observed that several factors had an influence, besides temperature and concentration of solution, with a significant effect, measured by test F ($P < 0.01$), for the following interactions: time × cultivar, temperature × cultivar, temperature × degree of agitation and cultivar × degree of agitation.

Table 3. Experimental values obtained in the essays of factorial design.

Essay	Independent variables			Responses			
	T (°C)	C (g·100 g^{-1})	F:OS	WL (g·100 g^{-1})	ML (g·100 g^{-1})	GS (g·100 g^{-1})	GS/WL
1	30 (−1)	40 (−1)	1:20 (−1)	16.94	10.62	6.32	0.37
2	30 (−1)	40 (−1)	3:20 (+1)	14.50	9.03	5.47	0.38
3	30 (−1)	60 (+1)	3:20 (+1)	23.69	14.88	8.81	0.37
4	30 (−1)	60 (+1)	1:20 (−1)	23.54	13.54	10.00	0.43
5	50 (+1)	40 (−1)	1:20 (−1)	25.72	18.75	6.97	0.27
6	50 (+1)	40 (−1)	3:20 (+1)	24.51	18.06	6.45	0.26
7	50 (+1)	60 (+1)	3:20 (+1)	37.02	26.40	10.62	0.29
8	50 (+1)	60 (+1)	1:20 (-1)	36.38	25.43	10.95	0.30
9	40 (0)	50 (0)	1:10 (0)	25.44	16.20	9.24	0.36
10	40 (0)	50 (0)	1:10 (0)	27.55	17.50	10.05	0.36
11	40 (0)	50 (0)	1:10 (0)	29.21	19.74	9.47	0.32
12	40 (0)	50 (0)	1:10 (0)	26.54	17.24	9.31	0.35

In Figure 1(b) it can be observed that ML followed a pattern similar to that of WL, as noted by [11] in their experiments, but the concentration was an effect with less relevance than temperature. That is due to the solids incorporated into the product during dehydration, as predicted by [16]. But [17] found out in their experiments of osmotic dehydration of peaches that the values of mass loss were less than those of water loss.

Figure 1(c) clearly shows that the effect having the most influence on GS was concentration, as ML is function of WL and GS, concentration must be less significant to ML, because of the incorporation of solids, even though a higher concentration of dehydrating solution generates a higher potential of mass transfer, which would favor ML.

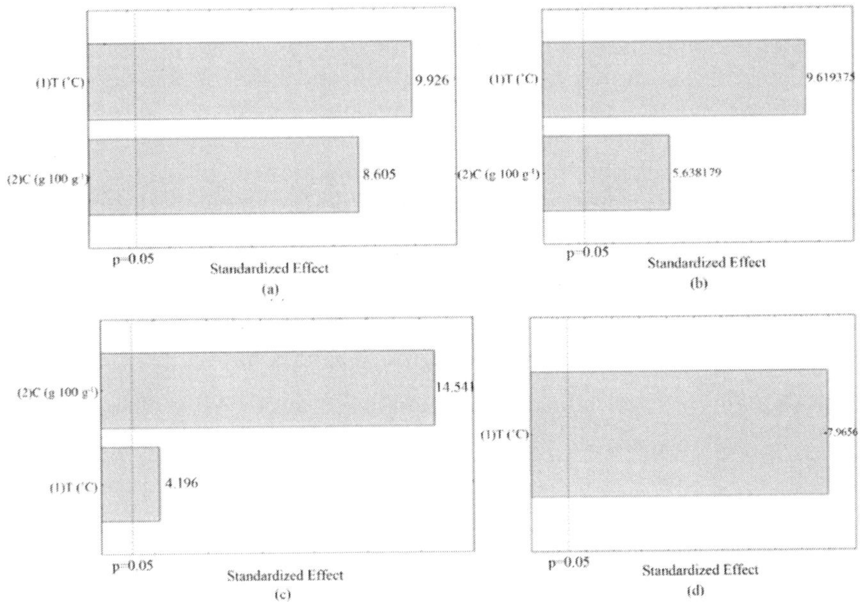

Figure 1. Pareto diagrams showing the significant effects for (a) loss of water (WL); (b) loss of mass; (c) gain of solids and (d) ratio gain of solids and loss of water.

Figure 1(d) shows that the only significant effect for the optimization variable GS/WL ratio was temperature, which had a negative effect on the parameter. That is normal, since the ratio GS/WL tends to zero when the process tends to its optimum and the goal is to achieve a minimum of solids incorporation and a maximum of water loss in the product during the osmotic process, as shown by [10].

Lastly, it can be perceived that the variable ratio of fruit and dehydration solution (F:OS) was not significant ($p < 0.05$) for any of the studied responses. With that is possible to affirm that can use the highest ratio of fruit in relation to the dehydrating solution, which in this case was 3 parts of fruit to 20 of solution (3:20). Such authors as [18] [19] used in their studies of dehydration the fixed ratio of 1:10 in order to maintain the osmotic potential during the processes, but the content of solute used in osmotic dehydration has a great influence on the costs of the final product, but the 3:20 ratio proved effective, thus it can be used, in the conditions of this study, and generate an acceptable product.

Equations (5)-(8) in Table 4 represent the linear models fitted to the experimental points of factorial design. Equations (5) and (7) represent WL and ML; the two models proved predictive, because the F_c/F_{tab} ratios were greater than 4 in both cases. They are also well fitted, with $R^2 = 0.94$ and the part representing the lack of fit of non-significant models by the Fischer test with F_{tab}values greater than F_c. Therefore, it is possible to affirm that the fitted linear models reliably represent the WL and ML phenomena in the stem of pineapples treated osmotically, in the studied range. Temperature and concentration increments positively influenced WL and ML. These results were expected, according to [20] , who studied osmotic dehydration in apples and realized that WL and ML were strongly influenced when the temperature was raised from 25°C to 55°C. [21] showed that an increase in the concentration of the sucrose osmotic solution also raised WL in pineapples, due to an increase in the gradient of osmotic pressure in the solution.

Equation (6) represents the model fitted to GS experimental points, which proved to be predictive, with F_c over 5 times higher than T_{tab}.; analysis of the model also showed a reasonable R_2 and a slight mismatch.

Thus, the GS model was taken into account in the determination of the optimization variable GS/WL shown in Table 5 by Equation (8).

As it has been shown, the F:OS ratio is not a variable that influences any of the responses studied, and concentration does not influence the GS/WL optimization in the conditions studied and statistical model chosen (the linear model). The optimization conditions for osmotic dehydration of sliced pineapple stems are understood to be those displayed in Table 5.

Table 4. Analysis of linear models fitted experimental points obtained in factorial design.

			Regression			Mismatch		
	Equation	R^2	F_c	$^*F_{tab}$	F_c/F_{tab}	F_c	$^*F_{tab}$	F_c/F_{tab}
WL	$WL = 25.92 + 5.62T + 4.87C$ (5)	0.94	71.04	4.26	16.68	1.32	6.94	0.19
GS	$GS = 8.64 + 0.57T + 1.90C$ (6)	0.83	23.10	4.26	5.42	7.39	6.94	1.06
ML	$ML = 17.28 + 5.07T + 2.97C$ (7)	0.94	81.12	4.26	19.04	0.65	6.94	0.09
GS/WL	$GS/WL = 0.34 - 0.05T$ (8)	0.83	49.12	4.96	9.90	1.46	8.89	0.16

Table 5. Optimized value for osmotic dehydration.

Process variable	Coded value	Real value
T (°C)	+1	50
C (g·100 g^{-1})	−1	40
F:OS	+1	3:20

From the values of temperature, concentration and ratio of fruit and solution shown in Table 5 and using fitted values, the loss of water was estimated at 26.37 g·100 g^{-1}, loss of mass at 19.38 g·100 g^{-1}, gain of solids at 7.31 g·100 g^{-1} and ratio of gain of solids and loss of water at 0.29. These values approach the average experimental values for the same conditions shown in Table 4. The relative errors were 7.00% for WL, 6.81% for ML, 11.76% for GS and 10.34% for GS/WL; the error values reflect the quality of the model fits.

Table 6 presents the results of the centesimal composition of the pineapple and the product after osmotic dehydration, utilizing the process conditions set in the optimization. In relation with the raw material, the moisture (84.20 g·100 g^{-1}) and sugar contents (14.50 g·100 g^{-1}) are in agreement with those found by [4] and those found at [22] , which were 86.07% e 86.30% for the sugars and 13.86% e 12.3% for moisture, respectively. The values found for the others components showed great disparities like, for example, the values found for proteins by [4] was 0.59%, while [22] found it to be 0.9%. That variation in results can be explained by some factors, such as climate conditions and the composition of soil, which have a direct bearing on the composition of plants, as shown by [23]. Alternatively, according to [24] , these possible discrepancies can be due to the structure of the fruit tissue, resulting in different behaviors during processing.

About the product after osmotic dehydration, a variation of components was perceived in comparison to the raw material. As expected, the moisture levels decreased (76.31 g·100 g^{-1}), since the objective of the process is to dehydrate the material. An important increase in the content of total sugars (19.31 g·100 g^{-1}) can also be observed in Table 6. This phenomenon is predictable and to be expected, according to [25] , because in an osmotic dehydration process occur three basic mass flows, namely: the water exit of the material into the hypertonic solution, solutes are incorporated from the solution into the material and to a lesser extent-solutes leave the product into the dehydration solution [16] [26] [27].

Table 6. Centesimal composition of raw material and of the post-osmotic dehydration product.

	Contents (g·100 g^{-1})	
Component	Raw material	Post-dehydration product.
Moisture	84.20	76.31
Ashes	0.53	1.02
Lipids	0.82	1.00
Proteins	0.45	0.32
Crude fiber	1.15	2.94
Total sugars	14.50	19.31

Since all the contents of components were calculated in wet basis, the increase in concentration was due to the loss of moisture in the product.

4. CONCLUSIONS

Statistical modeling of factorial design has shown that the variable F:OS was not significant to any of the responses studied, proving that in the process of osmotic dehydration, the costs with solutes can be lowered.

The models fitted for WL and ML were very predictive and fitted, with errors lower than 10%, relative to the experimental points. At the same time, the models for GS and GS/WL were predictive and reasonably fitted, showing relative errors slightly above 10%.

Temperature was the independent variable studied which had a significant impact on the optimization of the process, leading to these values: T = 50°C, C = 40 g·100 g^{-1} e F:OS = 3:20.

Chemical composition of the raw material showed values of moisture and sugars in agreement with the literature, but disparities in relation with the contents of the others components studied.

The product obtained through dehydration had moisture levels lower than the raw material, as was expected; the remaining components underwent a rise in concentration.

The product must go through complementary drying in order to reach the maximum moisture content allowed by legislation.

REFERENCES

1. Alves, R.E., Brito, E.A., Rufino, M.S.M. and Sampaio, C.G. (2008) Antioxidant Activity Measurement in Tropical Fruits: A Case Study with Acerola. Acta Horticulturae, 773, 299-305.http://dx.doi.org/10.17660/ActaHortic.2008.773.45
2. FAO (2011) Food and Agricultural Commodities Production.http: //faostat.fao.org/site/339/default.aspx
3. Sidra (2009) IBGE System Auto Recovery http://www.sidra.ibge.br
4. Valente, P.P.S.S. (2007) Desidrataç&atildeo osmótica e secagem de abacaxi (Ananas comosus L., Merril), variedade pérola. PhD Thesis, Universidade Estadual de Campinas, Campinas,107 p.
5. Carvalho, L.M.J., Castro, I.M. and Silva, C.A.B. (2008) A Study of Retention of Sugars in the Process of Clarification of Pineapple Juice (Ananas comosus, L. Merril) y Micro and Ultrafiltration. Journal of Food Engineering, 87, 447-454.http://dx.doi.org/10.1016/j.jfoodeng.2007.12.015
6. Karim, M.A. and Hawlader, M.N.A. (2005) Mathematical Modelling and Experimental Investigation of Tropical Fruits Drying. International Journal of Heat and Mass

Transfer, 48, 4914-4925.http://dx.doi.org/10.1016/ j.ijheatmasstransfer. 2005.04.035

7. Togrul, I.T. and Dursul, P. (2004) Modelling of Thin Layer Drying Kinetics of Some Fruits under Open-Air Sun Drying Process. Journal of Food Engineering, 65, 413-425.http://dx.doi.org/10.1016/j.jfoodeng.2004.02.001
8. Myers, R.H. and Montgomery, D.C. (2002) Response Surface Methodology: Process and Product Optimization Using Designed Experiments. 2nd Edition, Wiley, New York.
9. Almeida, F.A.C., Ribeiro, C.F., Tobinaga, S. and Gomes, J.P. (2005) Otimizaç&atildeo do processo de secagem osmótica na obtenç&atildeo de produtos secos da manga Tommy Atkins. Revista Brasileira de Engenharia Agrícola e Ambiental, 9, 576-584.http://dx.doi.org/10.1590/S1415-43662005000400022
10. El-Aouar, A.A., Azoubel, PM., Barbosa Jr., J.L. and Murr, F.E.X. (2006) Influence of the Osmotic Agent on the Osmotic Dehydration of Papaya (Carica papaya L.). Journal of Food Engineering, 75, 267-274.http://dx.doi.org/10.1016/j.jfoodeng.2005.04.016
11. BRASIL (2008) Ministério da Saúde. Agência de Vigilancia Sanitária (ANVISA). Instituto Adolfo Lutz. Métodos Físico-Químicos para Análise de Alimentos. Brasília.
12. Dionello, R.G., Berbert, P.A., Molina, M.A.B., Pereira, R.C., Viana, A.P. and Carlesso, V.O. (2009) Desidrataç&atildeo osmótica de frutos de duas cultivares de abacaxi em xarope de açúcar invertido. Revista Brasileira de Engenharia Agrícola e Ambiental, 13, 596-605.
13. Atarés, L., Chiralt, A. and González-Martínez, C. (2008) Effect of Solute on Osmotic Dehydration and Rehydration of Vacuum Impregnated Apple Cylinders (cv. Granny Smith). Journal of Food Engineering, 89, 49-56.http://dx.doi:10.1016 /j.jfoodeng.2008.04.002
14. Tonon, R.V., Baroni, A.F. and Hubinger, M.D. (2004) Osmotic Dehydration of Tomato in Ternary Solutions: Influence of Process Variables on Mass Transfer Kinetics and an Evaluation of the Retention of Carotenoids. Journal of Food Engineering, 65, 413-425.http://dx.doi:10.1016/j.jfoodeng.2007.03.008
15. Lombard, G.E., Oliveira, J.C., Fito, P. and Andrés, A. (2008) Osmotic Dehydration of Pineapple as a Pre-Treatment for Further Drying. Journal of Food Engineering, 85, 277-284.http://dx.doi:10.1016/j.jfoodeng.2007.07.009
16. Raoult-Wack, A.L., Lenart, A. and Guilbert, S. (1994) Recent Advances in the Osmotic Dehydration of Foods. Trends in Food Science & Technology, 5, 225-260.
17. Conceiç&atildeo Silva, M.A., Silva, Z.E., Mariani, V.C. and Darche, S. (2012) Mass Transfer during the Osmotic Dehydration of West Indian Cherry. LWT—Food Science and Technology, 45, 246-252.http://dx.doi:10.1016/j.lwt.2011.07.032
18. Azoubel, P.M. and Murr, F.E.X. (2004) Mass Transfer Kinetics of Osmotic Dehydration of Cherry Tomato. Journal of Food Engineering, 61, 291-295.http://dx.doi:10.1016/S0260-8774(03)00132-8
19. Ferrari, C.C. and Hubinger, M.D. (2008) Evaluation of the Mechanical Properties and Diffusion Coefficients of Osmo-Dehydrated Melon Cubes. International Journal of Food Science & Technology, 43, 2065-2074 http://dx.doi:10.1111/j.1365-2621.2008.01824.x
20. Khin, M.M., Zhou, W. and Pereza, C.O. (2007) Impact of Process Conditions and Coatings on the Dehydration Efficiency and Cellular Structure of Apple Tissue during Osmotic Dehydration. Journal of Food Engineering, 79, 817-827.http://dx.doi:10.1016/j.jfoodeng.2006.02.046
21. Saputra, D. (2001) Osmotic Dehydration of Pineapple. Drying Technology, 19, 415-425.
22. TACO (2006) Tabela Brasileira De Composiç&atildeo De Alimentos/NEPA-UNICAMP. Vers&atildeo II, NEPA-UNICAMP, Campinas, 113 p.
23. Battholomew, D.P., Paull, R.E. and Rohbach, K.G. (2003) The Pineapple Botany, Production and Uses. Volume 1, CABI Publishing, London, 281-288.
24. Lo Scalzo, R., Papadimitriu, C., Bertolo, G., Maestrelli, A. and Torreggiani, D. (2001) Influence of Cultivar and Osmotic Dehydration Time on Aroma Profiles of

Muskmelon (Cucumis melo, cv. Reticulatus Naud.) Spheres. Journal of Food Engineering, 49, 261-264.http://dx.doi.org/10.1016/S0260-8774(00)00209-0
25. Silva, W.P., Silva, C.M.D.P.S., Lins, M.A.A. and Gomes, J.P. (2013) Osmotic Dehydration of Pineapple (Ananas comosus) Pieces in Cubical Shape Described by Diffusion Models. LWT—Food Science and Technology, 55, 1-8.http://dx.doi.org/10.1016/j.lwt.2013.08.016
26. Giangiacomo, R., Torreggiani, D. and Abbo, E. (1987) Osmotic Dehydration of Fruit: Part 1. Sugars Exchange between Fruit and Extracting Syrups. Journal of Food Process Preservation, 11, 183-195.
27. Corzo, O. and Gomez, E.R. (2004) Optimization of Osmotic Dehydration of Cantaloupe Using Desired Function Methodology. Journal of Food Engineering, 64, 213-219.http://dx.doi.org/10.1016/j.jfoodeng.2003.09.035

CHAPTER 19

Antimicrobial Performance of Two Different Packaging Materials on the Microbiological Quality of Fresh Salmon

Manuela Rollini[1,*], Tim Nielsen[2], Alida Musatti[1], Sara Limbo[1], Luciano Piergiovanni[1], Pilar Hernandez Munoz[3] and Rafael Gavara[3]

[1]*DeFENS, Department of Food, Environmental and Nutritional Sciences, Università degli Studi di Milano, Via G. Celoria 2, Milano 20133, Italy*

[2]*SP Food and Bioscience, Ideon, Lund 22370, Sweden; Tim.Nielsen@sp.se*

[3]*Packaging Lab., Instituto de Agroquímica y Tecnología de Alimentos, IATA-CSIC, Paterna 46980, Spain*

ABSTRACT

In the present research the antimicrobial activity of two active packaging materials on the spoilage microbiota of fresh salmon fillets was tested. A PET-coated film (PET: Polyethylene Terephthalate) containing lysozyme and lactoferrin was tested in parallel with a carvacrol-coextruded multilayer film. Salmon fillet samples were stored up to four days at 0 and 5 °C, comparatively. The carvacrol multilayer film was found effective in preventing mesophiles and psychrotrophs at shorter storage time and at lower temperature (4.0 compared to 5.0 log CFU/g in the control sample—CFU: Colony Forming Units). Lysozyme/lactoferrin-coated PET was instead efficient in decreasing H_2S-producing bacteria at longer storage time and higher temperature (2.7 instead of 4.7 log CFU/g in the control sample). Even if is not intended as a way to "clean" a contaminated food product, an active package solution can indeed contribute to reducing the microbial population in food items, thus lowering the risk of food-related diseases.

Keywords: active packaging; carvacrol; coextrusion; lysozyme; lactoferrin; coating; salmon

1. INTRODUCTION

Salmon is an important product of aquaculture: 1,400,000 ton were produced in 2010 with a value of more than seven billion US dollars. In 2009, the main producers of Atlantic salmon were Norway, Chile, the EU and Canada [1]. In the EU the main farmed species is Atlantic salmon, accounting for 93% of the total aquaculture production. The EU is very dependent on the rest of the world for salmon since it imports 80% of its supply from other countries, and 80% of that from Norway.

Fresh seafood is characterized by a relatively short shelf-life and is typically spoiled by aerobic Gram-negative bacteria. In fish, the spoilage process is well documented and consists of autolytic degradation by fish enzymes and the production of unpleasant odors and flavors as a result of microbial action [2]. Typically, in the chilled seafood supply chain, microbial-mediated changes dominate the spoilage process [3]. The bacteria responsible for spoilage in marine fish vary according to the harvest environment, the degree of cross-contamination and the preservation methods applied post-harvest. The primary spoilage bacteria in aerobically packed fish are Gram-negatives from the genera *Pseudomonas* and *Shewanella* while in modified atmospheres, they are *Photobacterium* as well as lactic acid bacteria (LAB), such as *Lactobacillus* and *Carnobacterium* [3,4]. *Shewanella putrefaciens* and *Pseudomonas* spp. become the main producers of the volatile compounds associated with spoilage, such as trimethylamine (TMA), ammonia and sulphides. TMA is particularly responsible for the unpleasant odor of spoiled fish, and is a common index of seafood quality. However, the changes in sensory attributes often occur before products are hygienically spoiled [5].

During storage, spoilage bacteria are selected primarily as a result of the physical and chemical condition in the products; however, seafood spoilage obviously involves growth of the microorganisms to a high amount (>10^6–10^7 CFU/g) and the interaction between groups of microorganisms may influence their growth and metabolism [6]. In particular the high iron-binding capacity of the *Pseudomonas* siderophores may cause this bacterial group to be positively selected, as well as LAB inhibit the growth of other bacteria due to the formation of lactic acid and/or bacteriocins or by competition for nutrients [7].

A significant support to the fight against microbial spoilage may derive from food packaging, which not only acts as a barrier against moisture, water vapor, gases and solutes, but may also serve as a carrier of active substances, such as antimicrobials, in active packaging. Active packaging is defined as an integrated system in which the package, product and environment interact to prolong shelf-life or to enhance safety and/or quality of food products [8].

The antimicrobial features of food packaging materials can be achieved by different strategies: among others, the incorporation in the bulky polymer of migrating compounds, grafting of antimicrobial moieties, and immobilization of antimicrobial agents on the surface of the material in direct contact with the food are the most widely adopted routes. However, direct incorporation in the plastic polymer matrix is not a feasible approach when dealing with antimicrobial agents that are highly sensitive to package production conditions; indeed, high processing pressure and high temperature, or incompatibility with the packaging material, can inactivate the active agents [9]. Alternative production methods have recently been considered. In particular, coating technology has gained increasing attention due to its promising potential as a valid route to generate antimicrobial packaging materials [10].

The aim of the present research was to evaluate the antimicrobial effectiveness of two active packaging materials on the spoilage microbiota of fresh salmon fillets. In particular, a PET-coated film containing lysozyme and lactoferrin was tested in parallel with a carvacrol-coextruded multilayer film. These natural compounds have been selected for their interesting antimicrobial performance evidenced in the frame of the European funded project NAFISPACK—Natural Antimicrobials for Innovative and Safe Packaging (EU212544). Salmon samples after packaging were stored up to four days at 0 and 5 °C, comparatively, to show any significant effect of the developed materials on spoilage microbiota. These two temperatures were chosen as the first represents the one applied by the company to deliver salmon samples to the market, while the second is commonly used by consumers for home storage.

2. EXPERIMENTAL SECTION

2.1. Antimicrobial Films

Lysozyme-Lactoferrin (LZ-LF) water solutions were coated together onto PET: the formulation included gelatin as the main biopolymer network, glycerol as plasticizer, and a lipid phase (a monoglyceride acetylate) as slipping agent, in order to avoid blocking of the film coils during the unwinding operations on the reels. The total dry matter was 19.1 wt %, where LZ and LF accounted for 3 wt % (on the total weight). The procedure has been up-scaled in a pilot plant coating machine at a speed of 10 m/min. The removal of the solvent (water) was achieved by the combined effect of IR lamps and a flux of mild air. Right before the coating deposition, the plastic web underwent an in-line corona treatment to improve surface wettability.

Carvacrol (4.8%) was instead incorporated in a coextruded multilayer film 73 μm thick, made up of two external layers of polypropylene, two tie layers, and an internal layer of ethylene–vinyl alcohol copolymer with a 29% ethylene molar content (EVOH-29), prepared as reported elsewhere [11]. Briefly, films were obtained by extrusion processes using a flat sheet die 500 mm wide and three single screw extruders (Dr. Collin GmbH, Edersberg, Germany) with 30 mm of screw diameter and L/D ratio of 30. Die, feeding block, and transfer lines were maintained at 250 °C. Temperature of chill roll was set at 45 °C and an additional airknife was used. Line speed was 9.5 m/min.

2.2. Sample Preparation

Fresh salmon fillets cut to cubes (*ca.* 120 g) were packaged in laminated material made up of polyethylene/polyamide/polyethylene in Bremerhaven (Germany) and then transported on ice to the University of Milan, where they arrived the next day (Figure 1). Fillets were repackaged into pouches (20 cm × 12 cm) made from a high barrier multilayer film after being wrapped in sheets of the antimicrobial films. The transfer of the samples from the original package to the experimental packages was made in sterile conditions under a safety cabinet. Samples were all packaged under vacuum. Reference samples were also prepared in an identical way, without wrapping in antimicrobial films.

Figure 1. Sample of salmon fillets, cut into cubes, employed in shelf-life trials.

After packaging, samples were all stored at 0 °C and analyses were performed at days 0, 1 and 4. At day 1, some samples were moved to 5 °C, where they were left until the microbiological analysis.

2.3. Microbiological Analysis

Microbial analyses were performed at days 0, 1 and 4 after packaging. Samples were taken from the surface of the samples (2–3 mm) and transferred aseptically and weighed in a sterile Stomacher bag, diluted with peptone water (PW) (Scharlab, Barcelona, Spain) and blended in Stomacher (IUL S.L., Barcelona, Spain) for 6 min. Ten-fold dilution series in PW of the obtained suspensions were made and plated on selective solid media: TSA (Merck, Germany) for mesophiles and psychrotrophs, MEA (Sigma Aldrich, St. Louis, MO, USA) for yeasts and fungi, *Pseudomonas* agar base (Himedia, India) for *Pseudomonas* spp., VRBLA (Violet Red Bile Agar, Merck, Germany) for coliforms, MRS agar (Scharlab, Barcelona, Spain) for lactic acid bacteria, and Lyngby Iron agar (Oxoid, UK) for H_2S-producing bacteria. Colonies were counted after incubation at 30 °C for 24 h for mesophiles, 10 °C for 10 days for psychrotrophs, 30 °C for five days for yeasts and fungi, 25 °C for 24 h for *Pseudomonas*, 37 °C for 24 h for coliforms, 25 °C for five days for lactic acid bacteria and 20 °C for three days for H_2S-producing bacteria. Counts were performed in triplicate and reported as logarithms of the number of colony forming units (log CFU/g salmon), and means and standard deviations were calculated.

3. RESULTS AND DISCUSSION

Aerobic mesophiles and psychrotrophs as well as H_2S-producing bacteria were the prevalent population in the salmon at time of packaging (3.3–3.6 log CFU/g), followed by LAB (2.2 log CFU/g and *Enterobacteriaceae* (1.5 log CFU/g) (Table 1). According to the International Commission on Microbiological Specifications for Foods, most aquatic animals at the time of harvest have microbial counts in the range of 2 to 5 log CFU/g [12]. In the present study the initial value of microbial load was in the same range, as also reported by other authors [13]. After one day of storage at 0 °C, the psychrotroph population increased up to 5.5 log CFU/g in control samples. Salmon packed in carvacrol active films was characterized by a reduced psychrotroph population (3.9 log CFU/g), while LZ-LF–coated films were ineffective (5.8 log CFU/g) (Table 2). Gram-negative psychrotrophic bacteria are the major group of microorganisms responsible for the spoilage of fresh fish at chilled temperatures [2]. In this study, mesophiles in the first 24 h remained in the range 3.9–4.1 log CFU/g, without significant difference between the control sample and those stored in the active packages.

Table 1. Microbial population (log CFU/g) present in salmon at time of packaging.

Microorganisms	Microbial Count (log CFU/g)
Aerobic mesophiles	3.6 ± 0.2
Aerobic psychrotrophs	3.5 ± 0.1
Enterobacteriacae	1.5 ± 0.4
Lactic acid bacteria	2.2 ± 0.2
Pseudomonas	<2
H_2S-producing bacteria	3.3 ± 0.2

Table 2. Microbial population (log CFU/g) in salmon packed in two different antimicrobial materials and in one reference material (control) after storage for one day at 0 °C.

Microorganisms	Control	Carvacrol	LZ LF
Aerobic mesophiles	3.9 ± 0.7	4.1 ± 0.6	4.0 ± 0.5
Aerobic psychrotrophs	5.5 ± 0.4	3.8 ± 0.5	5.8 ± 0.2
Enterobacteriaceae	1.6 ± 0.2	1.6 ± 0.5	1.6 ± 0.2
Lactic acid bacteria	2.2 ± 0.3	2.2 ± 0.4	3.1 ± 0.1
Pseudomonas	<2	<2	<2
H_2S-prod. Bacteria	2.7 ± 0.6	3.6 ± 0.4	2.1 ± 0.2

At day 1 some samples were moved to 5 °C while, in parallel, other samples were kept at 0 °C. All samples were then stored for up to four days, one day more than the normal shelf-life suggested by the company.

After four days at 0 °C, salmon packed in carvacrol films maintained psychrotroph and mesophile populations at low levels, *i.e.*, around 4.0 log CFU/g, compared to 5.0 log CFU/g in control samples (Table 3). Also, LZ-LF–coated films showed an interesting performance, reducing the populations up to 3.0 and 3.7 log CFU/g, respectively. Note also, at this storage temperature, the efficacy of the two active films in reducing the *Pseudomonas* population of approximately 1 log cycle compared to control, from 3.2 to 2.2–2.3 log CFU/g.

Table 3. Microbial population (log CFU/g) in salmon packed in two different antimicrobial materials and in one reference material (control) after storage for four days at 0 °C and 5 °C, comparatively.

Temperature	Microorganisms	Control	Carvacrol	LZ LF
0 °C	Aerobic mesophiles	5.1 ± 0.2	3.9 ± 0.2	3.7 ±0.3
	Aerobic psychrotrophs	5.0 ± 0.3	4.0 ± 0.1	3.0 ± 0.1
	Enterobacteriaceae	2.0 ± 0.0	1.9 ± 0.3	2.2 ± 0.2
	Lactic acid bacteria	2.9 ± 0.2	2.6 ± 0.1	2.0 ± 0.1
	Pseudomonas	3.2 ± 0.5	2.2 ± 0.2	2.3 ± 0.6
	H_2S-prod. bacteria	3.3 ± 0.2	3.4 ± 0.4	3.4 ± 0.4
5 °C	Aerobic mesophiles	5.3 ± 0.6	4.6 ± 0.6	4.5 ± 0.3
	Aerobic psychrotrophs	5.3 ± 0.6	6.9 ± 0.4	3.8 ± 0.2
	Enterobacteriaceae	1.9 ± 0.0	2.5 ± 0.5	2.2 ± 0.2
	Lactic acid bacteria	2.8 ± 0.5	3.2 ± 0.4	3.1 ± 0.2
	Pseudomonas	2.5 ± 0.1	2.8 ± 0.3	2.2 ± 0.2
	H_2S-prod. bacteria	4.7 ± 0.6	4.0 ± 0.7	2.7 ± 0.5

Salmon samples stored for four days at 5 °C without active packaging were characterized by mesophile and psychrotroph population of 5.3 log CFU/g. Carvacrol active films were effective in maintaining only the mesophile population at a low level of 4.6 log CFU/g, while LZ-LF–coated PET reduced both the populations down to 4.5 and 3.8 log CFU/g, respectively. It must be noted, also, that in this last case there was a reduction of H_2S-producing bacteria to 2.7 instead of 4.7 log CFU/g as seen in the control samples. This positive observation that LZ/LF–coated PET films could prevent growth of H_2S-producing bacteria is of actual relevance since this type of microorganism has been identified as the most potent in causing rejection of fresh salmon fillets, due to the production of off-odors during growth [4].

Enterobacteriaceae were also found to be members of the microbial association implicated in the spoilage of fresh sliced salmon during refrigerated storage. This finding is in agreement with results reported for different fish species, including fresh Atlantic salmon [14] as well as rainbow trout [15], in which *Enterobacteriaceae* were determined as a part of the microbial population at the end of the product shelf-life under refrigerated storage. In this study, *Enterobacteriaceae* were always found in low levels, less than 3 log CFU/g; although *Enterobacteriaceae* can grow at low temperatures, their proliferation was slow during refrigerated storage, possibly because their growth rate is lower than that of other Gram-negative psychrotrophic spoilers [16].

The LAB population was always found lower than 3.2 log CFU/g, and was not influenced by the active packaging employed. The low LAB count in this study was expected since they tend to grow slowly at refrigeration temperatures [13].

The obtained data satisfied the recommended microbiological limits for fresh and frozen fish reported in [12], which are defined only for aerobic plate counts performed at 20–25 °C (5.5 to 7 log CFU/g) and for *E. coli* (1 to 2.7 log CFU/g). The first data reflect handling practices in the fish industry, from shipboard to market delivery, while the second parameter is considered an indicator of contamination and, when present in high numbers, suggests temperature abuse in product handling.

The present research was aimed at investigating the efficacy of two antimicrobial food packaging materials on microbial spoilage of fresh salmon fillets. Incorporation of natural antimicrobials into food packaging materials to control the growth of spoilage and pathogenic organisms has been researched for the last decades. As regards fish products and shelf-life, the vast majority of data relates to the possibility of applying antimicrobial edible films in contact with the fish surface. Jasour *et al.* [17] evaluated the effect of an edible coating based on chitosan coated with the lactoperoxidase system (LPS) on the quality and shelf-life extension of rainbow trout during refrigerated storage at 4 °C. Results indicated that antimicrobial coating was found efficient in reducing *Shewanella putrefaciens*, *Pseudomonas fluorescens* as well as psychrotrophic and mesophilic bacterial populations compared to the control sample.

Few authors have investigated the performance of an antimicrobial package prepared by applying a coating procedure. Gomez Estaca *et al.* [18] produced a complex gelatin-chitosan film incorporating clove essential oil which was applied to fish during chilled storage: the growth of microorganisms was drastically reduced in Gram-negative bacteria, especially *Enterobacteriaceae*, while LAB remained constant for much of the storage period. Neetoo and Mahomoodally [19] compared the antimicrobial efficacy against *Listeria monocytogenes* in smoked salmon fillets of films or direct coatings incorporating nisin (Nis) and sodiumlactate (SL), sodiumdiacetate (SD), potassiumsorbate (PS), and/or sodium benzoate (SB) in binary or ternary combinations on cold smoked salmon. Surface treatments incorporating Nis (25,000 IU/mL) in combination with PS (0.3%) and SB (0.1%) had the highest inhibitory activity, reducing the population of *L. monocytogenes* by a maximum of 3.3 log CFU/cm^2 (films) and 2.9 log CFU/cm^2 (coatings) relative to control samples after 10 days of storage at 21 °C. During refrigerated storage, coatings were more effective in inhibiting growth of *L. monocytogenes* than their film counterparts. Cellulose-based coatings incorporating Nis, PS, and SB reduced the population of *L. monocytogenes*, and anaerobic and aerobic spoilage microbiota by a maximum of 4.2, 4.8, and 4.9 log CFU/cm^2, respectively, after four weeks of refrigerated storage.

In the present study the carvacrol-incorporated multilayer film was found effective in preventing mesophiles and psychrotrophs at shorter storage time and at lower temperature, while lysozyme-lactoferrin–coated PET was mostly efficient in decreasing H_2S-producing bacteria at longer storage time and higher temperature. Even if it is not intended as a way to "clean" a contaminated food product, an antimicrobial package solution can indeed contribute to reducing the microbial population in food items, thus lowering the decay of organoleptic features and increasing shelf life.

4. CONCLUSIONS

This work provides examples of active food packages, in which the antimicrobial compounds, *i.e.*, the volatile carvacrol and the soluble association lysozyme-lactoferrin, were applied, the first incorporated in a coextruded multilayer film while the second association was coated onto PET. Application of these materials on actual salmon samples under conditions similar to those foreseeable for a future practical use gave positive results, in particular for the reduction of H_2S-producing bacteria in coated films. Future work will indicate whether the antimicrobial-loaded films used here may find other practical uses (e.g., liners or wraps) and whether they will be effective to improve the safety and to extend the shelf-life of other food products.

ACKNOWLEDGMENTS

Work supported by funds of the EU project FP7-NAFISPACK 212544.

REFERENCES

1. Food and Agriculture Organization (FAO), Fisheries and Aquaculture Department. *The State of World Fisheries and Aquaculture 2012*; Food and Agriculture Organization of the United Nations: Rome, Italy, 2012.
2. Gram, L.; Huss, H.H. Microbiological spoilage of fish and fish products. *Int. J. Food Microbiol.* **1996**, *33*, 121–137.
3. Emborg, J.; Laursen, B.G.; Rathjen, T.; Dalgaard, P. Microbial spoilage and formation of biogenic amines in fresh and thawed modified atmosphere-packed salmon (*Salmo salar*) at 2 °C. *J. Appl. Microbiol.* **2002**, *92*, 790–799.
4. Dalgaard, P.; Gram, L.; Huss, H.H. Spoilage and shelf-life of cod fillets packed in vacuum or modified atmospheres. *Int. J. Food Microbiol.* **1993**, *19*, 283–294.
5. Franzetti, L.; Martinoli, S.; Piergiovanni, L.; Galli, A. Influence of active packaging on the shelf life of minimally processed fish products in a modified atmosphere. *Packag. Technol. Sci.* **2002**, *14*, 267–274.
6. Gram, L.; Dalgaard, P. Fish spoilage bacteria: Problems and solutions. *Curr. Opin. Microbiol.* **2002**, *13*, 262–266.
7. Gram, L.; Ravn, L.; Rasch, M.; Bruhn, J.B.; Christensen, A.B.; Givskov, M. Food spoilage-interactions between food spoilage bacteria. *Int. J. Food Microbiol.* **2002**, *78*, 79–97.
8. Suppakul, P.; Miltz, J.; Sonneveld, K.; Bigger, S.W. Active packaging technologies with an emphasis on antimicrobial packaging and its application. *J. Food Sci.* **2003**, *68*, 408–420.
9. Perez-Perez, C.; Regalado-González, C.; Rodríguez-Rodríguez, C.; Barbosa-Rodríguez, J.R.; Villaseñor-Ortega, F. Incorporation of antimicrobial agents in food packaging films and coatings. In *Advances in Agricultural and Food Biotechnology*; Guevara-Gonzalez, R.G., Torres-Pacheco, I., Eds.; Research Signpost: Trivandrum, Kerala, India, 2006; pp. 193–216.
10. Farris, S.; Piergiovanni, L. Emerging coating technologies for food and beverage packaging materials. In *Emerging Food Packaging Technologies: Principles and Practice*; Yam, K., Lee, D.S., Eds.; Woodhead Publishing Ltd.: Oxford, UK, 2012; pp. 274–302.
11. Cerisuelo, J.P.; Bermudez, J.M.; Aucejo, S.; Català, R.; Gavara, R.; Hernandez-Munoz, P. Describing and modeling the release of an antimicrobial agent from an active PP/EVOH/PP package for salmon. *J. Food Eng.* **2013**, *116*, 352–361.
12. International Commission of Microbiological Specifications for Foods (ICMSF). Sampling plans for fish and shellfish. In *Microorganisms in Foods 2. Sampling for Microbiological Analysis: Principles and Specific Application*; ICMSF: Toronto, ON, Canada, 1986; pp. 181–196.
13. Sallam, K.I. Antimicrobial and antioxidant effects of sodium acetate, sodium lactate and sodium citrate in refrigerated sliced salmon. *Food Cont.* **2007**, *18*, 566–575.
14. Amanatidou, A.; Schluter, O.; Lemkau, K.; Gorris, L.G.M.; Smid, E.J.; Knorr, D. Effect of combined application of high pressure treatment and modified atmospheres on the shelf life of fresh Atlantic salmon. *Innov. Food Sci. Emerg. Technol.* **2000**, *1*, 87–98.
15. Chytiri, S.; Chouliara, I.; Savvaidis, I.N.; Kontominas, M.G. Microbiological, chemical, and sensory assessment of iced whole and filleted aquacultured rainbow trout. *Food Microbiol.* **2004**, *21*, 157–165.
16. Papadopoulos, V.; Chouliara, I.; Badeka, A.; Savvaidis, I.N.; Kontominas, M.G. Effect of gutting on microbiological, chemical, and sensory properties of aquacultured sea bass (*Dicentrarchus. labrax*) stored in ice. *Food Microbiol.* **2003**, *20*, 411–420.

17. Jasour, M.S.; Ehsani, A.; Mehryar, L.; Naghibi, S.S. Chitosan coating incorporated with the lactoperoxidase system: An active edible coating for fish preservation. *J. Sci. Food Agric.* **2014**, *95*, 1373–1378.
18. Gómez-Estaca, J.; López de Lacey, A.; López-Caballero, M.E.; Gómez-Guillén, M.C.; Montero, P. Biodegradable gelatin-chitosan films incorporated with essential oils as antimicrobial agents for fish preservation. *Food Microbiol.***2010**, *27*, 889–896.
19. Neeto, H.; Mahomoodally, F. Use of antimicrobial films and edible coatings incorporating chemical and biological preservatives to control growth of *Listeria monocytogenes* on cold smoked salmon. *BioMed Res. Int.* **2014**, *2014*.

CHAPTER 20

Thermal, Structural, and Physical Properties of Freeze Dried Tropical Fruit Powder

K. A. Athmaselvi[1] C. Kumar[2], M. Balasubramanian[3], and Ishita Roy[1]

[1]*Department of Food Process Engineering, Faculty of Engineering and Technology, SRM University, Kattankulathur, Tamil Nadu 603203, India*

[2]*Department of Physics, Government Arts College, Nandanam, Chennai, Tamil Nadu 600035, India*

[3]*Valliammai Engineering College, Kattankulathur, Tamil Nadu 603203, India*

ABSTRACT

This study evaluates the physical properties of freeze dried tropical (guava, sapota, and papaya) fruit powders. Thermal stability and weight loss were evaluated using TGA-DSC and IR, which showed pectin as the main solid constituent. LCR meter measured electrical conductivity, dielectric constant, and dielectric loss factor. Functional groups assessed by FTIR showed presence of chlorides, and O–H and N–H bonds in guava, chloride and C–H bond in papaya, and chlorides, and C=O and C–H bonds in sapota. Particle size and type of starch were evaluated by X-ray diffraction and microstructure through scanning electronic microscopy. A semicrystalline profile and average particle size of the fruit powders were evidenced by X-ray diffraction and lamellar/spherical morphologies by SEM. Presence of A-type starch was observed in all three fruits. Dependence of electric and dielectric properties on frequency and temperature was observed.

1. INTRODUCTION

Guava (*Psidium guajava* L.) and sapota (*Manilkara zapota*) are native of tropical America and grow well in tropical and subtropical regions and are widely cultivated in the tropics including India, Guatemala, and Mexico [1, 2]. Papaya (*Carica papaya* L.) is also an important fruit crop grown widely in tropical and subtropical countries in the world. The relative humidity and temperature are the more critical variables to extend the shelf-life of the fruit and to preserve its sensory, physical, and chemical properties. Thus, dehydration processes may be an efficient alternative for fruit storage, because the reduction of water activity is related to the decline of chemical and enzymatic reactions responsible for the deterioration of foods. Dehydration is based on the extraction of water contained in foods up to a minimal level which is enough to their conservation for long time. The hot air-drying is one of the most used methods for preservation of fruits since long time ago; however the main problem is the possibility to lose or change some of their properties due to the thermal treatment. Hence, gentle drying techniques, such as osmodehydration or lyophilisation (freeze-drying), have also been used in fruits. With the aim to find new alternatives for the commercialization of tropical fruits and to reduce the difficulties for its handling and transport, freeze-drying techniques were used to obtain added-value solids from tropical fruits.

The International Confederation for Thermal Analysis and Calorimetry (ICTAC) has defined thermal analysis (TA) as a group of techniques that allow us to study the relationship between material properties of a sample and its temperature. The use of TG analyses in these applications also allows us to determine at what temperatures and after what times drying is completed and at what point thermal degradation of the food material will occur. DSC is also used for the evaluation of quality parameters of vegetable oil and assessing oxidative deterioration of oil [3–5]. de Rosa et al. [6] studied the thermal stability of okra fibres by thermogravimetric analysis and reported that TG curve of okra fibres shows three weight loss steps, whilst their decomposition occurs in two main stages. Osorio et al. [1] studied thermal stability of guava (*Psidium guajavaK.*) powders obtained by two dehydration methods, hot air-drying and freeze-drying, and thermal analysis curves showed some differences between hot air dried and freeze dried powder.

A lot of work has been done for studying the electrical and dielectric properties of various food items and food ingredients at microwave frequencies by Sipahioglu and Barringer [7], Sipahioglu et al. [8], Soltani et al. [9], Sharma and Prasad [10], and many others, but there is no significant amount of work on the study of these properties for tropical fruits at very low frequency levels (0.001 MHz to 10 MHz). Li et al. [11] measured moisture content of cookies using dielectric spectroscopy. Nelson and Datta [12] measured the dielectric constant and loss factor with an open-ended coaxial line probe and an impedance analyzer on external surfaces and internal tissue of four cultivars of miniature watermelons provided new permittivity data over a range of maturities at frequencies from 10 MHz to 1.8 GHz at 24°C. They reported that both the dielectric constant and loss factor of internal tissues decreased monotonically with increasing frequency showing the dominance of ionic conduction at lower frequencies and dipolar losses at the higher frequencies. Mizukami et al. [13] measured moisture content of tea leaves using electrical impedance and capacitance method. This study attempts to develop relationship between the frequency and electrical and dielectric properties of spray-dried, pelletized powder of guava, sapota, and papaya.

The structural studies of the fruit powders were done using X-ray diffraction (XRD), scanning electron microscope (SEM), and Fourier transformer IR-spectroscopy (FTIR). XRD was used to reveal the presence and characteristics of crystalline structure of starch granules. Conventional laboratory analysis of fruit juices such as chemical testing is generally laborious and time consuming. But FTIR can be employed as a rapid and nondestructive method to identify different functional groups and molecules which can fingerprint a food.

Theivasanthi and Alagar [14] studied the XRD of Jackfruit Seed and reported diffraction peaks at around 15°, 17°, 23°, 31°, and 38°. Deraman et al. [15] have used XRD to study fiber of oil palm empty fruit bunch and rubberwood for medium-density fiberboard. Rengsutthi and Charoenrein [16] analyzed the jack fruit seed starch using XRD and reported that the XRD was used to reveal the presence and characteristics of crystalline structures of starch granules. Leopold et al. [17] used FTIR for quantitative determination of carbohydrates, such as glucose, fructose, and sucrose, in 28 processed commercial fruit juices and 5 genuine juices obtained from squeezed fruits. The objective of this study was to determine the thermal, dielectric, and structural properties of freeze dried fruit powders of guava, sapota, and papaya fruit.

2. MATERIALS AND METHODS

2.1. Preparation of Samples

Fruits of similar size, color, and ripeness were selected for dehydration studies. Selected fruits were washed with tap water, peeled, destoned, and crushed to pulp. The pulp was transferred to the freeze dryer plate and it is spread uniformly and kept in the deep freezer for pretreatment (12 h) and then it is freeze dried for 12 h at −52°C and −3 mbar pressure.

2.2. Differential Scanning Calorimetry (DSC) and Thermogravimetric Analysis (TGA)

The DSC and TGA studies were conducted for the fruit powder samples. In TGA, the weight loss of fruit powder was measured at different temperature. The DSC measures the temperatures and heat flows associated with test sample as a function of time and temperature in a controlled atmosphere. These measurements provide quantitative and qualitative information about physical and chemical changes that involve endothermic or exothermic processes or changes in heat capacity. In the present study, the DSC and TGA measurements were made in the temperature range 0° to 1000°C with heating range 20°C/min using SDT Q600 V8.0 Build 95 instrument. This differential scanning calorimeter has a temperature accuracy of ±0.2°C temperature reproducibility ±0.1°C. Since the DSC was interfaced with computer, the recorded spectrum is obtained from the computer [18].

2.3. Dielectric Constant and Electrical Conductivity

Pellets were prepared from the freeze dried tropical fruit powder. To perform dielectric study, the surface of the pellets was smoothened. The polished surface was maintained as highly smooth as possible for a good electrode contact. Platinum paint was used as the electrode and it was pasted on both sides of the test sample (pellet disc). The test sample was placed in copper conducting sample holder and the electrode connections were taken from the surfaces of the test sample and connected to the analyzer. The whole setup was placed inside a tubular furnace in which the temperature was gradually increased by a periodical adjustment of dimmer stat. The dielectric measurements were carried out using a two-probe LCR meter of ANDO AG4311B type analyzer up to 160°C for various frequency ranges [19].

2.4. FTIR Analysis

Fourier transformed infrared spectra absorption for each sample was obtained using a Bruker Alpha Model. The samples gently mixed with micronized KBr powder and compressed into pellets at a force of 10 KN for 2 min using a manual tablet presser at room temperature. The data were recorded at room temperature in the wavelength range of 4000–500 cm^{-1}.

2.5. X-Ray Diffraction

The XRD analysis of the powder samples was done in PANalytical Xpert Pro XRD instrument in powder mode with the source being Cu-Kα radiation at 1.540 Å. The spectra were scanned at a diffraction angle (2) range of 5–80° at a step size of 0.05°/step and 2 sec/step [20].

2.6. SEM Analysis

The SEM analysis was done on FEI Quanta-FEG 200 SEM. The experiment was done at low vacuum at 80 torr, each fruit powder sample was mounted on a stub using a carbon-tape, and field emission was done at 5 kV [1].

3. RESULT AND DISCUSSION

3.1. Differential Scanning Calorimetry and Thermogravimetric Analysis

A thermogravimetric measuring unit consists mainly of a sensitive weighing balance that is placed within an oven. During a temperature-time program in the oven, the weight of the sample is measured and recorded over time. The oven atmosphere is nitrogen. At high temperatures, volatile substances are driven off, and a decrease in sample mass is observed and measured. DSC plot is sometimes called a thermogram. In thermodynamic systems analysis, heat taken up by a system (gain) is considered positive, while heat given up (loss) is considered negative. It is likewise with heat flow, in which heat flowing into a system is positive (endothermic) and heat flowing out is negative (exothermic).

Thermal stability of freeze dried fruit powder was investigated by thermogravimetric analysis. The TGA-DSC curve of guava, sapota, and papaya fruit powder is shown in Figure 1.

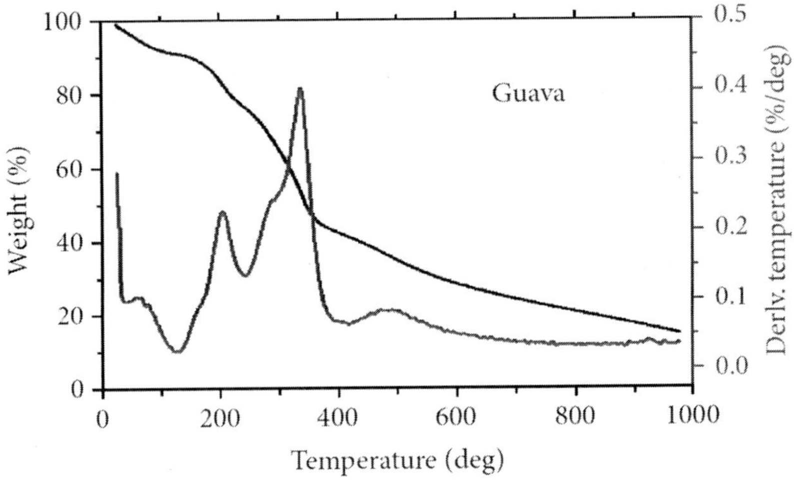
Guava
Weight (%)
Derlv. temperature (%/deg)
Temperature (deg)
100
80
60
40
20
0
0.5
0.4
0.3
0.2
0.1
0.0
0
200
400
600
800
1000

(a)

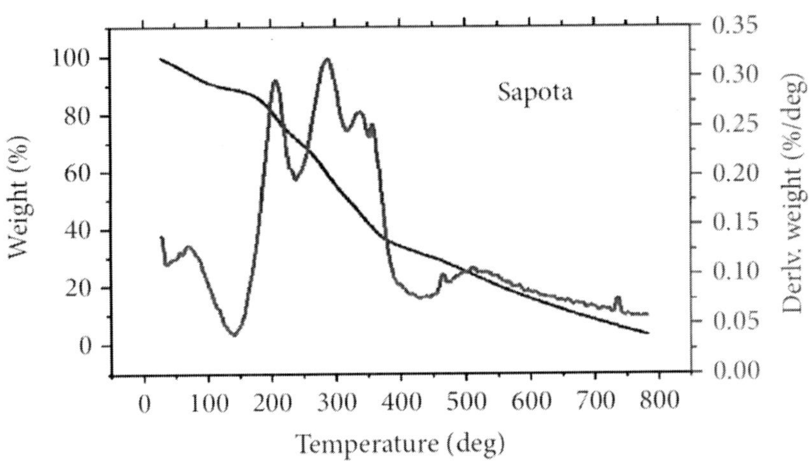
Sapota
Weight (%)
Derlv. weight (%/deg)
Temperature (deg)
100
80
60
40
20
0
0.35
0.30
0.25
0.20
0.15
0.10
0.05
0.00
0
100
200
300
400
500
600
700
800

(b)

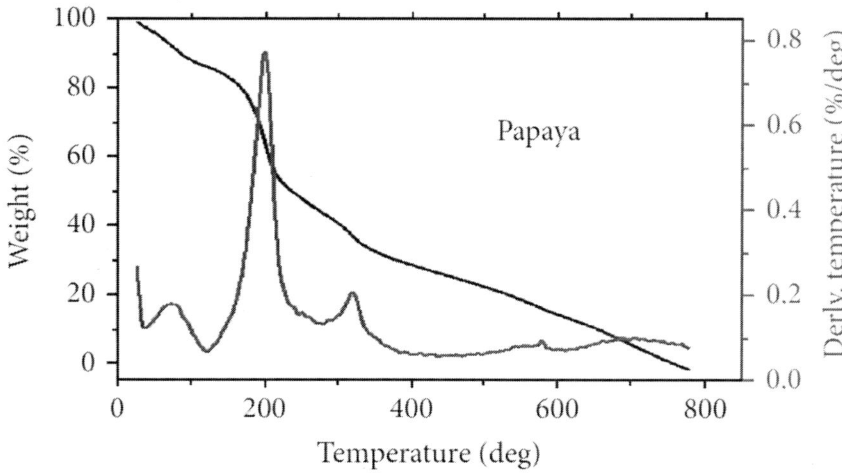

(c)

Figure 1: TGA-DSC of guava (a), sapota (b), and papaya (c).

The TGA curves of guava powder show three general regions of weight loss (54.37–178.23, 178.23–305.28, and 305.28–360.96°C). The first region (54.37–178.23°C) reveals a weight loss around 6.714%. This drop of mass can be assumed as water loss during the temperature rise. A double stage of mass drop for the guava powder in the second mentioned region (178.23–305°C) related to a possible more complex volatilization of matter. The third region (305.28–360.96°C) in the TGA analyses shows a more stable stage after volatilization of water and other compounds; between 150 and 350°C dehydroxylation, demethoxylation, and decarboxylation reactions should be occurring on pectin and similar polysaccharides (pectin is the main component of dehydrated guava) [21–23]. The total loss of weight for samples is 69.52% at 545°C; the residue left after incineration at 980°C is 14.32%, indicating the thermal stability of guava powder. On the other hand, DSC curves show that endothermic peak, centered at 68.64°C, 208.66°C, 341.16°C, and 500.75°C, involves the volatilization processes of water, melting of pectin contained in guava, and a possible demethoxylation, dehydroxylation, and decarboxylation of pectin and other polysaccharides [1].

The TGA curves of sapota powder show three general regions of weight loss (54.85–180.20, 180.20–332.76, and 332.76–372.17°C). The first region (54.85–180.20°C) reveals a weight loss of 10.13%. This drop of mass can be assumed as water loss during the temperature rise. The second region (180.20–332.76°C) shows a weight loss of 38.17%; the third region (332.76–372.17°C) shows a weight loss of 16.95%. The total weight loss of the sapota powder is 65.25% at 372.12°C; the residue left after incinerations at 784.37°C is 3.192%. The residual weight of sapota is much less when compared to guava powder. On the other hand, DSC curves show exothermic peak centered at 71.77°C, 209.09°C, 283.77°C, 341.58°C, and 358.44°C. Similar trend is also seen in okra fiber [24].

The TGA curves of papaya powder show three general regions of weight loss (63.08–184.93, 184.93–314.30, and 314.30–346.13°C). The first region (63.08–184.93°C) reveals a weight loss of 11.25%. This drop of mass can be assumed as water loss during the temperature rise. The second region (184.93–314.30°C) shows a weight loss of 41.45%; the third region (332.76–372.17°C) shows a weight loss of 14.26%. The total weight loss of the papaya powder is 73.45% at 589.55°C; the residue left after incinerations at 761.95°C is 0.4389%. The residual weight of papaya is much less when compared to guava powder. On the other hand, DSC curves show exothermic peak centered at 83.82°C, 205.47°C, 325.92°C, and 581.28°C.

3.2. Frequency Dependence of Electrical Conductivity

Electrical conductivity (S m^{-1}), dielectric constants, and dielectric loss factor values plotted versus different frequencies for guava, sapota, and papaya are given in Figure 2.

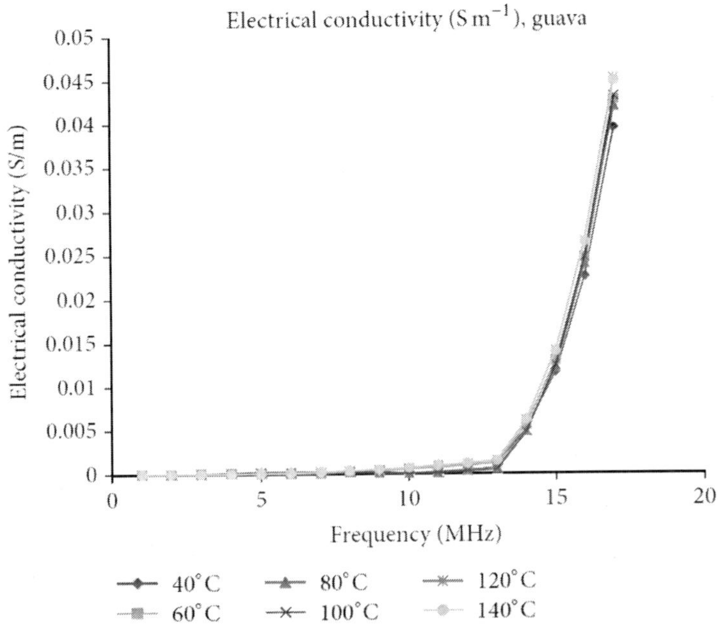

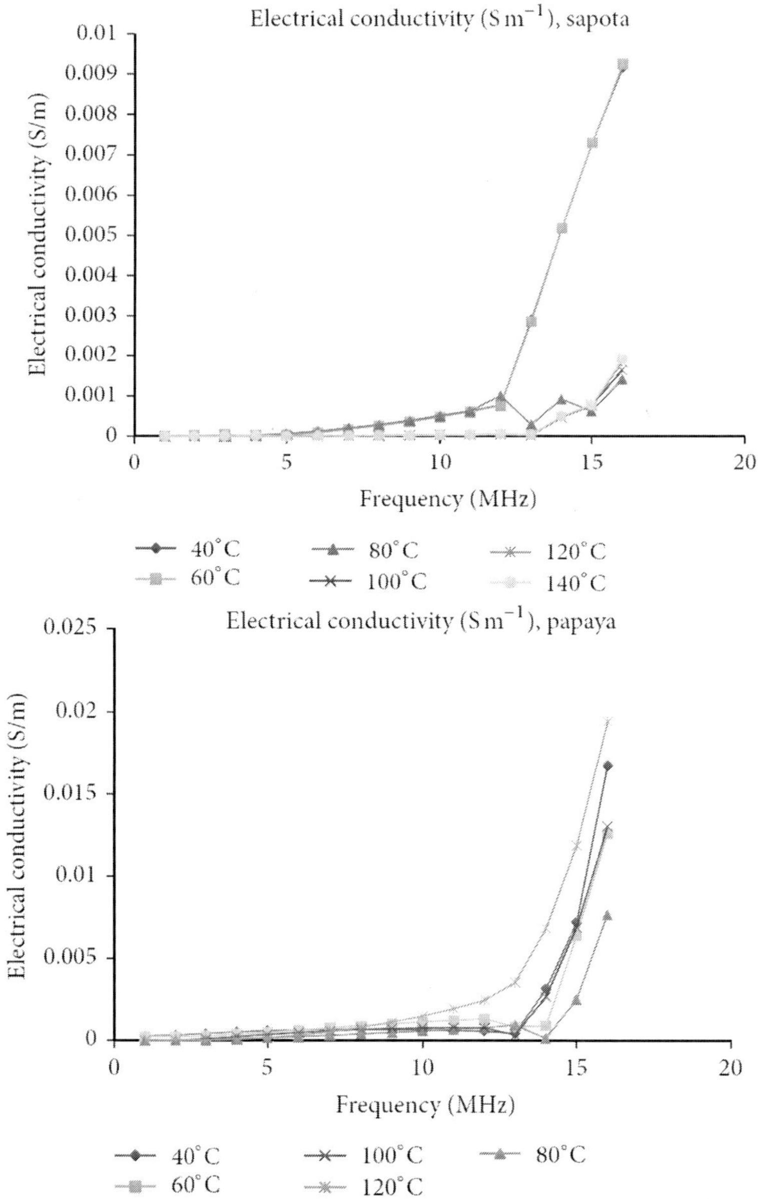

Figure 2: Dependence of electrical conductivity on frequency for guava, sapota, and papaya.

All the three fruits showed very low electrical conductivity at frequency between 0.001 MHz and 1 MHz. However, after 1 MHz, the electrical conductivities increased sharply. The maximum values achieved were 0.0449 (140°C), 0.0155 (60°C), and 0.025 (120°C) for guava, sapota, and papaya, at frequency of 5 MHz. This corresponds to an approximate increase by 6.3, 11.2, and 15.54 times in the electrical conductivities of guava, sapota, and papaya from 1 MHz to 5 MHz. From 0.001 MHz to 1 MHz, there was insignificant increase in electrical conductivities. It may be noted that the highest value of conductivity achieved was for guava followed by papaya and then for sapota which may be due to the difference in their acidity or moisture content, since electrical conductivity depends hugely on the moisture content, as stated by Sipahioglu and Barringer [7], Sharma and Prasad [10], Venkatesh and Raghavan [19], Feng et al. [24], Nelson and Datta [12], Mugdett and Goldblith [25], Funebo and Ohlsson [26], and many others. There was not much variation of electrical conductivity with temperature at very low frequencies, below 1 MHz, after which the conductivity showed higher values for higher temperature at 120°C and 140°C; the conductivity increases with increasing temperature, highest value for guava being 0.0449 (140°C) and that for papaya being 0.025 (120°C) as stated earlier. However, for sapota, the electrical conductivity increased with increasing temperature up to 60°C and then decreased with increasing temperature, with its maximum value being 0.0155 (S m^{-1}).

3.3. Frequency Dependence of Dielectric Properties

Dielectric properties, namely, dielectric constant and loss factor values, have previously been studied by Sipahioglu et al. [8], Ryynänen [27], and Lizhi et al. [28] and this study has shown similar trend for all three fruits. Dielectric constants for all three fruits are plotted in Figure 3 against the frequencies. It can be observed that there is a sharp decrease in their values from 0.001 MHz to 0.1 MHz. The maximum values obtained were 248.43 (at 80°C) for guava, 143.99 (at 40°C) for sapota, and 869.79 (at 40°C) for papaya, all at 0.001 MHz, which decreased by 2.8, 2.68, and 14.07 times when 0.1 MHz was reached. As the frequency increased, the dielectric constant values did not show much variation till 1 MHz. From 1 MHz to 5 MHz, as the frequency increased, the dielectric constant gradually decreased up to 49% for papaya. However, for guava and sapota, dielectric constant increased by 1.41 (60°C) and 1.42 (40°C) times, respectively. Thus, the overall curves for sapota and guava were roughly "U" shaped, whereas that for papaya was a curve with sharply decreasing values, followed by a lag phase and then a phase of gradual decrease. Effect of temperature on dielectric constant values was the same for guava and papaya where the constants increased with increasing temperature up to 80°C and then decreased with the increasing temperature. However, for sapota, the constants decreased steadily with increasing temperature. This was similar to the observation made by Lawrence et al. [29] for hard red winter wheat and Guo et al. [30] for chickpea flour.

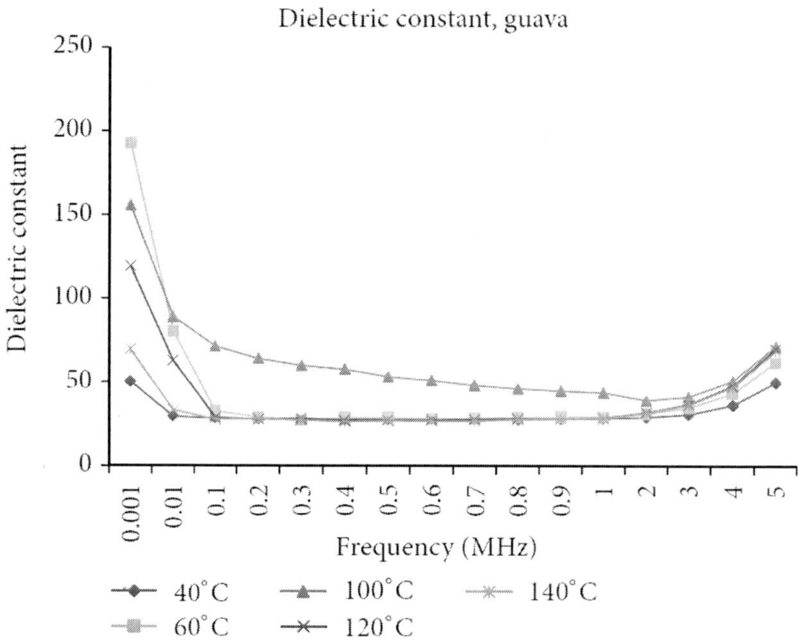

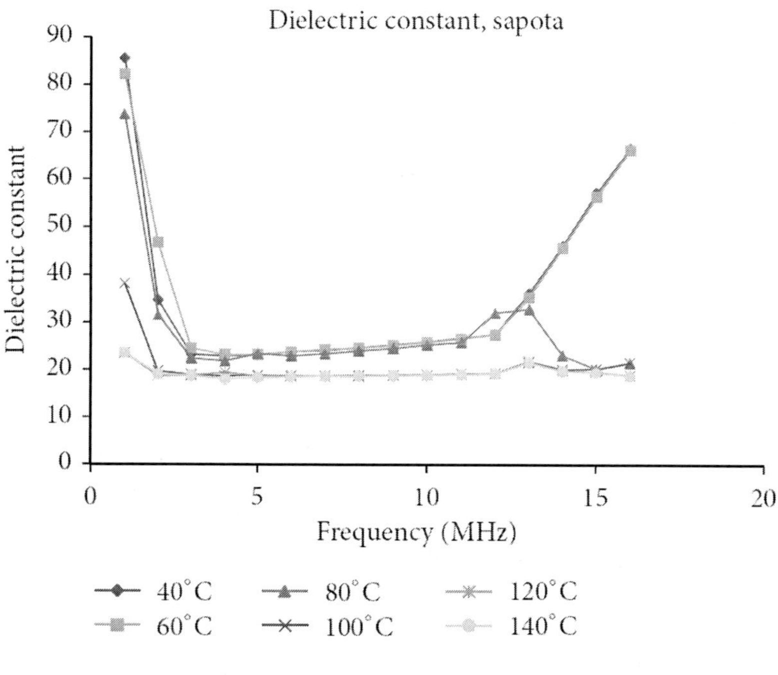

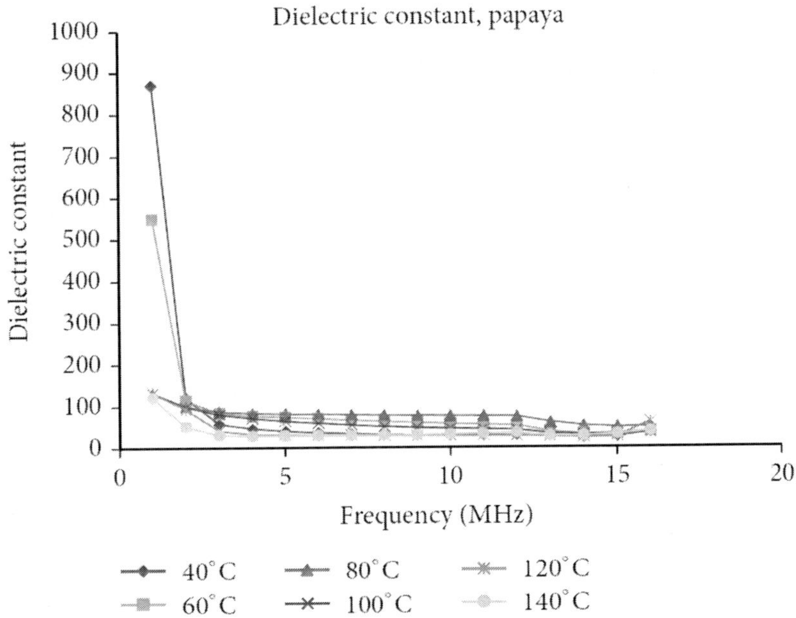

Figure 3: Dependence of dielectric constant on frequency for guava, sapota, and papaya.

Dielectric properties have previously been measured by Jiao et al. [18] and Nelson and Bartley [31] observed that loss factor decreased with increasing frequency but increased with increasing temperature for cowpea weevil, black eyed peas, and mung. Figure 4 represents the dielectric loss factors for the three fruits plotted versus frequency. Much similarly, for all the three fruits, there was an initial phase of sharp decrease but was followed by a gradual and then a sharp increase. For guava, the highest value of loss factor was 9.6859 at 40°C for 0.001 MHz and it decreased rapidly to 0.14428 at 0.1 MHz. For sapota, the decrease was from 1.54962 to 0.04296 at 100°C and for papaya it was from 7.01956 to 0.55029 at 60°C and for both fruits between 0.001 MHz and 0.1 MHz. After this, loss factors gradually increased for guava and papaya, but for sapota the dielectric loss was increased to great level at 40°C, 60°C, and 80°C. The final phase for all three fruits showed an upward curve, maximum values of the curve being at 4 MHz, with values finally decreasing at 5 MHz. For all three fruits, the dielectric loss was more at lower temperatures and decreased as the temperature increased, but, after 1 MHz, the loss increased with the increasing temperature. Such results were also reported by Sharma and Prasad [10] for garlic powder.

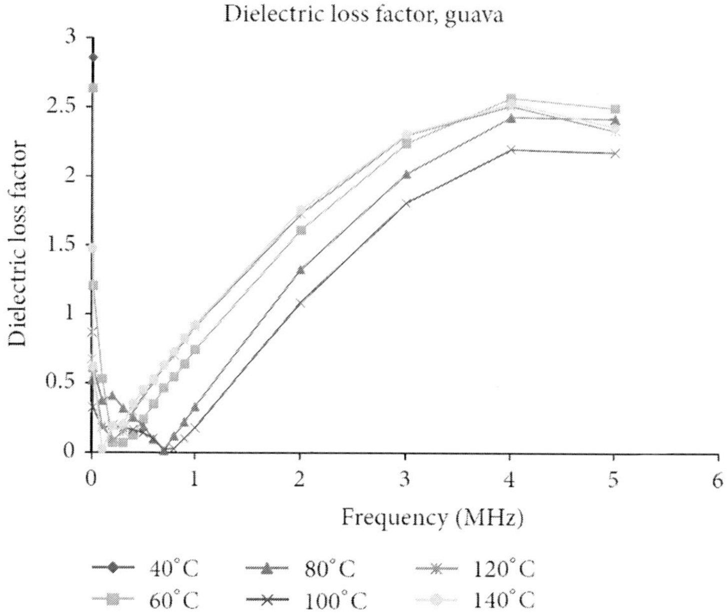
Dielectric loss factor, guava
Dielectric loss factor
Frequency (MHz)
40°C
80°C
120°C
60°C
100°C
140°C

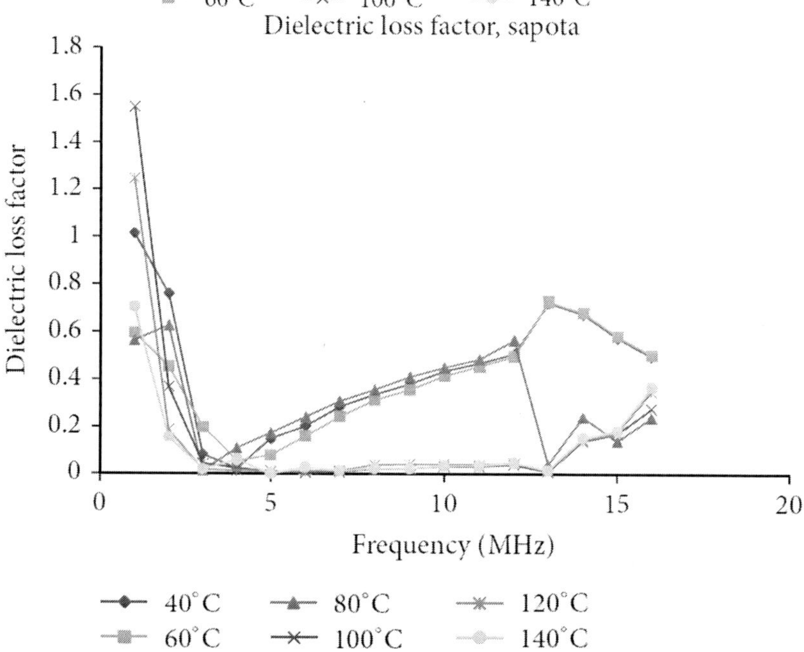
Dielectric loss factor, sapota
Dielectric loss factor
Frequency (MHz)
40°C
80°C
120°C
60°C
100°C
140°C

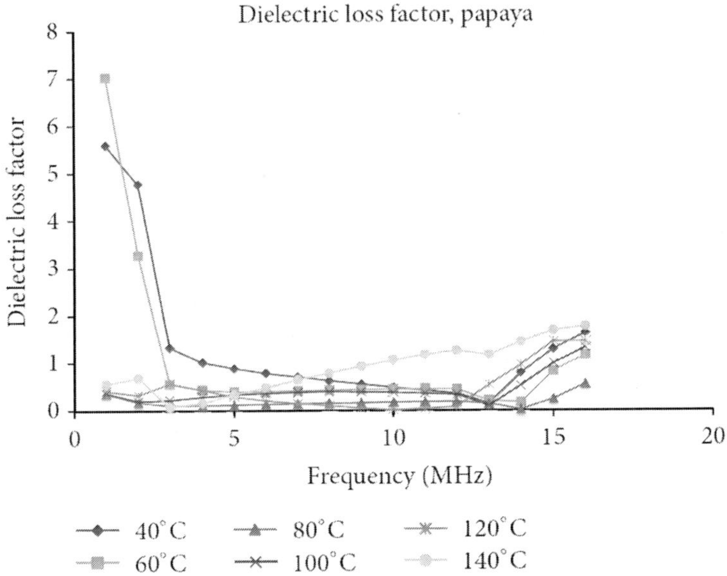

Figure 4: Dependence of dielectric loss factor on frequency for guava, sapota, and papaya.

3.4. FTIR Analysis of Freeze Dried Tropical Fruits Powder

The FTIR analysis of guava, sapota, and papaya powder is shown in Figure 5.

Peaks in the range 650–510 cm^{-1} are assigned to C–Br stretch—alkylhalides; sharp peak is absorbed at 546 and 586 cm^{-1} in sapota, 536, 579, and 637 cm^{-1} in guava, and 584 cm^{-1} in papaya; the peak may also be due to C–I stretch or C–Cl stretch. Peak at 772 cm^{-1} in guava may be assigned to ethyl branch–ethylene butane. Peak at 816 cm^{-1} in all the three fruits may be assigned to N–H to broad loop bending or CH_3 rocking. Peak at 1061 cm^{-1} in guava, 1063 cm^{-1} in sapota, and 1070 cm^{-1} is assigned to C–N stretch, aliphatic amines or aryl bromides, or aryl chloride or C–O stretch. Peak at 1256 cm^{-1} that is noticed in all three fruits may be assigned to P=O stretch or aryl fluorides. C=C stretch appears in the range of 1750–1735 cm^{-1}; there appears a peak at 1746 cm^{-1} for sapota and 1745 cm^{-1} for guava. Peak around 2900 cm^{-1} which is noticed in all three fruits may be due to asymmetric aliphatic C–H stretching vibrations of methylene ($-CH_2$).

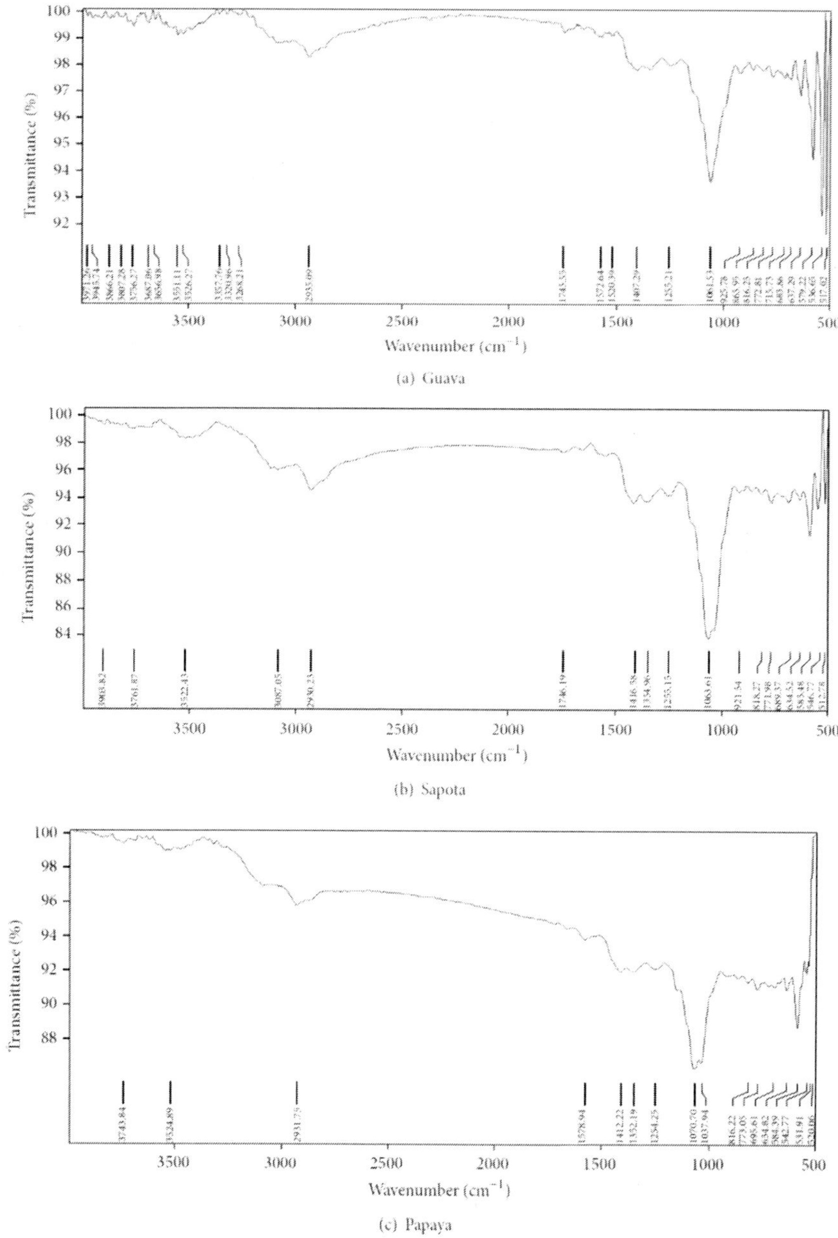

Figure 5: FTIR spectra of guava (a), sapota (b), and papaya (c).

Peak in the range of 3500–3000 cm^{-1} is assigned to N–H stretch; peak in this region is noticed in all the three fruits. The bands in the region 900–1153 cm^{-1} are assigned to C–O and C–C stretching modes, while those in the 1400–1199 cm^{-1}region are due to O–C–H, C–C–H, and C–O–H bending vibrational modes of the carbohydrates; similar results are also reported by Leopold et al. [17] in fruit juices. Fructose has a specific maximum peak at 1063 cm^{-1}, which is noticed in all three fruits. Peaks in the region from 3700 to 3000 cm^{-1} correspond to water and OH absorption frequencies [32]. The IR spectra of dehydrated fruit powders revealed the presence of carboxylic acids. In case of papaya there was broad absorption at 3400 cm^{-1} which is assigned to –OH stretching vibrations. In guava a peak at 3369 cm^{-1} was observed which is also assigned to –OH stretching. In both papaya and guava the peaks at 2926 cm^{-1} revealed the possible presence of –CH bonds. The peaks that appear near 1630 cm^{-1} in papaya powder represent C=O stretch and its conjugation with C=C. The peak marked at 1620 cm^{-1} in guava powder can be assigned to N–H bending for primary and secondary amides. The broad absorption of C=O at 1680–1630 cm^{-1} partially overlaps with N–H absorption band at 1640–1620 cm^{-1}. The peak at 1054 cm^{-1} in both papaya and guava powder is because of C–O absorptions indicating the presence of esters. This band is characteristic of carbohydrates, which shows high absorbance between 1250 and 950 cm^{-1}wavenumber values. The carboxylic acids present may be polygalacturonic acid or related acids [1].

3.5. X-Ray Diffraction Analysis

Crystallization is of great importance for the stability of powdered juice and can be determined by means of X-ray diffraction analysis. The presence of diffuse and large peaks in X-ray diffraction containing amorphous material is due to the fact that, in the amorphous state, the molecules are disorderly displayed producing disperse bands whereas crystalline materials yield sharp and defined peaks since they are presented in a highly ordered state. The X-ray diffraction patterns of papaya, guava, and sapota are shown in Figure 6, respectively. The diffraction peaks observed in the graph are broad, which indicate that the crystallite size is small which was similar to the results obtained by Theivasanthi and Alagar [14]. The particle size of each fruit powder has been determined using Debye-Scherrer formula:

$$\tau = \frac{K\lambda}{\beta \cos\theta} \tag{1}$$

is the shape factor, is the X-ray wavelength, is the line broadening at half the maximum intensity in radians, and is the Bragg angle; is the mean size of the ordered (crystalline) domains, which may be smaller than or equal to the grain size. The average particle sizes of papaya, guava, and sapota were found to be 0.411 μm, 0.339 μm, and 0.309 μm, respectively, and were represented in Table 1. Starch is a semicrystalline material. Actually the crystalline order of starch granules is lost during gelatinization. The semicrystalline structure can be identified at the light microscope level and through characteristic X-ray diffraction patterns. There are four major types of X-ray diffraction patterns of native starches, namely, A, B, C, and V [32]. The A form starch is mainly present in cereal starches, such as maize starch and wheat starch. The XRD patterns of these starches give the stronger diffraction peaks at around 15, 17, 18, and 23°. The B form starch is usually available in tuber starch such as potato and this type of starch gives the strongest diffraction peak at 17° 2. There were also few small peaks at around 2 values of 20, 22, and 24°. The C pattern starch is a mixture of both A and B types, such as smooth seeded pea starch and various bean starches [33]. In papaya, the peaks were found at 18° and 28° at 2 angle indicating the presence of A-type starch. A peak was also recorded at 20° showing the possibility of C-type starch. In case of guava, a sharp peak at 18° was observed at 2 angle. This shows the presence of type A starch. Other peaks were also noted at 19° and 25°. Similarly, in the case of sapota, a sharp peak was observed at 15° which again stresses the presence of A-type starch. The presence of A-type starch indicates that the semicrystalline nature of starch present in the fruit is not destructed after dehydration process.

Table 1: Particle size of tropical fruit powder.

Pos. [°2 Th.]	Height [cts]	FWHM left [°2 Th.]	*d*-spacing [Å]	Particle size (μm)
		Particle size of papaya fruit powder		
18.71	0.07	0.84	4.74	0.17
20.09	38.29	0.30	4.41	0.46
28.61	28.44	0.24	3.11	0.59
		Particle size of guava fruit powder		
18.60	8.99	0.39	4.76	0.37
19.68	8.38	0.49	4.50	0.28
25.23	13.90	0.39	3.52	0.36
		Particle size of sapota fruit powder		
15.61	35.17	0.72	5.67	0.19
50.17	14.61	0.36	1.81	0.42

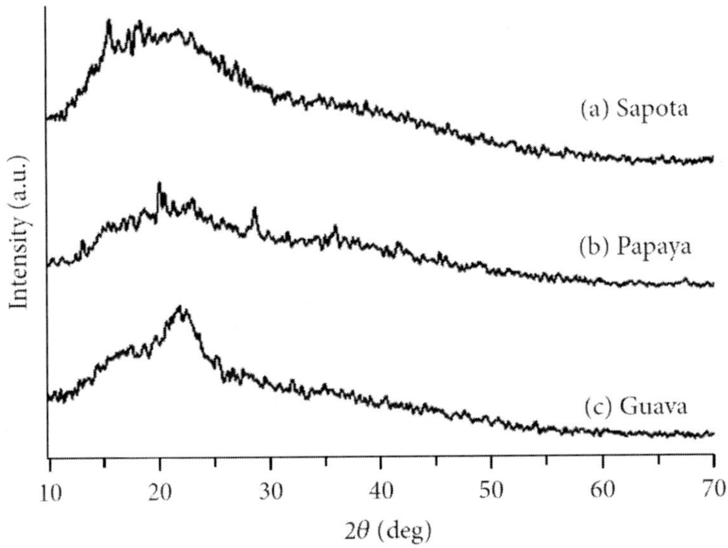

Figure 6: XRD of sapota (a), papaya (b), and guava (c).

3.6. SEM Image Analysis

The freeze dried powder of guava and sapota is free flowing, whereas the papaya powder is sticky, which may be due to the presence of pectin in papaya. SEM technique was considered to be a powerful tool for determining and observing the caking phenomenon on the powder surface. SEM micrographs for papaya, guava, and sapota powder were shown in Figures 7(a), 7(b), and 7(c). In morphological terms, two phases were observed in both guava and sapota solids: a lamellar surface structure (with flake form) and a spherical arrangement. These two morphologies indicate a possible separation of phases in the solids as a result of the dehydration processes. The interactions between the structural components on the surface of these materials can be understood as a surface tension increase, which lead to the separation and formation of spherical particles from the largely containing-pectin layered surface. However, papaya powder formed sponge platelets upon freeze-drying, which is also shown in the SEM image of papaya.

Guava
3/28/2012 5:25:14 PM
det LFD
HV 5.00 kV
mag 280 x
WD 11.8 mm
mode SE
300 μm
Quanta FEG

(a)

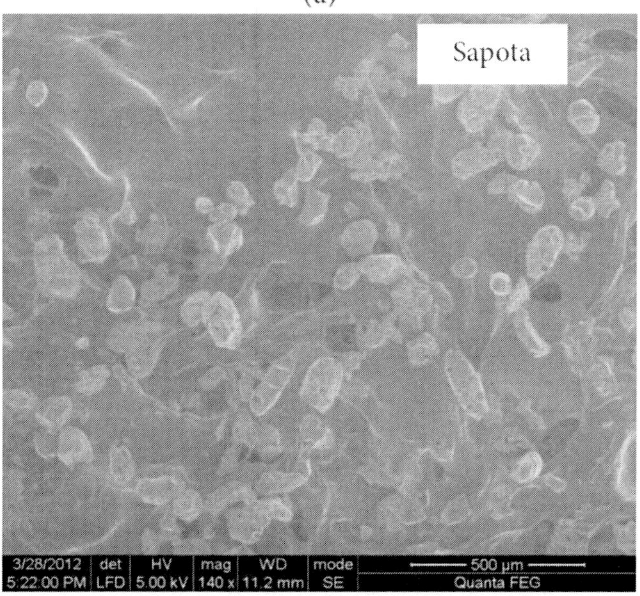
Sapota
3/28/2012 5:22:00 PM
det LFD
HV 5.00 kV
mag 140 x
WD 11.2 mm
mode SE
500 μm
Quanta FEG

(b)

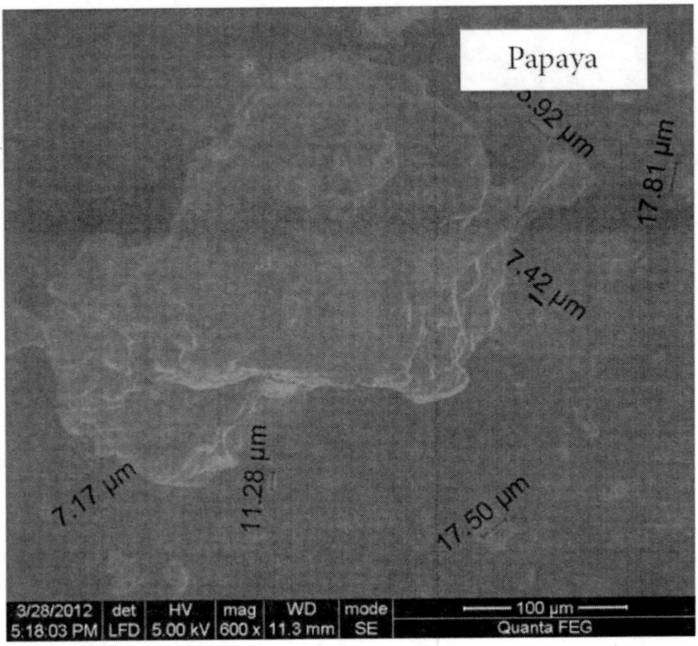

(c)

Figure 7: SEM images of guava (a), sapota (b), and papaya (c).

4. CONCLUSION

The TGA-DSC showed the thermal stability of guava powder, when compared to sapota and papaya powder. All the three fruits showed very low electrical conductivity at frequency between 0.001 MHz and 1 MHz. However, after 1 MHz, the electrical conductivities increased sharply. Effect of temperature on dielectric constant values was the same for guava and papaya where the dielectric constants increased with increasing temperature up to 80°C and then decreased with the increasing temperature. However, for sapota, the constants decreased steadily with increasing temperature. FTIR showed presence of chlorides and O–H and N–H bonds in guava, chloride and CH bond in papaya, and chlorides and C=O and C–H bonds in sapota. The presence of A-type starch in all the three fruits indicates that the semicrystalline nature of starch present in the fruit is not destructed after dehydration process.

REFERENCES

1. C. Osorio, J. G. Carriazo, and H. Barbosa, "Thermal and structural study of guava (Psidium Guajava L) powders obtained by two dehydration methods," Química Nova, vol. 34, no. 4, pp. 636–640, 2011
2. R. S. Chaughule, P. C. Mali, R. S. Patil, G. D. Joshi, and Y. M. Lo, "Magnetic resonance spectroscopy study of sapota fruits at various growth stages," Innovative Food Science and Emerging Technologies, vol. 3, no. 2, pp. 185–190, 2002
3. R. M. Maggio, L. Cerretani, C. Barnaba, and E. Chiavaro, "Application of differential scanning calorimetry-chemometric coupled procedure to the evaluation of thermo-oxidation on extra virgin olive oil," Food Biophysics, vol. 7, no. 2, pp. 114–123, 2012
4. C. P. Tan and Y. B. Che Man, "Recent developments in differential scanning calorimetry for assessing oxidative deterioration of vegetable oils," Trends in Food Science and Technology, vol. 13, no. 9-10, pp. 312–318, 2002
5. B. Kowalski, "Thermal-oxidative decomposition of edible oils and fats. DSC studies," Thermochimica Acta, vol. 184, no. 1, pp. 49–57, 1991
6. M. de Rosa, J. M. Kenny, D. Puglia, C. Santulli, and F. Sarasini, "Morphological, thermal and mechanical characterization of okra (Abelmoschus esculentus) fibres as potential reinforcement in polymer composites," Composites Science and Technology, vol. 70, no. 1, pp. 116–122, 2010
7. O. Sipahioglu and S. A. Barringer, "Dielectric properties of vegetables and fruits as a function of temperature, ash, and moisture content," Journal of Food Science, vol. 68, no. 1, pp. 234–239, 2003
8. O. Sipahioglu, S. A. Barringer, I. Taub, and A. Prakash, "Modeling the dielectric properties of ham as a function of temperature and composition," Journal of Food Science, vol. 68, no. 3, pp. 904–909, 2003
9. M. Soltani, R. Alimardani, and M. Omid, "Evaluating banana ripening status from measuring dielectric properties," Journal of Food Engineering, vol. 105, no. 4, pp. 625–631, 2011
10. G. P. Sharma and S. Prasad, "Dielectric properties of garlic (Allium sativum L.) at 2450 MHz as function of temperature and moisture content," Journal of Food Engineering, vol. 52, no. 4, pp. 343–348, 2002
11. X. Li, A. S. Zyuzin, and A. V. Mamishev, "Measuring moisture content in cookies using dielectric spectroscopy," in Proceedings of the IEEE Conference on Electrical Insulation and Dieletric Phenomena, pp. 459–462, Albuquerque, NM, USA, October 2003
12. S. O. Nelson and A. K. Datta, "Dielectric properties of food materials and electric field interactions," in Handbook of Microwave Technology for Food Application, pp. 69–114, Marcel Dekker, New York, NY, USA, 2001
13. Y. Mizukami, Y. Sawai, and Y. Yamaguchi, "Moisture content measurement of tea leaves by electrical impedance and capacitance," Biosystems Engineering, vol. 93, no. 3, pp. 293–299, 2006
14. T. Theivasanthi and M. Alagar, "An insight analysis of nano sized powder of jackfruit seed," Nano Biomedicine and Engineering, vol. 3, no. 3, pp. 163–168, 2011
15. M. Deraman, S. Zakaria, M. Husin et al., "X-ray diffraction studies on fiber of oil palm empty fruit bunch and rubberwood for medium-density fiberboard," Journal of Materials Science Letters, vol. 18, no. 3, pp. 249–253, 1999
16. K. Rengsutthi and S. Charoenrein, "Physico-chemical properties of jackfruit seed starch (Artocarpus heterophyllus) and its application as a thickener and stabilizer in chilli sauce," LWT—Food Science and Technology, vol. 44, no. 5, pp. 1309–1313, 2011
17. L. F. Leopold, N. Leopold, H.-A. Diehl, and C. Socaciu, "Quantification of carbohydrates in fruit juices using FTIR spectroscopy and multivariate analysis," Spectroscopy, vol. 26, no. 2, pp. 93–104, 2011

18. S. Jiao, J. A. Johnson, J. Tang, G. Tiwari, and S. Wang, "Dielectric properties of cowpea weevil, black-eyed peas and mung beans with respect to the development of radio frequency heat treatments," Biosystems Engineering, vol. 108, no. 3, pp. 280–291, 2011
19. M. S. Venkatesh and G. S. V. Raghavan, "An overview of microwave processing and dielectric properties of agri-food materials," Biosystems Engineering, vol. 88, no. 1, pp. 1–18, 2004
20. S. N. Jha and S. Gunasekaran, "Authentication of sweetness of mango juice using Fourier transform infrared-attenuated total reflection spectroscopy," Journal of Food Engineering, vol. 101, no. 3, pp. 337–342, 2010
21. L. M. Pereira, S. M. Carmello-Guerreiro, and M. D. Hubinger, "Microscopic features, mechanical and thermal properties of osmotically dehydrated guavas," LWT-Food Science and Technology, vol. 42, no. 1, pp. 378–384, 2009
22. U. Einhorn-Stoll and H. Kunzek, "Thermoanalytical characterisation of processing-dependent structural changes and state transitions of citrus pectin," Food Hydrocolloids, vol. 23, no. 1, pp. 40–52, 2009
23. M. Iijima, K. Nakamura, T. Hatakeyama, and H. Hatakeyama, "Phase transition of pectin with sorbed water," Carbohydrate Polymers, vol. 41, no. 1, pp. 101–106, 2000
24. H. Feng, J. Tang, and R. P. Cavalieri, "Dielectric properties of dehydrated apples as affected by moisture and temperature," Transactions of the American Society of Agricultural Engineers, vol. 45, no. 1, pp. 129–135, 2002
25. R. E. Mugdett and S. Goldblith, "Dielectric behavior of a semi-solid food at low, intermediate and high moisture contents," Journal of Microwave Power, vol. 15, pp. 27–36, 1980
26. T. Funebo and T. Ohlsson, "Dielectric properties of fruits and vegetables as a function of temperature and moisture content," Journal of Microwave Power and Electromagnetic Energy, vol. 34, no. 1, pp. 42–54, 1999
27. S. Ryynänen, "The electromagnetic properties of food materials: a review of the basic principles," Journal of Food Engineering, vol. 26, no. 4, pp. 409–429, 1995
28. H. Lizhi, K. Toyoda, and I. Ihara, "Dielectric properties of edible oils and fatty acids as a function of frequency, temperature, moisture and composition," Journal of Food Engineering, vol. 88, no. 2, pp. 151–158, 2008
29. K. C. Lawrence, S. O. Nelson, and P. G. Bartley Jr., "Measuring dielectric properties of hard red winter wheat from 1 to 350 MHz with a flow-through coaxial sample holder," Transactions of the American Society of Agricultural Engineers, vol. 41, no. 1, pp. 143–150, 1998
30. W. Guo, G. Tiwari, J. Tang, and S. Wang, "Frequency, moisture and temperature-dependent dielectric properties of chickpea flour," Biosystems Engineering, vol. 101, no. 2, pp. 217–224, 2008
31. S. O. Nelson and P. G. Bartley Jr., "Measuring frequency- and temperature-dependent dielectric properties of food materials," Transactions of the American Society of Agricultural Engineers, vol. 43, no. 6, pp. 1733–1736, 2000
32. C. Zhang, X. Liu, H. Qiang et al., "Inhibitory effects of rosa roxburghii tratt juice on in vitro oxidative modification of low density lipoprotein and on the macrophage growth and cellular cholesteryl ester accumulation induced by oxidized low density lipoprotein," Clinica Chimica Acta, vol. 313, no. 1-2, pp. 37–43, 2001
33. C. Adina, F. Florinela, T. Abdelmoumen, and S. Carmen, "Application of FTIR spectroscopy for a rapid determination of some hydrolytic enzymes activity on sea buckthorn substrate," Romanian Biotechnological Letters, vol. 15, no. 6, pp. 5738–5744, 2010

CHAPTER 21

Natural Pectin Polysaccharides as Edible Coatings

Arantzazu Valdés, Nuria Burgos, Alfonso Jiménez and María Carmen Garrigós *

Department of Analytical Chemistry, Nutrition & Food Sciences, University of Alicante, Campus San Vicente, 03690 San Vicente del Raspeig (Alicante), Spain

ABSTRACT

The most fashionable trends in food packaging research are targeted towards improvements in food quality and safety by increasing the use of environmentally-friendly materials, ideally those able to be obtained from bio-based resources and presenting biodegradable characteristics. Edible films represent a key area of development in new multifunctional materials by their character and properties to effectively protect food with no waste production. The use of edible films should be considered as a clean and elegant solution to problems related with waste disposal in packaging materials. In particular, pectin has been reported as one of the main raw materials to obtain edible films by its natural abundance, low cost and renewable character. The latest innovations in food packaging by the use of pectin-based edible films are reviewed in this paper, with special focus on the use of pectin as base material for edible coatings. The structure, properties related to the intended use in food packaging and main applications of pectins are herein reported.

Keywords: pectin; edible films; biopolymer coatings; fruits; vegetables; agricultural wastes; revalorisation

1. INTRODUCTION

Food packaging systems have been traditionally considered as simple containers to transport food from the place where they have been produced to the retail outlet and then to the consumer with no alteration of the food nutritional and organoleptic characteristics. Nevertheless, these systems are often unable to increase the shelf-life of fresh food resulting in a problem to producers, retailers and consumers. Since the main function of packaging is preservation of food from external contamination, other important features such as retardation of deterioration, extension of shelf-life, protection from transport impacts and maintenance of the food quality should be taken into account. Packaging materials should protect food from environmental influences such as heat, moisture, oxygen, enzymes, loss of aromas and unpleasant odor components, as well as from the attack from micro and macro-organisms. Furthermore, the global market is becoming more demanding and is continuously in need of novel and stable products which, at the same time, could retain the natural properties of food. In summary, the demand for new packaging materials and food packaging functionalities is increasing.

Among materials currently used in food packaging, polymers have taken a major share because of their versatility and advantageous performance/cost ratio. However, the most important polymers used for food packaging are obtained from non-renewable resources and they are not biodegradable or compostable, representing a global environmental problem. These drawbacks in the use of common polymers in food packaging are becoming an important issue in the design of attractive systems for distribution, making consumers aware of the problems related to the waste disposal. Although the stability of food packaging materials during the food shelf-life is an advantage, it turns into a disadvantage when the package enters the post-use phase. In summary, the use of polymers obtained from non-renewable resources and non-biodegradable in food packaging applications represents an important environmental impact and waste generation issue. In fact, packaging waste accounted for 32.5 million tons or 17.7% of the total municipal solid waste (MSW) in 2013 in the USA [1] and 19.3 million tons or around 25% in 2014 in Europe [2]. Currently, landfilling is the dominant method of packaging waste disposal, followed by recycling, incineration and composting. Even though recovery methods such as reuse, recycling and/or composting are encouraged as a way of reducing packaging waste disposal, there is still much work to do to substantially reduce the quantity of plastics present in MSW [3].

Some alternative packaging materials obtained from renewable resources, such as poly(lactic acid), PLA, poly(hydroxyalkanoates) (PHAs), starch or proteins, have been proposed as alternatives to non-biodegradable polymers in food packaging applications [4,5,6,7]. However, polysaccharides are gaining some space in their use as innovative packaging materials by their ubiquitous presence in Nature, as well as by their relative low cost compared to other biopolymers and the possibilities they offer to be used not only as polymer matrices but also in coatings. Polysaccharides are therefore strong environmentally-sound contenders in the food packaging market that fulfill all the environmental concerns (*i.e.*, derived from renewable raw materials and biodegradable) while being even possible to be metabolized by human body together with food, making them able to be used in edible films.

Polysaccharides are the main component of biomass, being the most abundant renewable polymer resources available from Nature. Among them, pectin is one of the most significant since the pectin world market demand is increasing, reaching a total production capacity around 45–50 Mton per year while the demand was around 140–160 Mton per year in 2011, showing that the industry interest on this complex polysaccharide is rising day by day.

One of the most important features in the still incipient commercial application in food packaging of bio-based polymers is the use of some of these materials (e.g., PLA, PHAs) for rigid containers [8], but flexible morphologies are still dependent on the use of additives. In addition, commercial biopolymers show some limitations in terms of performance (thermal resistance, poor barrier and brittleness) as well as high costs. In fact, the present situation is such that the development of polysaccharides-based materials in fields where unique performance or raw material characteristics show an added technological value (for instance, sourced from renewable materials, biodegradability, water barrier, antimicrobial properties, aroma barrier) or marketing-wise (e.g., green or sustainability image) is both a challenge and an opportunity for the food packaging industries.

The term edible coatings in food applications correspond to thin layers of edible materials applied onto surfaces of highly perishable foodstuff, such as fresh-cut fruits and vegetables. The aim of this paper is the full review of the properties and possible applications of pectin in the manufacture of edible films for food packaging. The most recent scientific and technological developments in this field will be highlighted and our goal would be to permit the reader getting a complete survey of the increasing use of this biopolymer in food industries.

2. STRUCTURE AND CLASSIFICATION OF PECTIC SUBSTANCES

Pectin is a white, amorphous and colloidal carbohydrate of high molecular weight occurring in ripe fruits, especially in apples, currants, *etc.*, and used in fruit jellies, pharmaceuticals and cosmetics for its thickening and emulsifying properties and ability to solidify to a gel. All these properties and applications have put pectin in the market of the biopolymers with great potential and possibilities for future developments. The structure of these polysaccharides is discussed in this section.

Pectic substances are present in the primary cell walls and middle lamellae of many plants and fruits, and they are frequently associated with cellulose, hemicellulose and lignin structures [9]. Their presence in the cell is important for some essential functions: (a) adhesion between cells; (b) mechanical strength of the cell wall; (c) ability to form stabilizing gels; and (d) they play a significant role in the growth of plant cells [10]. Pectin forms the most complex class of polysaccharides, mainly composed by high molecular weight heterogeneous groups of glycanogalacturonans and acidic structural polysaccharides with diverse structures. Pectin backbone consists of (1→4)-α-D-galacturonic acid molecules linked to a small number of rhamnose residues in the main chain and arabinose, galactose and xylose in the side chains [11]. Several authors stated that pectin polysaccharides can be classified in three types (Figure 1) [12,13,14].

Homogalacturonan (HG) is a linear polymer formed by D-galacturonic acid and it can be classified into three different families depending on the acetylation or methylation reactions suffered during polymerization: (i) pectin with more than 75% of methylated carboxyl groups; (ii) pectinic acid with less than 75% of methylation; and, finally, (iii) pectic acid or polygalacturonic acid without methyl-esterified carboxyl groups [12]. Is the bold/italics necessary?

Rhamnogalacturonan I (RGI) is composed by the repeating disaccharide rhamnose–galacturonic acid groups acetylated and linked to side chains of neutral sugars, such as galactose, arabinose and xylose. Finally, rhamnogalacturonan II (RGII) is also formed by homogalacturonan chains, but with complex side groups attached to 12 different types of glycosyl residues, such as rare sugar species (2-O-methyl xylose, 2-O-methyl fucose, aceric acid, 2-keto-3-deoxy-D-lyxo heptulosaric acid, and 2-keto-3-deoxy-D-manno octulosonic acid). RGI and RGII are called hairy regions in the pectin structure whereas HG is the smooth part of the molecule [13,14].

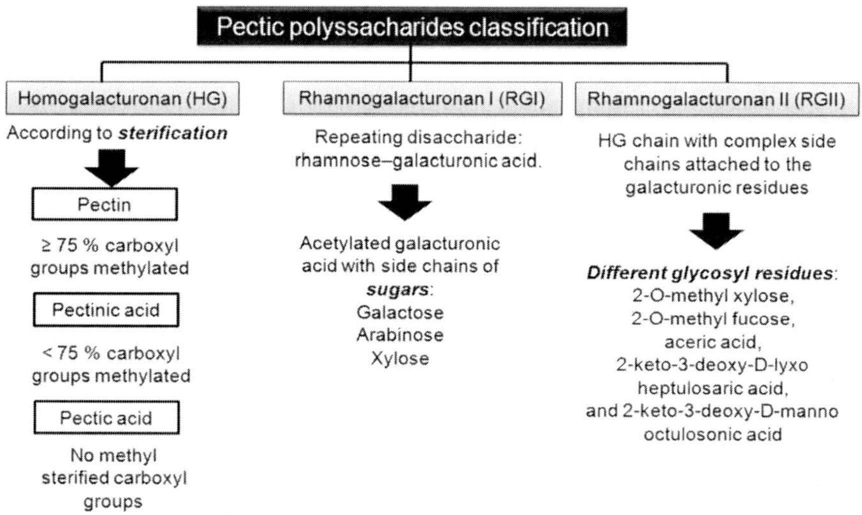

Figure 1. Classification of pectic polyssacharides based on D-galacturonic acid [12,13,14].

3. PROPERTIES AND CURRENT APPLICATIONS IN FOOD INDUSTRY

In general terms, pectins are used in food industry as stabilizers, thickening and gelling agents, crystallization inhibitors and encapsulating agents. But, in all cases, galacturonic residues should be previously modified by the addition of methyl groups to yield methoxides. This chemical modification is essential to improve pectins physical properties, which are dependent on their molecular masses and, primarily, on the degree of methyl-esterification (DM), calculated as moles of methanol per 100 moles of galacturonic acid [15]. In this context, two different DM degrees can be achieved resulting in different applications [10]:

- Pectins with high DM (HM pectins) containing 50% or higher of galacturonic residues. They are used as gelling component in heat-resistant bakery jams, fruit preservatives and juices, confectionaries jellies, milky products, glazing cakes and soft drinks. HM pectins form gels by hydrophobic interactions under acidic conditions in aqueous media and high sugar content.
- Pectins with low DM (LM pectins) are obtained by de-esterification of HM pectins under controlled pH, temperature and time. LM

pectins show DM degrees between 20% and 30% [16]. They are used to prepare gels at low pH values in the presence of divalent calcium. Applications in food industry include jams and jellies with low-sugar content, dairy desserts, ice cream with fruit gels, thickening agents of syrups for fruit and vegetable canning and food coatings.

In this context, edible coatings obtained from pectin and derivatives (pectate and amidated pectin) have been recently proposed in food-related applications by their excellent barrier to oxygen, aroma preservation, barrier to oil and good mechanical properties, but they are not effective against moisture transfer through films by their hydrophilic nature [17]. They are currently used in fresh and minimally processed fruits and vegetables, such as avocado, apple, apricot, chestnuts, berries, guava, melon, papaya, peach, walnuts, carrot and tomato [17]. Depending on the product, three different coating methods can be used, as shown in Figure 2 [18].

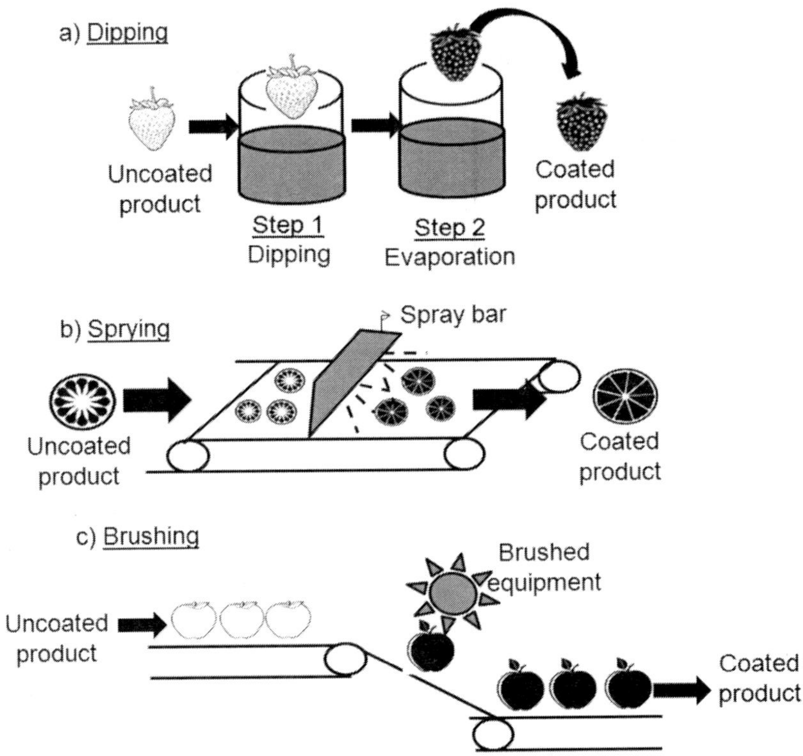

Figure 2. Main coating methods used in food industry. (Adapted from ref. [18]).

According to Dhanapal *et al* [18], dipping is the most common method to apply coatings to fruits and vegetables when the coating solution is highly viscous (Figure 2a). It is carried out by introducing the product for a time between 5 and 30 s in a coating solution under controlled conditions of density and surface tension.

However, when the coating solution is not highly viscous, spraying can be used (Figure 2b). The food product is introduced into the coating system and it is sprayed by controlling the final drop size of the spray solution, which depends on the thickness of the spray gun, nozzle temperature, air and liquid flow rates, humidity of incoming air and polymer solution, drying time and temperature [18]. The usual spraying instrument in this application can produce fine sprays, with drop-size distributions up to 20 μm. Finally, the brushing method is used in some products, such as fresh beans and strawberries, when the reduction of the moisture loss is an issue (Figure 2c). Thin coatings onto the surface of the product are obtained in all cases and they could act as semi-permeable membranes reducing gas transfer rates and creating new packaging materials to extend food shelf-life.

4. EXTRACTION METHODS FOR PECTINS: NEW TRENDS FOR THE REVALORIZATION OF AGRICULTURAL RESIDUES

Pomace from apples, citric albedo, sugar beet pulp and sunflower rinds have been proposed as by-products from the agricultural and food industries to obtain pectic substances with contents over 15% on dry basis [19]. In fact, apple pomace (15% to 20% of pectin) and citrus peel (30% to 35% of pectin), obtained in fruit juice industries, are the two main raw materials used for the industrial extraction of pectins. However, other natural sources with potential in the pectins extraction have been described in literature [15]. For example, plants of *Lupinus* genus (pectins content 1.5%–7% depending on the variety, year, and climatic conditions), burdock from the *Arctium* genus (pectins content higher than 21%) or peach (*Persica vulgaris L.)* with pectins content up to 10% have been proposed.

The diverse nature of the pectic substances has resulted in different extraction methods. Depending on the raw material the extraction methods and further chemical purification treatments are different. This section focuses on natural pectin sources and the associated extraction methods used to produce high added-value products. The most frequent extraction techniques are: (i) conventional solvent extraction based on stirring and heating; (ii) microwave assisted extraction (MAE); (iii) electromagnetic induction (EMI); (iv) ultrasound-assisted extraction (UAE); and (v) enzymatic extraction.

In general terms, pectins are commonly extracted by solvent-liquid methods in acidic aqueous media (nitric, sulfuric, phosphoric or citric acid) at high temperatures due to their availability, applicability and ease of use [20]. Pectins are usually extracted by hot dilute mineral acids at pH 1.5–2.5, taking 2–4 h with further precipitation with ethanol or isopropyl alcohol, separation to remove impurities, drying, grinding and blending with other additives. The extracted pectins can be categorized into two major types, depending on their DM degree. Factors like pH, temperature, time and solvent:liquid (S:L) ratio are usually studied to optimize the extraction conditions. Table 1 shows the main extraction methods for natural pectins reported in literature.

Table 1. Natural pectins extracted by solvent extraction methods.

Natural Source	Extraction Parameters	Pectin Yield Extraction (%)	Ref.
Okra pods	Citric and phosphate buffer, pH 2 and 6, 60 min, 80 °C, 1:15 g·mL^{-1}	13.3 (pH 2.0); 15.7 (pH 6.0)	[21]
Durian rinds	HCl, pH 2.8, 43 min, 86 °C, S:L 1:10 g·mL^{-1}	9.1	[22]
Mango peel	Sulfuric acid in water, pH 1.5, 2.5 h, 90 °C	>70	[23]
Banana peel	Citric acid solution, pH 2.0, 160 min, 87 °C, 1:20 g·mL^{-1}	13.89	[24]
	Citric acid and HCl, pH 1.5, 4 h, 90 °C	16.54	[25]
Tropical fruit peels	Citric acid, pH of 2, 3.3 and 4.5, 120 min, 70 °C	12.56–14.24	[26]
Sugar beet pulp	Citric acid, pH 1, 166 min, 99 °C, 1:20 g·mL^{-1}	23.95	[27]
Bagasse and pomace lime fruit	Citric acid, 60 min, 90 °C, 1:20 g·mL^{-1}	13.31 (Bagasse); 15.1 (Pomace)	[28]
Valencia orange peel	Citric acid extraction, pH 1.5, 75 min, 90 °C	16.7	[29]
Watermelon seed	HCl, pH 2, 60 min, 85 °C, 1:15 g·mL^{-1}	19.75	[30]
Faba bean hulls	HCl, pH 1.5, 80 min, 85 °C, 1:25 g·mL^{-1}	15.75	[31]
Honeydew melon seeds and damaged skin	HCl at pH 1, 80 °C for 4 h	7.9	[32]
Tomato peel	Ammonium oxalate and oxalic acid, 90 °C in two extraction steps (24 and 12 h)	32.0	[33]
Sunflower head	Sodium citrate, 85 °C, 3.5 h, 1:40 g/mL	16.90	[34]

Alba, Laws and Kontogiorgos extracted okra pod pectins at different pH values (2.0 and 6.0 with 100 mM citric and phosphate buffer, respectively) and examined the effect of the extraction conditions on their molecular and compositional characteristics [21]. Pectins obtained could not be quantitatively recovered in a single extraction step and second extraction was required. They concluded that pectins extracted with citric buffer (pH 2.0) resulted in slightly lower yields (13.3% ±

0.3%) when compared to extraction at pH 6.0 (15.7% ± 0.2%). The highest extraction yields were obtained at high temperatures and low pH values due to the cleavage of pectin and other cell wall materials. Furthermore, okra pectins were found to be LM with DM 40.0% ± 1.6% and 24.6% ± 1.0% for pH 2.0 and 6.0, respectively. These results could be attributed to de-esterification process caused by the β-elimination of the esterified homogalacturonan backbone, removing methyl ester groups and resulting in pectins with lower DM and molecular mass. These pectins can be used as functional ingredients (thickeners, viscosity enhancers, gelling agents and texture modifiers) in the food industry.

Durian (*Durio zibethinus*) is considered as a high-value fruit due to its distinct flavor and unique taste, but its processing generates large amounts of waste and by-products rich in polysaccharides. A second-order polynomial equation using a three-level Box-Behnken response design to optimize the extraction conditions of pectins from Durian residues was recently proposed [22]. Those authors studied the influence of four variables: S:L ratio (5–15 g·mL^{-1}), pH (2–3), extraction time (20–60 min) and extraction temperature (75–95 °C) in the pectins extraction yield. The optimum selected conditions were pH 2.8, 43 min, 86 °C and 1:10 g·mL^{-1} S:L ratio, yielding 9.1% for the extraction efficiency.

Hubert *et al.* studied the impact of different Mango (*M. indica* L.) cultivars, technological procedures and ripeness degree on the pectin quality of their peels [23]. They concluded that grinding of mango peels (with particle sizes of 42 μm) significantly enhanced the pectins extraction yield (70%) without increasing the impurity contents.

Banana fruit has been considered an excellent source of pectins. A central composite design was used to determine the effect of pH (2.0–4.5), extraction temperature (70–90 °C) and time (120–240 min) in the extraction yield of pectins in citric acid solution (1:20 w/v) under stirring (150 rpm) [24]. High temperatures and pH values resulted in higher extraction yields with a decrease in DM from 79% to 43%. The optimum conditions for pectins extraction were 87 °C, 160 min and pH 2.0 with extraction yield 13.9%. Castillo-Israel *et al.* proved that banana peel waste could be a promising source of pectins with extraction yield 17.0% and DM 75.0% by using HCl (0.5 N, pH 1.5) and citric acid (0.5 N, pH 1.7) at 90 °C for 4 h [25]. Other agricultural residues used as pectins sources are peels of Yellow passion fruit (*Passiflora edulis* var. *flavicarpa*), red dragon fruit (*Hylocereus polyrhizus*) and soursop (*Annona muricata* L.) [26]. Peels of yellow passion fruit and dragon fruit gave pectins yield of 14.2% and 12.6% with DM 55.5% and 47.9% at pH = 2.4 and extraction times 58 and 65 min, respectively. It was concluded that the soursop peels are not suitable by their low pectins yield (lower than 6%). In contrast, the extraction of sugar beet pulp pectins in citric acid at 99 °C and 166 min at pH = 1.0 and S:L ratio 20 allowed the maximum extraction yield (23.9%) [27].

Bagasse and pomace of Mexican lime fruit have been proposed as pectins sources with S:L ratio 1:20, citric acid 1% at 90 °C for 60 min with yields of 13.3 and 15.1%, respectively [28]. Pectins yield extraction from Valencia orange peels was carried out by using citric acid solution at 90 °C for 75 min and pH 1.5 yielding acceptable pectins extraction (16.7%) [29]. Korish reported the use of the *Citrullus lanatus* var. Colocynthoides watermelon seeds as source of pectins [30]. The optimum yield was obtained with acidic water solution (HCl, pH 2.0, S:L ratio 1:15 and 85 °C), for 60 min yielding pectins extraction efficiency of 16.7%. This waste was proposed as raw material for the production of useful biomaterials and chemicals. These authors also studied the potential use of bean hulls as pectins source [31]. The maximum yield of the extracted pectins was not coincident with the highest esterification degree with the maximum yield (15.7%) recorded at pH 1.5 and 85 °C for 80 min and S:L ratio 1:25; while the highest degree of esterification (54.6%) occurred at pH 2.5 and 90 °C for 60 min. Therefore, it was concluded that the yield of pectins was variable with the extraction conditions.

Seeds and damaged skin of fresh honeydew melon (*Cucumis melo Inodorus*) were proposed as other pectins sources with acceptable efficiencies (7.9%) in HCl at pH = 1.0 [32]. Pectins from tomato (*Lycopersicum esculentum*) peels were also extracted under reflux at 90 °C with ammonium oxalate (16 $g{\cdot}L^{-1}$) and oxalic acid (4 $g{\cdot}L^{-1}$) in two extraction steps (24 and 12 h respectively) [33]. The highest pectins yields were 32.6% and 31.9% in two batches, but the pectin quality was low in the second extraction step. Kang *et al.* extracted pectins from sunflower heads (SFHP) with sodium citrate at 85 °C and S:L ratio 1:40 [34]. The content in uronic acid was used in this study as the indicator of the pectin extraction yield by using the carbazole colorimetric method with a maximum yield in uronic acid of 16.9%.

Giosafatto *et al.* recently reported the production of an edible hydrocolloid film made by using Citrus pectin and the protein phaseolin crosslinked by microbial transglutaminase, an enzyme able to covalently modify proteins by formation of isopeptide bonds between glutamine and lysine residues [35,36]. These authors included different biopolymer matrices in their formulations, including chitosan [37,38] and whey proteins [39].

In general terms, it can be concluded that the combination of high temperatures, times and solvents volume resulted in the recovery of pure pectins. However, two problems in the pectins extraction were described, since long heating times resulted in ruptures of the pectin chains leading to difficulties in pectin isolation, while acid extraction may result in corrosion and pollution whereas the alkaline extraction reduces the chain of galacturonic acid by beta-elimination [20]. Therefore, innovative methods such as MAE, EMI, UAE and enzymatic treatment have been proposed to counteract these effects. Table 2 summarizes the most recent trends on the application of these methods for pectins extraction.

Table 2. Innovative extraction methods for pectins.

Extraction Method	Natural Source	Extraction Parameters	Pectin Yield Extraction (%)	Ref.
MAE	Dragon fruit peel	400 W, 45 °C, 20 min, 24 g·mL^{-1}	7.5	[40]
		HCl, pH 2.07, 800 W, 65 s, 66.6 g·mL^{-1}	18.57	[41]
	Bagasse and pomace of Mexican lime fruit	Citric acid, 800 W, 120 °C, 5 min, 1:30 g·mL^{-1}	16.9 (Bagasse); 8.4 (Pomace)	[28]
	Pomelo peel	Tartaric acid solution, pH 1.5, 660 W, 9 min, 1:40 g·mL^{-1}	23.83	[42]
	Mango peel	Aqueous solution, pH 2.7, 413 W, 134 s, 1:18 g·mL^{-1}	28.86	[43]
	Papaya peel	Aqueous solution, 512 W, pH 1.8, 140 s, 1:15 g·mL^{-1}	25.41	[44]
EMI	Citrange albedos	Aqueous solution, pH 1.2 with H_2SO_4, 80 °C, 90 min	29	[45]
UAE	Pomegranate peel	Aqueous solution, pH 1.27, 61.9 °C, 28.31 min, 1:17.52 g·mL^{-1}	24.18	[46]
	Sisal fiber	Aqueous solution, 61 W, 50 °C, 26 min, 1:28 g·mL^{-1}	29.32	[47]
	Sugar beet pulp	Aqueous solution, 10.70 MPa, 120.72 °C, 30.49 min, 44.0 g·mL^{-1}	24.63	[48]
	Grapefruit peel	Deionized water, HCl, pH 1.5, 12.56 W/cm^2, 66.7 °C, 27.9 min, 1:50 g·mL^{-1}	27.34	[49]
Enzymes	Gold kiwifruit pulp, skin and seed	Purified water, pH 3.7, 25 °C, 30 min, Celluclast 1.5 L enzyme (1.05 mL·kg^{-1}), 1:3 g·mL^{-1}	2.14	[50]
	Gold kiwifruit pomace		4.5	[51]

MAE is an extraction technique that uses microwave power to heat samples in solution by enhancing the mass transfer rate of the solutes from the vegetable raw materials to solvents. Two main mechanisms allows the pectin extraction by MAE, *i.e.*, the rapid increase in temperature, which reduces the emulsion viscosity and breaks the chemical bonds in the raw material structure, and the molecular rotation due to the electrical charges of molecules, resulting in enhanced movement of ions which increase the efficiency of the extraction process [41]. MAE has been reported as the preferred extraction method of pectins from natural sources such as dragon fruit peels [40,41], bagasse and pomace obtained from Mexican lime fruit [28], pomelo [42], mango [43] and papaya peels [44] under different operation conditions. MAE conditions are dependent on different factors, such as microwave power, pH, time and S:L ratio. MAE methods show significant advantages over the conventional extraction techniques, such as the reduction in the amount of the extraction solvents, low energy consumption, high recoveries, good reproducibility, short extraction times (minutes rather than hours) and minimal sample manipulation.

Since these variables should be optimized for the development of every method, response surface methodologies (RSM) are used to study their interactive effects on pectins extraction with good correlation between experimental data and the predicted values. Table 2 summarizes the proposed MAE conditions to obtain maximum pectins extraction yields from natural sources. The most usual procedure includes the use of Box-Behnken experimental designs (BBD) or central composite designs (CCD). Good results were reported for the extraction of target compounds, such as antioxidants from almond skin [52], peanut skin [53] and batata leaves [54], highlighting the potential of MAE as a promising extraction method for pectins.

The use of EMI methods involving mixing, heating and maintenance of samples in an acidic medium to extract pectins has been also proposed. Zouambia *et al.* introduced a new approach for pectins extraction from citrange albedos [45]. In this study, the extraction yield, composition and physico-chemical characteristics of pectins extracted by electromagnetic induction heating were studied and compared with those obtained by direct boiling. EMI methods were also evaluated at different extraction times and power levels. A considerable reduction in time was observed by using EMI, since only 30 min were necessary to obtain pectins extraction yield (24%) similar to those obtained with conventional heating methods for 90 min. This innovative method can be suggested as an alternative for pectins extraction with short times, good yield and maintaining the main physico-chemical properties of pectins.

UAE is a process that uses acoustic energy and selected solvents to extract specific compounds from various matrices in relatively short times giving high yields. The mechanism of UAE is based on the interaction of the ultrasound waves and the solvent molecules allowing the acoustic cavitation and the disruption of cell walls with enhanced mass transfer across the cell membranes [20]. High efficiencies were recently reported for the UAE of pectins from several agricultural wastes, such as pomegranate peel [46] and sisal fiber [47]. In both cases, BBD was successfully used as valuable multivariate technique to evaluate the relationship between variables, helping to determine the accurate optimum values of all the experimental parameters with a reduced number of experiments [55]. In pomegranate peel, the effects of pH, S:L ratio, extraction time and temperature on the recovery of pectins were studied [46]. The maximum extraction yields were determined at pH = 1.27; 28.3 min 1:17.5 g·mL^{-1} S:L ratio and 61.9 °C resulting in yield extraction of 24.2%. In sisal fibers, the extraction conditions were also optimized [47]. The analysis of variance showed that the contribution of a quadratic model was significant for the pectins extraction yield. The optimal extraction condition were 61 W, 1:28 g·mL^{-1} S:L ratio, 26 min and 50 °C resulting in pectins yield of 29.3%. Therefore, UAE was proposed as an alternative method to extract pectins from sisal waste with the advantages of lower extraction temperatures and energies and shorter extraction times.

Several authors have combined UAE with other extraction techniques. Pectin-enriched materials obtained from sugar beet pulp wastes were extracted by UAE combined with subcritical water by using the RSM methodology [48]. The highest extraction yields (24.6%) were obtained at pressure 10.70 MPa, 120.7 °C, 30.5 min and S:L ratio 44.0 g·mL^{-1}.

The effect of the extraction temperature and sonication time is essential for UAE, but not much published work has focused on the optimization of UAE conditions and exploration of the combined effects of these variables in the pectins extraction. Wang *et al.* used RSM to study the extraction of grapefruit peel pectins by UAE, evaluating the combined effects of ultrasound power and extraction time and temperature on the extraction yield in HCl aqueous solutions (pH = 1.5) [49]. This study also compared UAE results with those obtained with conventional extraction methods using the same solvent at 80 °C for 90 min. UAE provided extraction yields of 16.3% at low temperatures and short times. In summary, UAE could be considered as an efficient and economic technique to extract pectins from plant resources.

All these methods have in common the use of diluted acid solutions, but in some cases non-solvent methods are preferred in food products. Enzymatic methods represent an innovative and clean alternative for the "green" extraction of pectins reducing water pollution and waste generation. These methods are based on the degradation of pectins by selective depolymerisation or their isolation by breaking the plants cell wall [20]. Only few works have reported the use of these methods by comparing the enzymatic extraction of pectins obtained from gold kiwifruit wastes at different stages of maturity with other extraction methods [50,51]. In both cases, the enzymatic treatment was conducted at 25 °C for 30 min with a Celluclast 1.5 L (1.05 mL·kg^{-1}) at pH 3.70 ± 0.05. These methods showed relative low yields (2.14% for peel, skin and seeds and 4.5% for pomace) when compared to acid/water extraction methods. Nevertheless, the advantages of enzymatic methods, *i.e.*, low temperatures that contribute to avoid degradation of pectins and low cost, make this method a promising alternative that deserves further study.

5. RECENT TRENDS IN THE USE OF PECTIN EDIBLE COATINGS

Edible coatings are currently used in highly perishable foodstuff to protect their nutritional and organoleptic properties by their potential to extend shelf-life and reducing the negative effects caused by processing, such as enzymatic browning, texture breakdown and off-flavors development [56,57]. In addition, edible coatings may contribute to the reduction of gas exchange, moisture and solute migration as well as to the reduction of the respiration and oxidation reaction rates and the decrease of the risk of pathogens growth on food surfaces during storage [58].

The selection of the most adequate edible materials for this purpose depends on different factors, such as cost, availability, mechanical properties, transparency, brightness, gas barrier effects and resistance to water and microorganisms. The selection is also influenced by the coating processing conditions (pH, type of solvent, temperature), type and concentration of additives (plasticizers, emulsifiers or cross-linking agents) and by the presence of active compounds in the edible matrix (antimicrobials, antioxidants, texture enhancers or nutraceuticals) [56,58].

Plasticizers, such as glycerol or sorbitol, in edible coating formulations play the role of improving coatings flexibility by reducing the number of internal hydrogen bonds between polymer chains and by increasing the free volume in the matrix to permit oxygen and water vapor diffusion through the coating film. Emulsifiers are usually added into coating formulations to improve wettability, stability and adhesion to the food surface, while cross-linking agents are necessary to react with carbohydrates forming a solid polymeric matrix upon the food. In particular, the gelation of low-methoxyl pectins takes place in the presence of multivalent metal cations, such as calcium, due to their electrostatic interactions with the opposite charged cavities formed by polymer chains [59,60,61].

The most recent trends in the field of pectins coating applications include the shelf-life extension of fresh-cut highly perishable food, the application of pectin coatings as pre-frying treatment to reduce the oil consumption in deep-fat fried products and the use as pre-dried treatments to improve the retention of nutrients and quality characteristics of dehydrated and lyophilized food.

5.1. Shelf-life Extension of Fresh-cut Fruits and Vegetables

Consumers usually accept fresh food on the basis of their appearance and organoleptic characteristics, but their shelf-life is usually low, representing a clear problem for their commercial distribution. The increasing interest for fresh-cut fruits consumption has resulted in the development of physical technologies to extend their shelf-life [62]. The traditional method of storage at low temperatures to slow down the fruits metabolism can be used in combination with other strategies, being modified atmosphere packaging (MAP), thermal treatments (blanching and heat-shock), irradiation, UV light and edible films and coatings the most common preservation methods [63].

Among all these methods, the direct application of edible coatings onto the surface of fresh-cut horticultural products is one of the most studied possibilities. In particular, pectin coatings have been widely evaluated for this purpose by their good oxygen and carbon dioxide barrier, their ability to retard lipid migration and moisture loss while maintaining the sensory and quality of food. Low-methoxyl pectins are mostly used as edible coatings by their ability to form firm gels at low pH in the presence of calcium cations, promoting higher firmness and structural integrity while reducing water vapor permeability [64,65]. Table 3 summarizes the most recent reports on applications of edible pectin-based coatings to extend shelf-life in fresh fruits and vegetables.

In this context, low methoxyl edible pectins coated onto fresh-cut melon improved water vapor resistance, preventing dehydration and maintaining the initial firmness during storage (15 days at 4 °C) by using calcium chloride as cross-linking agent and sunflower oil as lipid source in the preparation of the film-forming emulsion. This edible coating did not reduce the microbiological growth, but could help to decrease the wounding stress of fresh-cut melon, providing antioxidant properties and maintaining quality attributes [66].

Another study reported the use of pectin coatings to restrict the loss of moisture from avocados during storage, decreasing the O_2 absorption and slowing down the respiration rate [68]. The delay in changes on texture and color was observed, concluding that pectin coatings could extend the shelf-life of avocados to over a month at 10 °C. Similar results were reported by Moalemiyan *et al.* in fresh-cut mangoes coated with pectins, which remained in acceptable quality conditions for over two weeks by applying the optimized coating formulation: pectin (1.3% in water), sorbitol (28%) as plasticizer and beeswax (23%) based on pectins dry weight [67]. Authors reported that some optimization of the coating formulation was necessary to prevent the off-flavor development and the anaerobic respiration. A balance between gas and water vapor permeability should be achieved, but it is dependent on the respiration rate of each fruit or vegetable.

Table 3. Pectin-based coatings to extend shelf-life of fresh fruits and vegetables.

Coating	Active Agent	Food Matrix	Effect	Ref.
Low methoxyl pectin (2%), glycerol, sunflower oil, $CaCl_2$	--	Melons (*Cucumis melon L.*)	Antioxidant properties maintaining quality attributes (4 °C, 15 days)	[66]
Pectin (3%), sorbitol, beeswax	--	Mangoes (*cv. Ataulfo*)	Reduction of physiological changes, respiration rates and weight loss (15 days, room temperature)	[67]
Pectin (3%), sorbitol, beeswax	--	Avocados	Reduction of firmness, color, respiration rates and moisture loss (10 °C, 1 month)	[68]
Osmotic dehydration (with 0.5% calcium lactate) + pectin coating (1%)	--	Ripe Melons (*Cucumis melon cv. inodorus*)	Reduction of respiration rate maintaining sensory properties and quality parameters (5 °C, 80% RH, 14 days)	[69]
Pectin (2%) + glycerol combined with MAP	--	Nectarines (*Prunus persica L. cv. Babygold*)	Texture, color and hygienic quality (3 °C, 7 days)	[62]
Multi-layered alginate (1%)-β-CD-trans (2%)/pectin (2%) Calcium lactate	trans-cinnamaldehyde encapsulated in β-Cyclodextrins	Watermelon (*C. lanatus*)	Antimicrobial activity maintaining quality and sensory attributes (4 °C, 12 days)	[70]
Multi-layered chitosan (2%)-β-CD-trans (1%)/pectin (1%) $CaCl_2$	trans-cinnamaldehyde encapsulated in β-Cyclodextrins	Cantaloupe melon	Antimicrobial activity maintaining quality and sensory attributes (4 °C, 9 days)	[71]
Multi-layered chitosan (2%)-β-CD-trans (1%)/pectin (2%) $CaCl_2$	trans-cinnamaldehyde encapsulated in β-Cyclodextrins	Papaya fruits (*Carica papaya L. cv Maradol*)	Antimicrobial activity maintaining quality and sensory attributes (4 °C, 15 days)	[72]
Pectin (2%), glycerol, $CaCl_2$	*N*-acetylcysteine Glutathione	Pears (*Pyrus communis L.*)	Antibrowning, antimicrobial, antioxidant maintaining sensory attributes (4 °C, 14 days)	[73]
Pectin (3.5%), glycerol	Sodium benzoate; Potassium sorbate	Strawberries (*Fragaria ananassa cv. Albion*)	Antimicrobial activity maintaining physico-chemical and sensory attributes (4 °C, 15 days)	[74]
Pectin (2%), glycerol, $CaCl_2$	Apple fiber; Inulin; Ascorbic acid	"Golden delicious" apples	Nutritional value maintaining firmness, color and antioxidant activity (4 °C, 7 days)	[75]
Pectin (2%) $CaCl_2$	Eugenol Citral; Ascorbic acid	Strawberries	Antimicrobial, antioxidant, antibrowning activities maintaining sensory properties (0.5 °C, 7 days)	[76]
Pectin (1%) $CaCl_2$	Eugenol Citral; Ascorbic acid	Raspberries	Antimicrobial, antioxidant, antibrowning activities maintaining sensory properties (0.5 °C, 7 days)	[77]
Pectin (3%) + glycerol	Cinnamon leaf essential oil	Peach (*Prunus persica*)	Antimicrobial, antioxidant activity. Odor acceptability up to 10 days (5 °C)	[78]

A recent report highlighted that the combination of osmotic dehydration (in 40 °Bx sucrose solution containing 0.5% calcium lactate) and pectins coating (1%) was a good preservation alternative for fresh-cut melon by reducing the fruit respiration rate and maintaining the organoleptic properties and quality parameters for 14 days of storage at 5 °C and 80% RH [69].

MAP is another popular technique to extend the shelf-life of fresh-cut fruits by decreasing O_2 and increasing CO_2 levels in the headspace of closed packages, reducing the respiration rate during storage with the positive effect of delaying ageing and senescence [63]. However, anaerobic respiration should be avoided to limit the proliferation of hazardous microorganisms and, consequently, some increase in ethanol production and off-flavors can be observed. In this sense, the combination of low-temperature storage, MAP and edible coatings have shown to be a promising alternative to control respiration of fruits [62]. It was reported that nectarine sections can be stored at 3 °C for 7 days maintaining texture, color and hygienic quality by using pectin, chitosan and sodium caseinate coatings combined with MAP.

Another current trend in food packaging research is the development of active systems with materials containing active compounds in their composition with the possibilities of release at controlled rate to the food surface during storage, resulting in shelf-life extension [79]. It was stated that pectins edible matrices show excellent performance as carriers for active compounds, such as antioxidants, antimicrobials and texture enhancers. The incorporation of active compounds into edible films and coatings has the main goal to reduce, inhibit or stop the microbial growth in food surfaces. The effect on the permeability and mechanical properties of active agents in coating formulations need further studies [58,80]. Herbs and spices are the most common sources of active additives, in particular phenolics, flavonoids and terpenoids [79]. Essential oils derived from plants and their components are also interesting bioactive agents in active packaging formulations. However, their application to packaging of fresh fruits and vegetables is still under investigation by their high volatility, reactivity and strong aroma that could affect the organoleptic characteristics of food. For example, the incorporation of cinnamon leaf essential oil into pectin edible coatings resulted in an effective antimicrobial activity and the enhancing of the antioxidant status of fresh-cut peach, with good acceptation by consumers [78].

The main components of essential oils obtained from herbs and spices have been evaluated as active agents in food packaging applications [79]. Eugenol and citral are examples of active compounds obtained from plant-derived essential oils, since they have been reported as good antimicrobial agents in pectin coatings to extend fresh strawberries and raspberries shelf-life [76,77]. In this context, *N*-acetylcysteine (0.75% w/v) and glutathione (0.75% w/v) were incorporated into pectin coatings (in the calcium chloride solution) to help in the enzymatic browning control while also retarding the microbial growth in fresh-cut pears and contributing to maintain the high antioxidant potential of this fruit without affecting firmness for 14 days at 4 °C [73]. Sodium benzoate and potassium sorbate are other active compounds successfully incorporated in pectin solutions to form edible coatings [74]. Results showed that pectin active coatings improved the physico-chemical, microbiological and sensory quality (color, flavor, texture and acceptance) of strawberries from 6 (control) to 15 days at 4 °C and 90% RH.

A limited number of studies on the effect of the incorporation of nutraceutical compounds into edible coatings to extend food shelf-life while providing nutritional value have been reported. Moreira *et al.* introduced apple fiber and inulin into pectins solutions as edible coatings in fresh-cut "Golden delicious" apples evaluating the effect on the quality attributes [75]. Ascorbic acid was also added as antioxidant. Results demonstrated that these edible active formulations increased the nutritional value of fresh-cut apples while maintaining the firmness and color and helping to keep their antioxidant potential for seven days of storage at 4 °C.

Multilayer coatings have received recent interest to overcome the problem of the poor adhesion between the commonly hydrophobic coatings and the hydrophilic surface of fruits. In the layer-by-layer technique fruits and vegetables are dipped into different solutions with opposite charged polyelectrolytes with the aim to obtain strong chemical bonds between them [71]. In this context, pectins have been studied in combination with other edible matrices, such as alginate or chitosan. Multilayer coatings were successfully prepared with the layer-by-layer technique by using a solution of sodium alginate containing one antimicrobial agent (trans-cinnamaldehyde encapsulated in β-cyclodextrins (β-CD-trans)) in combination with pectins and calcium lactate solutions to obtain stable and homogeneous coatings [70]. Authors reported that multilayer coatings with 1% alginate, 2% β-CD-trans and 2% pectins maintained the quality and organoleptic attributes of fresh-cut watermelon while extending the shelf life at 4 °C from 7 (control) to 12 days. In addition, microbiological analysis demonstrated the effectiveness of the encapsulated agents in the multilayer coatings against microbial growth. The application of multilayer coatings formed by chitosan (2%), 2% β-CD-trans and pectins (1%) (optimum formulation) was reported with the extension of shelf-life of fresh-cut cantaloupe at 4 °C up to 9 days [71]. Similar multilayer coatings were also successful in prolonging the shelf-life of fresh-cut papaya [72]. In both cases, calcium chloride (2%) was used as cross-linking agent.

5.2. Pectin Coatings in Pre-frying Treatments

Fried foods usually contain a significant amount of saturated and unsaturated fats, in some cases up to 1/3 of the total product mass. This high fats content has been related to obesity and coronary heart diseases. During deep-frying, the internal moisture of food is evaporated while oil is confined to the external surface. These processes are influenced by factors such as oil quality, frying conditions (temperature and time), food composition and pre-frying treatments (blanching and/or drying). The increase in the consumer's interest for more healthy food products has led to the trend of reducing lipid uptake in fried-food products while maintaining acceptable organoleptic characteristics. Edible coatings prepared from food hydrocolloids, such as cellulose and derivatives, gums, alginate, whey protein, albumin or pectins have been studied for such purpose. The high water absorption capacity and retention of the original food firmness of these coatings decrease the moisture loss by evaporation, reducing lipid uptake [56].

In the case of pectin coatings, few investigations have been reported. One of them studied the influence of sunflower head pectin and calcium chloride on the oil absorption in fried-potato chips [16]. Authors concluded that coatings prepared with 1% (w/v) pectins solutions induced by $CaCl_2$ (< 0.1 mol·L^{-1}) effectively decreased the lipid uptake by about 30%. These authors also observed that the $CaCl_2$ concentration influenced significantly on the brittleness and crispiness of coated chips but the consumer's acceptation was still high.

In another study, the combination of carboxymethyl cellulose (CMC) and pectins (0.5 and 1% w/w respectively) was used as one layer in a double coating, formed by guar and xantham gums (1% w/w), or in a three-multilayer coating with CMC 1% (w/w) as the third layer [81]. Those authors recommended the CMC-pectin mixture for single-layer coatings in French-fries since this combination leads to the lowest lipid uptake. Further reductions in the oil absorption were observed in fried food coated with multilayer systems, but the high moisture content of the final products led to the decrease in the crunchiness and loss of acceptance by consumers.

Other authors observed that the application of pectin coatings (2% w/v) reduced significantly the acrylamide formation in banana chips (around 33%), while helping to slightly improve the organoleptic attributes and crunchiness of the final product [82]. In addition, a synergic effect of the combination of pectin edible coatings and blanching treatments at 90 or 100 °C for 1 min was observed, resulting in higher acrylamide reduction in fried-banana chips up to 91.9% and 90.8%, respectively. In summary, all these results indicate that the use of pectin coatings in pre-frying treatments is a promising process to reduce the lipid uptake and acrylamide content in potato and banana chips, respectively.

5.3. Pectin Coating as Pre-drying Treatment

Drying has been traditionally applied as a method of preservation in fruits and vegetables by reduction of their moisture content, decreasing the action of enzymes and preventing the microbial growth. However, this method implies important physical changes in color and texture, as well as chemical modifications related to the degradation of bioactive compounds sensitive to oxygen, light and heating, such as vitamins, antioxidants or minerals. Edible coatings have been proposed as a possibility to prevent the nutrients loss while improving the organoleptic quality of dried products after different pre-treatments, such as blanching and osmotic dehydration [83].

Some authors have recently reported that pectin coatings on papaya slices increased the moisture diffusion coefficient due to the hydrophilic character of pectins while enhancing the retention of vitamin C, resulting in a significant protective effect against nutritional losses [84]. Pectin coatings were also applied to osmotic dehydrated pineapple in sucrose (50%)/calcium lactate (4%)/ascorbic acid (2%) solutions before the application of the hot-air-drying method [85]. Results showed that pectin edible coatings were effective as oxygen barrier and did not affect the drying efficiency and the water diffusion coefficient. However, dried pineapples coated with pectins showed high retention of vitamin C and slightly changed their lightness. Pectin-based coatings (2% w/w) with and without vitamin C (1% w/w) were also applied to papaya slices before air-drying without reducing the efficiency of the dehydration process [86]. Authors observed that the incorporation of vitamin C on pectin coatings reduced its retention in dried samples after 30 days of storage, while increased significantly the vitamin C content in the final product. In addition, sensorial analysis confirmed that the pectin + vitamin C coating can be used as pre-drying treatment of papaya slices.

A combination of pectins (1.5%), CMC (1.5%) and ascorbic acid (0.6%) was used as edible coating of quince slices prior to their osmotic dehydration, with a subsequent hot-air-drying process [87]. In a first step, the composition of the osmotic solution was optimized (fructose 47.7%, $CaCl_2$ 4% and citric acid 3.5%) and authors reported the important effect of the fructose concentration on water loss. Pectin-CMC-based coatings increased the porosity of the quince dried structure, while the osmotic dehydration increased the collapse of their porous structure. These authors concluded that the decrease observed on the final product firmness could be due to the plasticizing effect of low molecular mass components, which were absorbed from the coating and osmotic solutions.

In summary, all these results suggest that pectin-based coatings can be effectively used as a valid pre-drying treatment constituting a promising alternative to improve the retention of nutrients and quality characteristics of dehydrated food, also opening the possibilities to carry other active additives and nutrients.

6. CONCLUSIONS

Much work has been recently carried out to propose new formulations of edible coatings based on pectins to be applied in the protection of fresh food, improving organoleptic and nutritional characteristics and extending shelf-life. Despite the fact that recent studies have reported significant improvements in specific applications of these formulations, there is still a large amount of work to be performed since some of the most remarkable improvements are not yet experimentally attained or reproducible at large scale. In general terms, there is still a need for a better understanding of the composition-structure-processing-properties relationships in edible films based on pectins for food packaging, both at the laboratory and industrial scale. Moreover, since many of the studies related to this issue have been carried out using some basic pectins already in the market, there is still a lot of room for variation and maturation in the development of edible coatings for application in food packaging.

REFERENCES

1. Advancing Sustainable Materials Magement: Facts and Figures 2013. Available online: http://www.epa.gov/wastes/nonhaz/municipal/pubs/2013_advncng_smm_rpt.pdf (accessed on 25 September 2015).
2. Collection and transport. Waste Management World Homepage. Available online: http://www.waste-management-world.com/collection-transport (accessed on 25 September 2015).
3. Jamshidian, M.; Tehrany, E.A.; Imran, M.; Jacquot, M.; Desobry, S. Poly-lactic acid: Production, applications, nanocomposites, and release studies. *Compr. Rev. Food Sci. Food Saf.* **2010**, *9*, 552–571.

4. Martucci, J.F.; Ruseckaite, R.A. Biodegradable three-layer film derived from bovine gelatin. *J. Food Eng.* **2010**, *99*, 377–383.
5. Kechichian, V.; Ditchfield, C.; Veiga-Santos, P.; Tadini, C.C. Natural antimicrobial ingredients incorporated in biodegradable films based on cassava starch. *LWT-Food Sci. Technol.* **2010**, *43*, 1088–1094.
6. Arzu, A.B.; Tulay, O.; Oya, I.S.; Lutfiye, Y.E. The utilisation of microbial poly-hydroxy alkanoates (PHA) in food industry. *Res. J. Biotechnol.* **2010**, *5*, 76–79.
7. Ahmed, J.; Varshney, S.K. Polylactides—Chemistry, properties and green packaging technology: A review. *Int. J. Food Prop.* **2011**, *14*, 37–58.
8. Auras, R.; Harte, B.; Selke, S. An overview of polylactides as packaging materials. *Macromol. Biosci.* **2004**, *4*, 835–864.
9. Anuradha, K.; Padma, P.N.; Venkateshwar, S.; Reddy, G. Fungal isolates from natural pectic substrates for polygalacturonase and multienzyme production. *Indian J. Microbiol.* **2010**, *50*, 339–344.
10. Lopes da Silva, J.A.; Rao, M.A. *Food Polysaccharides and Their Applications*, 2nd ed.; Taylor & Francis: Abingdon, UK, 2006.
11. Kohli, P.; Gupta, R. Alkaline pectinases: A review. *Biocatal. Agric. Biotechnol.* **2015**, *4*, 279–285.
12. Pedrolli, D.B.; Monteiro, A.C.; Gomes, E.; Carmona, E.C. Pectin and pectinases: Production, characterization and industrial application of microbial pectinolytic enzymes. *Open Biotechnol. J.* **2009**, *3*, 9–18.
13. Ridley, B.L.; O'Neill, M.A.; Mohnen, D. Pectins: Structure, biosynthesis, and oligogalacturonide-related signaling. *Phytochemistry* **2001**, *57*, 929–967.
14. Caffall, K.H.; Mohnen, D. The structure, function, and biosynthesis of plant cell wall pectic polysaccharides. *Carbohydr. Res.* **2009**, *344*, 1879–1900.
15. Ovodov, Y.S. Current views on pectin substances. *Russ. J. Bioorg. Chem.* **2009**, *35*, 269–284.
16. Hua, X.; Wang, K.; Yang, R.; Kang, J.; Yang, H. Edible coatings from sunflower head pectin to reduce lipid uptake in fried potato chips. *LWT-Food Sci. Technol.* **2015**, *62*, 1220–1225.
17. Ciolacu, L.; Nicolau, A.I.; Hoorfar, J. *Global Safety of Fresh Produce. A Handbook of Best Practice, Innovative Commercial Solutions and Case Studies*; Woodhead Publishing Limited: Sawston, UK, 2014.
18. Dhanapal, A.; Sasikala, P.; Rajamani, L.; Kavitha, V.; Yazhini, G. Edible films from polysaccharides. *Food Sci. Qual. Manag.* **2012**, *3*, 9–18.
19. Giovanetti, M.H.; Nogueira, A.; de Oliveira, C.L.; Wosiacki, G. *Chromatography—The Most Versatile Method of Chemical Analysis*; InTech: Rijeka, Croatia, 2012.
20. Sundari, N. Extraction of pectin from waste peels: A review. *Res. J. Pharm. Biol.* **2015**, *6*, 1841–1848.
21. Alba, K.; Laws, A.P.; Kontogiorgos, V. Isolation and characterization of acetylated lm-pectins extracted from okra pods. *Food Hydrocoll.* **2015**, *43*, 726–735.
22. Maran, J.P. Statistical optimization of aqueous extraction of pectin from waste durian rinds. *Int. J. Biol. Macromol.***2015**, *73*, 92–98.
23. Geerkens, C.H.; Nagel, A.; Just, K.M.; Miller-Rostek, P.; Kammerer, D.R.; Schweiggert, R.M.; Carle, R. Mango pectin quality as influenced by cultivar, ripeness, peel particle size, blanching, drying, and irradiation. *Food Hydrocoll.* **2015**, *51*, 241–251.
24. Oliveira, T.Í.S.; Rosa, M.F.; Cavalcante, F.L.; Pereira, P.H.F.; Moates, G.K.; Wellner, N.; Mazzetto, S.E.; Waldron, K.W.; Azeredo, H.M.C. Optimization of pectin extraction from banana peels with citric acid by using response surface methodology. *Food Chem.* **2015**. in press.
25. Castillo-Israel, K.A.T.; Baguio, S.F.; Diasanta, M.D.B.; Lizardo, R.C.M.; Dizon, E.I.; Mejico, M.I.F. Extraction and characterization of pectin from Saba banana [*musa 'saba'(musa acuminata x musa balbisiana)*] peel wastes: A preliminary study. *Int. Food Res. J.* **2015**, *22*, 202–207.

26. Liew, S.Q.; Chin, N.L.; Yusof, Y.A.; Cheok, C.Y. Citric acid extraction of pectin from tropical fruit peels of passion fruit, dragon fruit and soursop. *J. Food Agric. Environ.* **2015**, *13*, 45–51.
27. Li, D.Q.; Du, G.M.; Jing, W.W.; Li, J.F.; Yan, J.Y.; Liu, Z.Y. Combined effects of independent variables on yield and protein content of pectin extracted from sugar beet pulp by citric acid. *Carbohyd. Polym.* **2015**, *129*, 108–114.
28. Sánchez Aldana, D.; Contreras-Esquivel, J.C.; Nevárez-Moorillón, G.V.; Aguilar, C.N. Characterization of edible films from pectic extracts and essential oil from mexican lime. *CyTA J. Food.* **2014**, *13*, 17–25.
29. Casas-Orozco, D.; Villa, A.L.; Bustamante, F.; González, L.M. Process development and simulation of pectin extraction from orange peels. *Food Bioprod. Process.* **2015**, *96*, 86–98.
30. Korish, M. Potential utilization of citrullus lanatus var. Colocynthoides waste as a novel source of pectin. *J. Food Sci. Technol.* **2015**, *52*, 2401–2407.
31. Korish, M. Faba bean hulls as a potential source of pectin. *J. Food Sci. Technol.* **2015**, *52*, 6061–6066.
32. Denman, L.J.; Morris, G.A. An experimental design approach to the chemical characterisation ofpectin polysaccharides extracted from cucumis melo inodorus. *Carbohydr. Polym.* **2015**, *117*, 364–369.
33. Grassino, A.N.; Halambek, J.; Djaković, S.; Rimac Brnčić, S.; Dent, M.; Grabarić, Z. Utilization of tomato peel waste from canning factory as a potential source for pectin production and application as tin corrosion inhibitor. *Food Hydrocoll.* **2016**, *52*, 265–274.
34. Kang, J.; Hua, X.; Yang, R.; Chen, Y.; Yang, H. Characterization of natural low-methoxyl pectin from sunflower head extracted by sodium citrate and purified by ultrafiltration. *Food Chem.* **2015**, *180*, 98–105.
35. Giosafatto, C.V.L.; di Pierro, P.; Gunning, P.; Mackie, A.; Porta, R.; Mariniello, L. Characterization of citrus pectin edible films containing transglutaminase-modified phaseolin. *Carbohydr. Polym.* **2014**, *106*, 200–208.
36. Giosafatto, C.V.L.; Di Pierro, P.; Gunning, A.P.; Mackie, A.; Porta, R.; Mariniello, L. Trehalose-containing hydrocolloid edible films prepared in the presence of transglutaminase. *Biopolymers* **2014**, *101*, 931–937.
37. Porta, R.; Mariniello, L.; di Pierro, P.; Sorrentino, A.; Giosafatto, C.V.L. Transglutaminase crosslinked pectin- and chitosan-based edible films: A review. *Crit. Rev. Food Sci. Nutr.* **2011**, *51*, 223–238.
38. Di Pierro, P.; Sorrentino, A.; Mariniello, L.; Giosafatto, C.V.L.; Porta, R. Chitosan/whey protein film as active coating to extend ricotta cheese shelf-life. *LWT Food Sci. Technol.* **2011**, *44*, 2324–2327.
39. Rossi Marquez, G.; di Pierro, P.; Esposito, M.; Mariniello, L.; Porta, R. Application of transglutaminase-crosslinked whey protein/pectin films as water barrier coatings in fried and baked foods. *Food Bioprocess Technol.* **2014**, *7*, 447–455.
40. Thirugnanasambandham, K.; Sivakumar, V.; Prakash Maran, J. Process optimization and analysis of microwave assisted extraction of pectin from dragon fruit peel. *Carbohyd. Polym.* **2014**, *112*, 622–626.
41. Rahmati, S.; Abdullah, A.; Momeny, E.; Kang, O.L. Optimization studies on microwave assisted extraction of dragon fruit (*Hylocereus polyrhizus*) peel pectin using response surface methodology. *Int. Food Res. J.* **2015**, *22*, 233–239.
42. Quoc, L.P.T.; Huyen, V.T.N.; Hue, L.T.N.; Hue, N.T.H.; Thuan, N.H.D.; Tam, N.T.T.; Thuan, N.N.; Duy, T.H. Extraction of pectin from pomelo (*Citrus maxima*) peels with the assistance of microwave and tartaric acid. *Int. Food Res. J.* **2015**, *22*, 1637–1641.
43. Maran, J.P.; Swathi, K.; Jeevitha, P.; Jayalakshmi, J.; Ashvini, G. Microwave-assisted extraction of pectic polysaccharide from waste mango peel. *Carbohyd. Polym.* **2015**, *123*, 67–71.

44. Maran, J.P.; Prakash, K.A. Process variables influence on microwave assisted extraction of pectin from waste *Carcia papaya* L. peel. *Int. J. Biol. Macromol.* **2015**, *73*, 202–206.
45. Zouambia, Y.; Youcef Ettoumi, K.; Krea, M.; Moulai-Mostefa, N. A new approach for pectin extraction: Electromagnetic induction heating. *Arabian J. Chem.* **2014**.
46. Moorthy, I.G.; Maran, J.P.; Surya, S.M.; Naganyashree, S.; Shivamathi, C.S. Response surface optimization of ultrasound assisted extraction of pectin from pomegranate peel. *Int. J. Biol. Macromol.* **2015**, *72*, 1323–1328.
47. Maran, J.P.; Priya, B. Ultrasound-assisted extraction of pectin from sisal waste. *Carbohyd. Polym.* **2015**, *115*, 732–738.
48. Chen, H.M.; Fu, X.; Luo, Z.G. Properties and extraction of pectin-enriched materials from sugar beet pulp by ultrasonic-assisted treatment combined with subcritical water. *Food Chem.* **2015**, *168*, 302–310.
49. Wang, W.; Ma, X.; Xu, Y.; Cao, Y.; Jiang, Z.; Ding, T.; Ye, X.; Liu, D. Ultrasound-assisted heating extraction of pectin from grapefruit peel: Optimization and comparison with the conventional method. *Food Chem.* **2015**, *178*, 106–114.
50. Yuliarti, O.; Matia-Merino, L.; Goh, K.K.T.; Mawson, J.; Williams, M.A.K.; Brennan, C. Characterization of gold kiwifruit pectin from fruit of different maturities and extraction methods. *Food Chem.* **2015**, *166*, 479–485.
51. Yuliarti, O.; Goh, K.K.T.; Matia-Merino, L.; Mawson, J.; Brennan, C. Extraction and characterisation of pomace pectin from gold kiwifruit (actinidia chinensis). *Food Chem.* **2015**, *187*, 290–296.
52. Valdés, A.; Vidal, L.; Beltrán, A.; Canals, A.; Garrigós, M.C. Microwave-assisted extraction of phenolic compounds from almond skin byproducts (prunus amygdalus): A multivariate analysis approach. *J. Agric. Food Chem.* **2015**, *63*, 5395–5402.
53. Ballard, T.S.; Mallikarjunan, P.; Zhou, K.; O'Keefe, S. Microwave-assisted extraction of phenolic antioxidant compounds from peanut skins. *Food Chem.* **2010**, *120*, 1185–1192.
54. Song, J.; Li, D.; Liu, C.; Zhang, Y. Optimized microwave-assisted extraction of total phenolics (TP) from ipomoea batatas leaves and its antioxidant activity. *Innov. Food Sci. Emerg. Technol.* **2011**, *12*, 282–287.
55. Montgomery, D.C. *Design and Analysis of Experiments*; Wiley: Hoboken, NJ, USA, 2009.
56. Falguera, V.; Quintero, J.P.; Jiménez, A.; Muñoz, J.A.; Ibarz, A. Edible films and coatings: Structures, active functions and trends in their use. *Trends Food Sci. Technol.* **2011**, *22*, 292–303.
57. Sánchez-Ortega, I.; García-Almendárez, B.E.; Santos-López, E.M.; Amaro-Reyes, A.; Barboza-Corona, J.E.; Regalado, C. Antimicrobial edible films and coatings for meat and meat products preservation. *Sci. World J.* **2014**.
58. Rojas-Graü, M.A.; Soliva-Fortuny, R.; Martín-Belloso, O. Edible coatings to incorporate active ingredients to fresh-cut fruits: A review. *Trends Food Sci. Technol.* **2009**, *20*, 438–447.
59. Rojas, M.A. Use of edible coatings for fresh-cut fruits and vegetables. In *Advances in Fresh-cut Fruits and Vegetables Processing*; CRC Press: Boca Raton, FL, USA, 2010; pp. 285–311.
60. Espitia, P.J.P.; Du, W.-X.; Avena-Bustillos, R.d.J.; Soares, N.d.F.F.; McHugh, T.H. Edible films from pectin: Physical-mechanical and antimicrobial properties—A review. *Food Hydrocolloid* **2014**, *35*, 287–296.
61. Martín-Diana, A.B.; Rico, D.; Frías, J.M.; Barat, J.M.; Henehan, G.T.M.; Barry-Ryan, C. Calcium for extending the shelf life of fresh whole and minimally processed fruits and vegetables: A review. *Trends Food Sci. Technol.* **2007**, *18*, 210–218.
62. Ramirez, M.E.; Timón, M.L.; Petrón, M.J.; Andrés, A.I. Effect of chitosan, pectin and sodium caseinate edible coatings on shelf life of fresh-cut *Prunus persica* var. Nectarine. *J. Food Process. Preserv.* **2015**.

63. Rico, D.; Martín-Diana, A.B.; Barat, J.M.; Barry-Ryan, C. Extending and measuring the quality of fresh-cut fruit and vegetables: A review. *Trends Food Sci. Technol.* **2007**, *18*, 373–386.
64. Zhang, Y.; Rempel, C.; McLaren, D. Edible coating and film materials: Carbohydrates. In *Innovations in Food Packaging*, 2nd ed.; Elsevier: Amsterdam, The Netherlands, 2014; pp. 305–323.
65. Ciolacu, L.; Nicolau, A.I.; Hoorfar, J. Edible coatings for fresh and minimally processed fruits and vegetables. In *Global Safety of Fresh Produce*; Hoorfar, J., Ed.; Woodhead Publishing: Sawston, UK, 2014; pp. 233–244.
66. Oms-Oliu, G.; Soliva-Fortuny, R.; Martín-Belloso, O. Using polysaccharide-based edible coatings to enhance quality and antioxidant properties of fresh-cut melon. *LWT Food Sci. Technol.* **2008**, *41*, 1862–1870.
67. Moalemiyan, M.; Ramaswamy, H.S.; Maftoonazad, N. Pectin-based edible coating for shelf-life extension of ataulfo mango. *J. Food Process Eng.* **2012**, *35*, 572–600.
68. Maftoonazad, N.; Ramaswamy, H.S. Effect of pectin-based coating on the kinetics of quality change associated with stored avocados. *J. Food Process. Preserv.* **2008**, *32*, 621–643.
69. Ferrari, C.C.; Sarantópoulos, C.I.G.L.; Carmello-Guerreiro, S.M.; Hubinger, M.D. Effect of osmotic dehydration and pectin edible coatings on quality and shelf life of fresh-cut melon. *Food Bioprocess Technol.* **2013**, *6*, 80–91.
70. Sipahi, R.E.; Castell-Perez, M.E.; Moreira, R.G.; Gomes, C.; Castillo, A. Improved multilayered antimicrobial alginate-based edible coating extends the shelf life of fresh-cut watermelon (citrullus lanatus). *LWT Food Sci. Technol.* **2013**, *51*, 9–15.
71. Martiñon, M.E.; Moreira, R.G.; Castell-Perez, M.E.; Gomes, C. Development of a multilayered antimicrobial edible coating for shelf-life extension of fresh-cut cantaloupe (*Cucumis melo* L.) stored at 4 °C. *LWT-Food Sci. Technol.* **2014**, *56*, 341–350.
72. Brasil, I.M.; Gomes, C.; Puerta-Gomez, A.; Castell-Perez, M.E.; Moreira, R.G. Polysaccharide-based multilayered antimicrobial edible coating enhances quality of fresh-cut papaya. *LWT-Food Sci. Technol.* **2012**, *47*, 39–45.
73. Oms-Oliu, G.; Soliva-Fortuny, R.; Martín-Belloso, O. Edible coatings with antibrowning agents to maintain sensory quality and antioxidant properties of fresh-cut pears. *Postharvest Biol. Technol.* **2008**, *50*, 87–94.
74. Treviño-Garza, M.Z.; García, S.; Flores-González, M.S.; Arévalo-Niño, K. Edible active coatings based on pectin, pullulan, and chitosan increase quality and shelf life of strawberries (*Fragaria ananassa*). *J. Food Sci.* **2015**, *80*, M1823–M1830.
75. Moreira, M.R.; Cassani, L.; Martín-Belloso, O.; Soliva-Fortuny, R. Effects of polysaccharide-based edible coatings enriched with dietary fiber on quality attributes of fresh-cut apples. *J. Food Sci. Technol.* **2015**.
76. Guerreiro, A.C.; Gago, C.M.L.; Faleiro, M.L.; Miguel, M.G.C.; Antunes, M.D.C. The use of polysaccharide-based edible coatings enriched with essential oils to improve shelf-life of strawberries. *Postharvest Biol. Technol.* **2015**, *110*, 51–60. [] []
77. Guerreiro, A.C.; Gago, C.M.L.; Faleiro, M.L.; Miguel, M.G.C.; Antunes, M.D.C. Raspberry fresh fruit quality as affected by pectin- and alginate-based edible coatings enriched with essential oils. *Sci. Hortic.* **2015**, *194*, 138–146.
78. Ayala-Zavala, J.F.; Silva-Espinoza, B.A.; Cruz-Valenzuela, M.R.; Leyva, J.M.; Ortega-Ramírez, L.A.; Carrazco-Lugo, D.K.; Pérez-Carlón, J.J.; Melgarejo-Flores, B.G.; González-Aguilar, G.A.; Miranda, M.R.A. Pectin–cinnamon leaf oil coatings add antioxidant and antibacterial properties to fresh-cut peach. *Flavour Fragr. J.* **2013**, *28*, 39–45.
79. Valdes, A.; Mellinas, A.C.; Ramos, M.; Burgos, N.; Jimenez, A.; Garrigos, M.C. Use of herbs, spices and their bioactive compounds in active food packaging. *RSC Adv.* **2015**, *5*, 40324–40335.
80. Mellinas, C.; Valdés, A.; Ramos, M.; Burgos, N.; del Carmen Garrigós, M.; Jiménez, A. Active edible films: Current state and future trends. *J. Appl. Polym. Sci.* **2015**.

81. Daraei Garmakhany, A.; Mirzaei, H.O.; Maghsudlo, Y.; Kashaninejad, M.; Jafari, S.M. Production of low fat french-fries with single and multi-layer hydrocolloid coatings. *J. Food Sci. Technol.* **2014**, *51*, 1334–1341.
82. Suyatma, N.E.; Ulfah, K.; Prangdimurti, E.; Ishikawa, Y. Effect of blanching and pectin coating as pre-frying treatments to reduce acrylamide formation in banana chips. *Int. Food Res. J.* **2015**, *22*, 936–942.
83. Oliveira, S.M.; Brandão, T.R.S.; Silva, C.L.M. Influence of drying processes and pretreatments on nutritional and bioactive characteristics of dried vegetables: A review. *Food Eng. Rev.* **2015**.
84. Garcia, C.C.; Caetano, L.C.; de Souza Silva, K.; Mauro, M.A. Influence of edible coating on the drying and quality of papaya (*Carica papaya*). *Food Bioprocess Technol.* **2014**, *7*, 2828–2839.
85. Silva, K.S.; Garcia, C.C.; Amado, L.R.; Mauro, M.A. Effects of edible coatings on convective drying and characteristics of the dried pineapple. *Food Bioprocess Technol.* **2015**, *8*, 1465–1475.
86. Canizares, D.; Mauro, M.A. Enhancement of quality and stability of dried papaya by pectin-based coatings as air-drying pretreatment. *Food Bioprocess Technol.* **2015**, *8*, 1187–1197.
87. Akbarian, M.; Ghanbarzadeh, B.; Sowti, M.; Dehghannya, J. Effects of pectin-CMC-based coating and osmotic dehydration pretreatments on microstructure and texture of the hot-air dried quince slices. *J. Food Process. Preserv.* **2015**, *39*, 260–269.

CHAPTER 22

Silk Fibroin as Edible Coating for Perishable Food Preservation

B. Marelli[†], M. A. Brenckle[‡], D. L. Kaplan & F. G. Omenetto

Department of Biomedical Engineering Tufts University 4 Colby St., Medford, MA, 02155, USA.

ABSTRACT

The regeneration of structural biopolymers into micelles or nanoparticles suspended in water has enabled the design of new materials with unique and compelling properties that can serve at the interface between the biotic and the abiotic worlds. In this study, we leveraged silk fibroin quintessential properties (i.e. polymorphism, conformability and hydrophobicity) to design a water-based protein suspension that self-assembles on the surface of food upon dip coating. The water-based post-processing control of the protein polymorphism enables the modulation of the diffusion of gases through the silk fibroin thin membranes (e.g. O_2 and CO_2 diffusion, water vapour permeability), which is a key parameter to manage food freshness. In particular, an increased beta-sheet content corresponds to a reduction in oxygen diffusion through silk fibroin thin films. By using the dip coating of strawberries and bananas as proof of principle, we have shown that the formation of micrometre-thin silk fibroin membranes around the fruits helps the management of postharvest physiology of the fruits. Thus, silk fibroin coatings enhance fruits' shelf life at room conditions by reducing cell respiration rate and water evaporation. The water-based processing and edible nature of silk fibroin makes this approach a promising alternative for food preservation with a naturally derived material.

INTRODUCTION

Food waste has an impact on food quality and safety, representing a loss of economic value and resources, and, a hindrance to economic development. The Food and Agriculture Organization (FAO) of the United Nations has estimated that one-third of the food produced for human consumption worldwide is annually lost or wasted along the chain that stretches from farms to postharvest treatments, processing, distribution and end-user consumption[1]. For fruit and vegetable commodities, the FAO has estimated a 50% loss of crops throughout the food supply chain, mainly concentrated in the postharvest, distribution and end-user consumption stages and mostly due to the premature deterioration of perishable crops. Many perishable fruit and vegetables possess in fact high metabolic activity and suffer from high possibility of microbial contamination, resulting in short shelf-life, fungal decay, colour change, and off-flavour[2]. To date, several treatments have been explored to extend the postharvest life of fruit and vegetables (e.g. cryopreservation, exposure to synthetic chemical fungicides, modified atmosphere packaging, osmotic treatments, hypobaric and heat treatments)[3]. In this scenario, the formation of an edible coating represents an alternative route to extend crop freshness by combining extended storage periods with ease of handling[4,5]. In general, food coatings should be mechanically robust matrices with hydrophobic groups to exert low permeability to oxygen and water vapours (i.e. control over fruit respiration rate and firmness retention). Biocompatibility, biodegradability, antibacterial and antifungal activities, membrane forming capacity and safety (i.e. edible and not allergenic) should also be compelling properties for an edible coating material[6]. To date, polysaccharides, proteins, resins, lipids and their combinations are the commonly used options for coating formulations[7,8,9,10,11,12,13]. Polysaccharides and proteins are known to form films with good mechanical properties, but with poor permeability, while lipids and resins form brittle films with improved permeability. Among all the perishable food, strawberries are considered one of the most difficult to preserve fresh along the food supply chain and have therefore been used as a model to test the efficacy of food preservation strategies and, in particular, of edible coatings[7,14,15].

Silk fibroin is an extensively investigated biomaterial for its potential in textile, biomedical, photonic, and electronic applications[16,17]. Silk fibroin is a structural protein, like collagen, but with a unique feature: it is produced from the extrusion of an amino-acidic suspension by a living complex organism (while collagen is produced in the extracellular space by self-assembly of cell-produced monomers). Silk fibroin properties are derived from its structure, which consists of hydrophobic blocks staggered by hydrophilic, acidic spacers[18]. In its natural state, silk fibroin is organized in beta-sheet crystals alternated by amorphous regions, which provide strength and resilience to the protein[18]. The multiplicities of forms in which regenerated silk fibroin can be processed at a high protein concentration and molecular weight make it attractive for several high-tech applications, as recently reported[17,19,20,21,22,23,24,25,26]. In addition, the amino-acidic nature of silk fibroin brings a diversity of side chain chemistries that allows for the incorporation of macromolecules useful in drug delivery applications or in providing cellular instructions[27]. The natural properties of silk fibroin have been exploited in the regenerated silk materials. The combination of beta-sheet structures with inter- and intra-molecular hydrogen bonds that provides high-flexibility in the natural fibre also guarantees extreme conformability in regenerated film format. Silk fibroin films have in fact been used as biocompatible and biodegradable substrates for implantable electrodes to be applied on the brain and as adhesive, edible sensors to monitor fruit ripening and cheese aging[21,22]. In addition, the gas (i.e. oxygen, carbon dioxide, water vapour) diffusion properties of silk materials present in the cocoon, which allows the *Bombyx mori* to survive through the pupal phase of its lifecycle, can be tailored in regenerated protein by regulating silk fibroin polymorphism[28,29,30]. This, for example, has ensured the correct gas exchange in silk fibroin materials engineered for corneal grafting[28,29,30]. In particular, silk fibroin with modular degrees of beta-sheet content can be obtained by regulating the time (minutes to hour range) and temperature (4–95 °C) to which the protein gets exposed to water vapour or edible polar solvents (i.e. ethanol)[26].

Remarkably, silk fibroin has also shown to possess features that are distinct of its regenerated form. For example, regenerated silk fibroin stabilizes heat labile molecules and compounds (e.g. enzymes and antibiotics)[31]. Silk is indeed considered a platform technology in biomaterials fabrication as its robustness and qualities bring needed assets to provide a portfolio of distinct features (*e.g.* nanopatterning, biochemical functionalization) for the final construct. Processing of regenerated fibroin generally involves the partial or total dehydration of a fibroin suspension (protein content of 1–15 wt%) to form films, sponges, gels, spheres (micron- to nano-sized) and foams with numerous techniques (*e.g.* solvent casting, freeze drying, salt leaching, sonication)[32]. The rationale beyond these fabrication processes is to manufacture a robust material that combines mechanical strength with *ad-hoc* biochemical properties.

RESULTS AND DISCUSSION

In this study, we report the use of silk fibroin suspension as an edible coating material for perishable fruit preservation. Silk fibroin is generally considered flavorless and odorless, which are compelling properties for food coating and packaging applications[33,34]. A Silk fibroin suspension was used to coat freshly picked strawberries through dip coating (Fig. 1a)[32]. The molecular weight of silk fibroin was tailored in the 170–90 kDa range by regulating the boiling time (i.e. 30 minutes) during fibroin extraction and the concentration of the protein in the dipping suspension was adjusted to 1 wt% to obtain a final suspension with rheological properties (i.e. viscosity and surface tension) similar to the ones previously used for biopolymer-based coatings[7,35.] Inductively coupled plasma mass spectrometry (ICP-MS) was used to investigate the presence of metal elements in the silk suspension, to assess possible toxicity deriving from the silk fibroin regeneration process. For all the elements considered, the detection values were significantly below the toxicity levels in drinking water (Table S1), as per World Health Organization guidelines[36]. The coating of strawberries with silk fibroin was achieved with a two-phase process. In the first phase, a multiple-step dip coating process (number of coating steps =D 1, 2 and 4) was used to expose strawberries to silk fibroin suspension. Secondly, silk fibroin-coated fruits were treated with water vapour under vacuum (namely water annealing post-processing) to explore the effects of increasing beta-sheet content in silk fibroin materials on the management of fruits' postharvest physiology[26]. The results suggested that silk fibroin coatings prolonged the freshness of perishable fruits by slowing fruit respiration, extending fruit firmness and preventing dehydration.

Silk fibroin as a coating to manage strawberries' postharvest physiology

The thickness of silk fibroin coatings, which was in the range of 27–35 µDm, was not statistically significantly influenced (p >D 0.05) by the number of dip coating steps (Table 1). Conversely, water annealing post-process strongly influenced the properties of the silk fibroin coatings by increasing the relative content of beta-sheet structures from 23 ±D 2% (for untreated coatings) to 58 ±D 5% (for coatings exposed to water vapour for 12 hours). The effects of water annealing time on the the relative beta-sheet content of silk fibroin coatings are reported in Table 1. No changes in the microscopic morphology of silk fibroin materials exposed to water annealing was visible with scanning electron

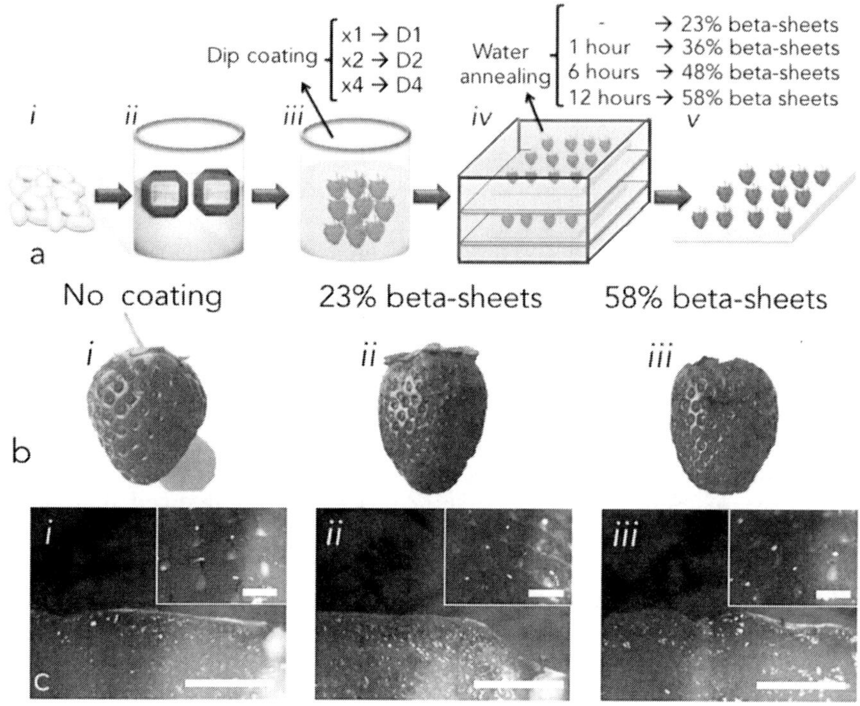

Figure 1: Coating of perishable fruits with edible silk fibroin. (**a**) The impact of silk fibroin coating was investigated on freshly picked strawberries. (*i*) Silk fibroin was extracted from *Bombyx Mori* cocoon fibers by dissolution in 9.3 M LiBr solution and (*ii*) dialysis in deionized water. The concentration of the protein in water was then adjusted to 1 wt%. (*iii*) Coating of strawberries was then achieved by dip coating process in silk fibroin suspension (1wt%). The dip coating process was repeated up to 4 times. *iv*) Beta-sheet content in silk fibroin edible coatings was modulated using water annealing post-processing. The longer the exposure to water vapour (up to 12 hours), the higher the beta-sheet content of the protein, as reported in Table 1. *v*) Silk fibroin-coated strawberries were then left at room conditions (T =D 22 °C, RH =D 38%) to investigate the impact of the coating on the quality of the fruit. Crystal violet dye was used to stain the silk fibroin coating. (**b**) Representative images of stained strawberries *i*) freshly picked, *ii*) coated silk fibroin edible coating (4 dip coating processes, 23% beta-sheet, *i.e.* no water annealing applied), *iii*) coated with silk fibroin (4 dip coating processes), 58% beta-sheet, *i.e.* exposed to water annealing post-processing. The crystal violet dye is barely visible on the surface of the coated strawberries (black dots) due to the few-micron thickness of the coating. (**c**) Stereoscopic images of the surface and of the cross-section (insets) of crystal violet-stained fresh strawberries (*i*) as picked, (*ii*) coated with silk fibroin with 23% beta-sheet content and (*iii*) coated with silk fibroin (4 dip coating processes) with 58% beta-sheet content. Scale bars: 2 mm.

microscopy analysis (Fig. S1). Crystal violet dye was used to stain the silk fibroin coating. Representative macroscopic images of stained strawberries uncoated and coated with silk fibroin edible materials with increasing beta-sheet content are presented in Fig. 1b *i*, *ii* and *iii*, respectively. Crystal violet staining is barely visible on the surface of the coated strawberries (black dots) due to the few-micron thickness of the coating. In addition, stereoscopic microscopy of the surface and of the cross-section of crystal violet-stained strawberries coated with silk fibroin materials showed no changes in the appearance of the fruit when compared to the uncoated control (Fig. 1c). The efficacy of silk fibroin edible coatings in managing strawberries' postharvest physiology was evaluated as a function of number of coating steps and of silk fibroin beta-sheet content. The external and internal analyses of the aging of representative strawberries as a function of dip coating steps and of silk beta-sheet content is presented in Fig. 2a and Figure S2. A prolonged preservation of strawberries tissues corresponded to increasing coating steps and to longer water annealing process (i.e. enhanced beta-sheet content). This was visible by time dependent reduction in shading from the original red colour and deformation from the original morphology. In addition, increasing coating steps and higher beta-sheet content possibly down-regulated microbial growth, as suggested by reduced fungal- and mould-induced decay. This was also confirmed by investigation of the time dependent decay of the strawberries' flesh. Higher beta-sheet content in silk fibroin coatings corresponded to an enhanced preservation of the internal tissue of strawberries, for the time points considered.

Table 1: Characterization of edible silk fibroin coating thickness as a function of dip coating steps and of relative beta-sheet content as a function of exposure time to water annealing.

Number of dip coating processes (x) in silk fibroin suspension [Dx]	**Thickness [μm]**
D1	27 ± 8
D2	32 ± 7
D4	35 ± 8
Exposure time to water annealing post-processing [hours]	Silk fibroin coating crystallinity [relative content of beta-sheet structures, %]
0	23 ± 2
1	36 ± 3
6	48 ± 4
12	58 ± 5

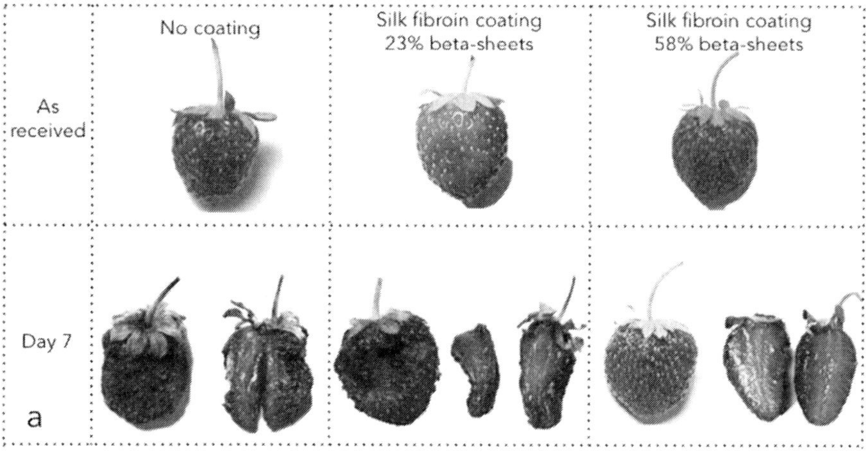

Figure 2: Ripening and weight loss of strawberries coated with edible silk fibroin coating. (**a**) Time-lapse of strawberries ripening. As picked strawberries were stored at 22 °C and 38% RH (no coating) or dip coated in silk fibroin suspension (23% beta-sheets coating). Water annealing was used as post-process to modulate the relative content of beta-sheets in silk fibroin. At day 7, silk fibroin coating showed to improve the quality of the stored strawberries. (**b**) Weight loss of strawberries stored for up to 14 days at 22 °C and 38% RH. Strawberries were stored as picked (i.e. *no coating*) or after coating with silk fibroin suspension (*Dx-xx%*). *Dx* stands for 'x' dip coating steps. *xx%* stands for relative amount of beta-sheet content. (e.g. D1-23% means that strawberries were dip coated with silk fibroin materials with 23% beta-sheets content). One-way ANOVA test with Tukey mean analysis was used to evaluate the weight loss data. Silk beta-sheet content (which is proportional to water-annealing time) but not number of dip coating steps affected the dehydration of the strawberries considered. *No coating* controls lost circa 70 wt% of their original weights in the 14 days considered (highlighted within the red rectangles). Strawberries coated silk fibroin retained more water than the *no coating* controls at day 3 (p <D 0.05). An increase in silk fibroin beta-sheets content via water annealing process further decreased fruit dehydration compared to non-water annealed counterparts (p <D 0.05) and to the *no coating* controls (p <D 0.05) but no statistically significant difference was found between samples with a beta-sheet content ≥D 36% (p >D 0.05).

Measurement of dehydration

Dehydration of strawberries is an indication of the breakdown of the red receptacle tissue, which results in off-flavouring, microbial decay, loss of turgor and water evaporation. Figure 2b shows the time dependent weight loss (express as fraction of original weight) of strawberries as a function of dip coating steps and of silk fibroin beta sheets content. Two-way ANOVA test with Tukey mean analysis was used to evaluate the data. Silk fibroin beta-sheet content but not number of coating steps affected the dehydration of the strawberries considered. Uncoated controls lost circa 70 wt% of their original weight in the 14 days considered. Strawberries simply coated (i.e. no water-annealing post-process) silk fibroin (Dx23%) retained more water compared to controls at day 5 (p <D 0.05). Strawberries coated with silk fibroin and further treated with water annealing process presented a decreased dehydration compared to strawberries non-treated with water-annealing (p <D 0.05) and to uncoated controls (p <D 0.05). However, water annealing time (i.e. silk fibroin beta-sheet content >36%) did not statistically significantly affect changes in the strawberries' original weight (p >D 0.05) for all the time considered. To investigate the phenomenon of fruit dehydration through thin silk fibroin coatings, the interplay between silk fibroin thin membranes and water have been investigated (Table 2 and Table S2). In particular, hydrodynamic permeability and water diffusivity phenomena were explored (Fig. 3a,b, respectively). Investigation of hydrodynamic permeability showed that silk fibroin beta-sheet content did not statistically significantly affect (p >D 0.05) water permeation (as per one-way ANOVA test with Tukey mean analysis). The study of water diffusivity revealed that the beta-sheet content of silk fibroin membranes slightly affected mass transport of water through the protein (as per one-way ANOVA test with Tukey mean analysis). Calculated diffusion coefficients for mass transport experiments are reported in Table S2. Surprisingly, there is an apparent inconsistency between calculated diffusion coefficients for water (Table S2) and for water vapor permeability (WVP, Table 2). When silk fibroin beta-sheet content increased from 36%–58%, WVP of the film containing 58% beta-sheet content decreased to 1/100 of that with 36% beta-sheet content. However, the measured diffusion coefficient for water was only five times smaller. While further analyses are necessary to better understand these results, the data may signify that an increase in beta-sheet content induces a water vapor barrier effect without significant change of water (molecule or cluster) diffusion in the silk fibroin film. This may be due to a combination of beta-sheet domain re-arrangement during the water-annealing process (i.e. beta-sheet crystal dimension, formation of inter- and intra-molecular beta-sheet, alignment of the crystals) and by the presence of wicking phenomena.

Table 2: Calculated water vapour permeability in silk fibroin films of increasing beta-sheet content ($\Delta RH = 75\%$).

Silk fibroin edible coating beta-sheets content	Water vapour permeability
($g\ m^{-1} s^{-1}\ Pa^{-1}$)	
36%	$7.93 \cdot 10^{-9}$
48%	$5.37 \cdot 10^{-10}$
58%	$6.49 \cdot 10^{-11}$

Gas diffusivity through silk membranes

Gas diffusivity through silk fibroin membranes plays a major role in fruit preservation as gases play a pivotal role in fruit stem cells metabolism (i.e. oxygen), are by-products of their metabolism (i.e. carbon dioxide) and may act as growth factors (i.e. ethylene in climacteric fruit). To evaluate the efficacy of silk fibroin coatings as a gas barrier, we measured oxygen diffusivity through silk fibroin membranes (t =D 70–80 μDm) with increasing beta-sheet content (Fig. 3c, Table S3). In particular, the effective oxygen diffusion coefficient was found to be modulated by protein beta-sheet content. A decrease of two orders of magnitude in the oxygen effective diffusivity coefficient was measured between silk fibroin materials with a relative beta-sheet content of 23 ±D 5% (no water annealing treatment) and 58 ±D 4% (12 h of water annealing).

Respiration rate is an important parameter to evaluate the metabolic activity of the stem cells present in the fruits. The higher the respiration rate, the higher the metabolic activity and the faster the fruit decay. In this study, the respiration rate of silk fibroin-coated strawberries was measured as a function of beta-sheet content in silk materials (Fig. 3d). To higher beta-sheet content corresponded a statistically significant decrease in the production of CO_2 (p <D 0.05), which indicated a decrease in the respiration rate of the fruits. It was evident that the beta-sheet content of the edible silk fibroin coatings controlled the gas exchange of between the strawberries and the environment due to the strong influence that silk polymorphism has on the permeability to gases like CO_2 and O_2. The increase in beta-sheet content, in fact, yields a less hydrated material (decrease in free water and freezing bound water) where the molecules are more packed and organized in nanoscopic beta-sheet structures26,37. O_2 and CO_2 are not soluble in silk proteins, while they are soluble in water. It is then possible that the dissolution of gas decreases for silk materials that have a decreased amount of water (i.e. higher beta-sheet content) in

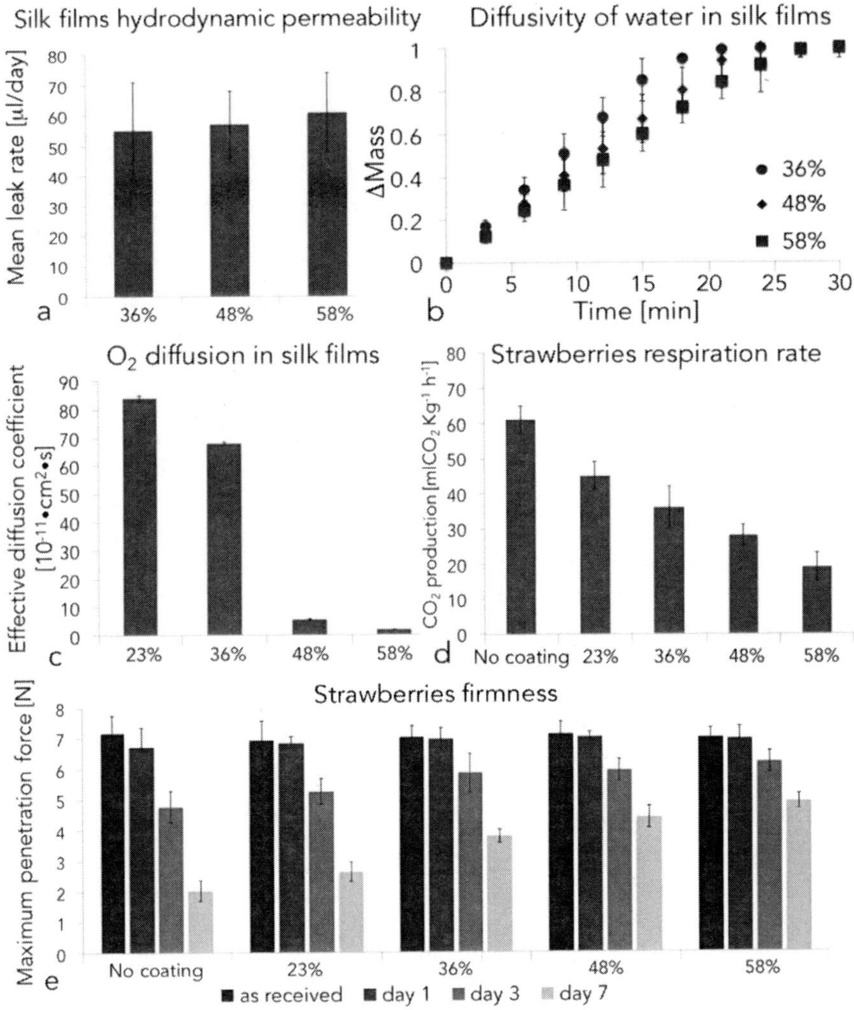

Figure 3: Permeability and diffusivity of water and oxygen in silk fibroin films as a function of protein crystallinity and their effects on the quality of the edible coating. (**a**) Hydrodynamic permeability of silk fibroin films. The beta-sheets content of the protein did not affect the leakage of water through the film (p <D 0.05). (**b**) Diffusivity of water in silk fibroin films. The beta-sheet content in silk fibroin materials affected the diffusivity of water in the transient state (0 <D t <D 25 minutes) but not in the steady state (t >D 25 minutes). (**c**) Oxygen permeability in silk fibroin films. Silk fibroin polymorphism strongly affected oxygen permeability as a 50 fold decreased in oxygen permeability was measured between silk fibroin with a relative beta-sheet content of 23% and 58%.

(**d**) Respiration rate of silk coated strawberries was measured as a function of beta-sheet content. To higher beta-sheet contents corresponded a statistically significant decrease in the production of CO_2 (p <D 0.05). (**e**) Effects of silk coating on the firmness of strawberries as a function of storage time and of beta-sheet content. Natural decay of the strawberries caused a decrease in the fruit firmness, as measure by a time-dependent decrease in the force required to penetrate the fruit (p <D 0.05). An increase in the coating beta-sheet content corresponded to a statistically significant delay in the decay of the fruit firmness at days 3 and 7.

their structure, resulting in a decreased diffusion of water-soluble gases through silk membranes.

Firmness rate in silk fibroin-coated strawberries

A puncture (or penetration) test was also used to evaluate the effects of silk coating on the firmness of strawberries as a function of storage time and of beta-sheet content (Fig. 3e). While natural decay of strawberries caused a decrease in the fruit firmness, as measured by a time-dependent decrease in the force required to penetrate the fruit (p <D 0.05), an increase in the coating beta-sheet content corresponded to a statistically significant delay of fruit firmness decay at days 3 and 7.

Silk fibroin as coating for climacteric fruits

Climacteric fruits, such as apples, bananas, tomatoes, ripen through ethylene production and increased cell respiration. The climacteric event is associated with changes in fruit colour and with the production of sugar in the extracellular space. To evaluate the efficacy of a silk-based climacteric fruit coating, the ripening of bananas was evaluated by comparing silk coated and non-coated fruits (Fig. 4). Fruits were stored at 22 °C and 38% relative humidity (RH) as received (*No coating*) and after coating with water-annealed treated silk fibroin materials (t_{WA} =D 12 h, *Silk fibroin coating*). Bananas were hung from their respective stem throughout the whole experiment. Time-lapse photography of banana ripening indicated that silk coating decreased the ripening rate. The beta-sheet content of the silk fibroin used to coat the climacteric fruit did not affect fruit ripening (data not shown). Investigation of silk-coated banana firmness showed that the coating decreased fruit ripening, when compared to uncoated control at day 9 after coating. In addition, morphological analysis of the flesh of non-coated and coated bananas at day 9 post silk coating-treatment revealed a more preserved fruit when the silk coating was applied. Flesh of non-coated bananas presented a brown colour, while silk-coated fruits preserved a tallow flesh, indication of a decreased ripening rate within the silk-coated sample.

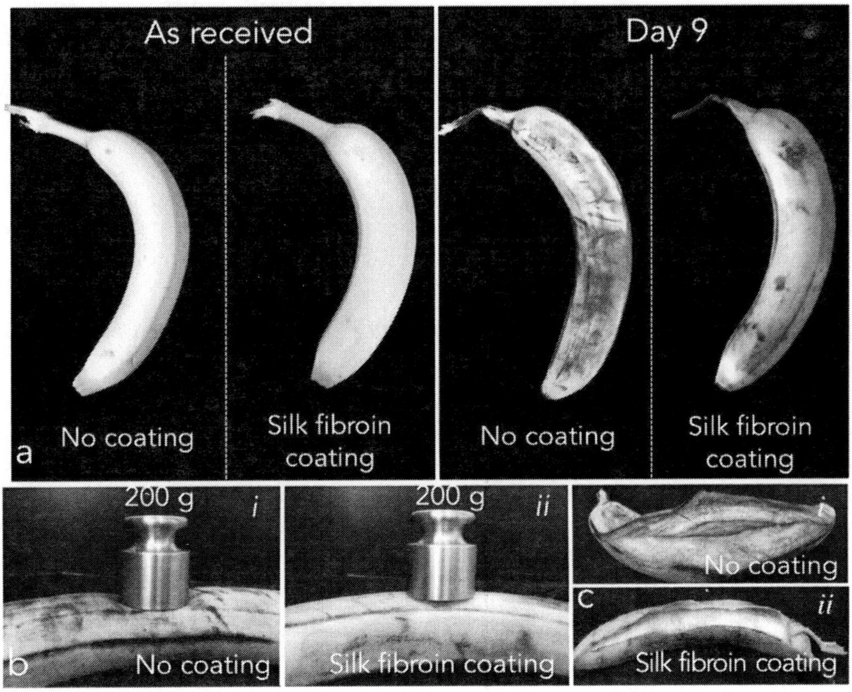

Figure 4: Evaluation of bananas ripening with and without crystalline silk coating.

Fruits were stored at 22 °C and 38% RH as received (*no coating*) and after coating with silk fibroin materials. Bananas were hanged from their respective stem throughout the whole experiment. (**a**) Time lapse photography of banana ripening indicating that silk fibroin coating decreased the ripening rate. (**b**) Investigation of silk fibroin-coated banana turgidity (i) when compared to uncoated controls. Turgidity was studied qualitatively by applying a dead load (200 g) on the surface of the fruit. (ii). The test was accomplished ad day 9 after coating. Silk fibroin-coated banana showed an increased firmness, when compared to uncoated control. (**c**) Images of the internal flesh of non-coated (i) and silk fibroin-coated (ii) bananas at day 9 post coating-treatment. Flesh of non-coated banana presented a brown colour, while silk fibroin-coated fruits preserved a tallow flesh, indication of a decreased ripening rate for the silk fibroin-coated samples.

In summary, we have demonstrated that silk fibroin is an effective, water-based coating to enhance freshness of perishable food. The water-based processing and comestible nature of this material makes the approach a promising alternative for preservation with a naturally derived material. Silk polymorphism may be used to tailor the properties of the coating, affecting the interplay between silk fibroin and water evaporation and food decay. The ability to consume silk or enhance coatings with stabile biomacromolecules further adds utility for functional coatings that could be envisioned to add therapeutic function to consumable goods without resorting to complex chemistries, while preforming the preservation function.

METHODS

Silk Fibroin Regeneration

Cocoon from *Bombyx Mori* were used as source of fibroin. Extraction of silk fibroin was achieved by standard degumming process, which involved boiling (t =D 30 minutes) 2.5 g of chopped silk cocoons per litre of 0.02 M sodium carbonate solution. SF was then solubilized in 9.3 M lithium bromide for 4 hours in a 60 °C oven. The chaotropic salt was subsequently removed through dialysis (3.5 kDa MWCO) against Milli-Q water for a total of 72 hours, yielding an 8% (w/v) silk fibroin suspension. The resulting SF suspension was then purified by centrifugation at 9000 rpm (~12,700 *g*) over two 25 minute-long periods, at a constant of 4 °C. The final concentration of SF suspension was then adjusted to 5 wt% by adding MilliQ water.

Strawberry dip coating

For experiments run during the summer, freshly picked New England native strawberries (Wilson Farms, Lexington, MA) were dip coated for 1, 2 and 4 times (namely D1, D2 and D4) in a 60 mm deep silk suspension at 4 °C ensuring that the whole surface of the strawberries and of their calyx and epicalyx were exposed to the suspension. The dipping step(s) last for 10 s each and then strawberries were dried by hanging them from the peduncle for 4 hours at 22 °C, 38% RH. For experiments conducted off-season, strawberries were bought from a local store (Whole Foods, Medford, MA).

Banana dip coating

Bananas (Del Monte) green in colour were bought from a local store (Whole Foods, Medford, MA) and then dip-coated with silk fibroin as aforementioned.

Coating crystallization

Silk fibroin coating crystallization (i.e. enhancement of beta-sheet content) was obtained through exposure of coated strawberries to water vapours under vacuum (namely water annealing) at 22 °C, according to previously developed protocols26. The choice of operating at room temperature has been dictated by the possible detrimental effects on the fruit quality upon exposure to high temperatures (T >D 40 °C) for prolonged time (t >D 30 minutes). Exposure time was set to 0 s, 1 hours, 6 hours and 12 hours. To longer water annealing post-processing corresponded an increase in the protein beta-sheet content (Table 1), as previously reported26.

Attenuated Total Reflectance Fourier Transform Infrared Spectroscopy (ATR-FTIR) was used to evaluate the relative content of beta-sheet structures in silk fibroin materials. FTIR spectroscopy scans were taken on a Jasco FTIR 6200(Easton, MD) spectrometer with attached ATR detector. A total of 64 scans at a resolution of 4.0 cm^{-1} were co-added to produce spectra ranging from 400–4000 cm^{-1}. A cosine apodization was simultaneously applied by the software. From these scans, the Amide III region (1200–1350 cm^{-1}) was selected for its sensitivity to protein secondary structure and lack of sensitive to water content. Amide III curves were normalized and baseline corrected, and then fit to 12 Gaussian curves. Bands corresponding to beta-sheet secondary structure motifs were then added to give a relative value for the beta-sheet content of the films.

Evaluation of strawberry freshness

The effect of different silk fibroin coating steps and of water annealing times (i.e. beta-sheet content) on strawberry freshness was evaluated morphologically and gravimetrically. Changes in strawberries colour and shape were evaluated through time-lapse photography. Gravimetric analysis of berries as received in the laboratory, after dip coating and at days 1, 3, 5, 7 and 14 was evaluated with a standard laboratory balance (Mettler Toledo MS204S). Berries weight was calculated as an average of three measurements.

Water vapour permeability of silk fibroin films

WVP values were determined using the method reported in ASTM E96-E96M38, with some modifications[39,40]. Circular silk fibroin films (Ø =D 16 mm, n =D 5) of increasing beta-sheet content (36 ±D 4%, 48 ±D 3, 58 ±D 4) were obtained by modulating the exposure to water-annealing vapours at 22 °C, following a previously published experimental procedure, to investigate silk fibroin materials with beta-sheet content comparable to the fibroin coatings[26]. Silk fibroin films were sealed with lubricant grease over the circular opening (Ø =D 14 mm) of custom-made plastic test cups that were stored in a desiccator at 22 °C. A sodium chloride saturated solution was placed in the desiccator (75% RH) while anhydrous calcium chloride pellets were positioned on the bottom of the test cups (0% RH). The difference in RH corresponded to a water vapour partial pressure of 1753.55 Pa. Water vapour permeability was then determined as:

$$WVP = \frac{WVTR \cdot d}{A \cdot S \cdot t \cdot (RH_2 - RH_1)} \quad \left[\frac{g}{m \cdot s \cdot Pa}\right] \tag{1}$$

where the water vapour transmission rate (WVTR) was calculated by the linear regression (r^2 >D 0.98) between the weight gain of the test cups and time, d is the thickness of the silk fibroin samples, A is the area of the circular opening in the test cups, S is the the saturation vapour pressure of water, 2645 Pa at 22 °C, RH_1 is RH in the desiccator and $RH_2$is RH in the test cup.

Measurement of oxygen diffusion in silk fibroin films

The effective oxygen diffusion coefficients were measured in silk fibroin films (n =D 5) of increasing crystallinity (23 ±D 5%, 36 ±D 4%, 48 ±D 3, 58 ±D 4) using a conventional diffusion system - Microx TX3 Microsensor Oxygen Meter (Presens, Germany) equipped with an Oxygen Microptode (Presens, Germany) and a PermeGear sealed, water-jacked, gas chamber, as previously described[41,42]. The system consisted of two compartments that contained known (measured) concentrations of oxygen, separated by the silk film of interest. The opening between the two chambers had an area of 2.75 cm^2. Silk fibroin film thickness was measured with a micrometer (n =D 7). Before use, a two-point calibration was performed according to manufacturer protocols using an oxygen-free environment (sodium sulphite) and an air-saturated environment (water vapour). The average of the initial and final oxygen concentration readings was used for diffusivity data analysis. Oxygen concentrations were measured at

10 minutes intervals and each silk fibroin film was tested 3 times consecutively.

The effective oxygen diffusion coefficient was calculated as previously reported41·42. In brief, under the assumptions of (i) well-mixed fluid in the diffusion system, (ii) no oxygen consumption, (iii) linear oxygen concentration across the tested silk films and (iv) instantaneous steady state, the mass of oxygen in the receiving chamber is described by:

$$V^R\frac{dc_R}{dt} = A_M j_M + Q_B \tag{2}$$

$$j_M = \frac{\mathcal{D}_{e,M}}{l}(c^* - c_R) \tag{3}$$

$$Q_B = D_B(c^* - c_R) \tag{4}$$

where Equation 2 describes the change of the oxygen concentration (c_R) as a function of time (t) in the receiving chamber of volume V_R through a membrane of area A_M and with a flux j_M. Q_B represents the experimental baseline input rate of oxygen. Equation 3 describes the flux of oxygen j_M through the semipermeable membrane, where $D_{e,M}$ is the effective membrane oxygen diffusion coefficient, l is the membrane thickness and C_D is the concentration of dissolved oxygen in the donor chamber, which is constant. Equation 4 describes the experimental baseline flux, which is assumed to be linearly proportional to D_B, the experimental baseline volumetric diffusion factor. The parameter c*, is the maximum amount of oxygen in solution, which is equal to $c_D$measured in the donor chamber. Equation 4 can then be re-written as:

$$Q_B = D_B(c_D - c_R) \tag{5}$$

The combination of Equations 2, 4 and 5 gives:

$$\frac{dc_R}{(c_D - c_R)} = \mathcal{D}'dt \tag{6}$$

in which

$$\mathcal{D}' = \beta'\mathcal{D}_{e,M} + \frac{D_B}{V_R} \tag{7}$$

and

$$\beta' = \frac{A_M}{lV_R} \tag{8}$$

where D' is the diffusion factor and β' is the effective membrane characteristic geometric constant.

The following initial conditions apply: at t =D 0, c_R =D c_{R0}. Thus, integrating Equation 6 yields:

$$ln\left(\frac{C_D - C_R}{C_D - C_{R0}}\right) = D't \tag{9}$$

Indeed, the diffusion factor D' was calculated for each dataset from the slope of the experimental logarithmic concentration data versus time. The baseline volumetric diffusion factor (D_B) is determined by replacing silk film samples with an oxygen impermeable barrier (i.e. rubber-made stopper). Because it is assumed that in this case $D_{e,M}$ =D 0, then Equation 9 becomes:

$$ln\left(\frac{C_{D,rs} - C_{R,rs}}{C_{D,rs} - C_{R0,rs}}\right) = \frac{D_B}{V_R}t \tag{10}$$

where the suffix *rs* indicates the use of a rubber-made stopper instead of a silk fibroin membrane. The effective silk film oxygen diffusion coefficient $D_{e,M}$ is then calculated as:

$$D_{e,M} = \frac{1}{\beta}\left(D' - \frac{D_B}{V_R}\right) \tag{11}$$

where D_B is obtained by applying Equation 10 when the baseline effect is investigated. As baseline effect is independent to the membrane used, the measured D_B and D' may be used to calculate $D_{e,M}$ from Equation 11.

Measurement of strawberry respiration rate

A previously published method was followed to measure the respiration rate of strawberries. In brief, strawberry samples (around 100 g, n =D 3) were placed in a 1 L hermetic glass jars with a septum in the lid for sampling gas at different sampling times, over period of 36 hours. The jars were stored at ambient temperature of 22 °C and RH =D 38%. Gas sampling was carried out every 30 minutes for the first 5 hours, every

90 minutes till the 12th hour and every 180 minutes for the remaining 24 hours with a needle probe. Three replicates were performed for each coating treatment. The respiration rate was calculated by using the following equation:

$$Respiration\ rate = \frac{\Delta CO_2}{100} \cdot V_{headspace} \cdot \frac{1000}{m} \cdot \frac{60}{t} \qquad \left[\frac{mlCO_2}{kg \cdot h}\right] \tag{12}$$

where m is mass of strawberry, $V_{headspace}$ is the empty volume of the hermetic jar [ml], ΔDCO_2 is the difference between the initial and final concentration of CO_2 and t is sampling time [min].

Measurement of strawberry firmness

The firmness of silk fibroin uncoated and coated strawberries was measured at 22 °C and RH =D 38% through a puncture test using an Instron uniaxial system equipped with a 10N load cell. Strawberry firmness was evaluated as a function of storage time (as received, days 1, 3 and 7) and silk fibroin crystallinity degree. The test was performed according to a previously reported protocol[43]. In brief, a 5 mm diameter stainless steel rod with a flat end was used as probe to penetrate the strawberries analysed. The maximum penetration force (N) was defined as the maximum force required pushing the probe into the strawberries (n =D 5) to a depth of 8 mm at a cross-head speed of 1 mm/s.

Statistics

All data were collected for n =D 5, unless differently specified in the text. Statistical significance of the measurements was evaluated via one-way ANOVA test with Tukey mean analysis.

ACKNOWLEDGEMENTS

The authors would like to acknowledge funding from the Office of Naval Research (N00014-13-1-0596) for this work.

REFERENCES

1. FAO. Global food losses and food waste. Available at http://www. fao.org/docrep /014/mb060e/mb060e.pdf(Accessed: 8th March 2016) (2011).

2. Falguera, V., Quintero, J. P., Jiménez, A., Muñoz, J. A. & Ibarz, A.Edible films and coatings: Structures, active functions and trends in their use. *Trends in Food Science & Technology* **22**, 292–303, http://dx.doi.org/10.1016/j.tifs. 2011.02. 004 (2011).
3. Guilbert, S. & Gontard, N. In *Innovations in Food Packaging* (ed. Han J. H.) 263–276 (Academic Press, 2005).
4. Vargas, M., Pastor, C., Chiralt, A., McClements, D. J. & González-Martínez, C. Recent advances in edible coatings for fresh and minimally processed fruits. *Critical Reviews in Food Science and Nutrition* **48**, 496–511, 10.1080/10408390 701537344 (2008).
5. Janjarasskul, T. & Krochta, J. M. Edible packaging materials. *Annual Review of Food Science and Technology* **1**, 415–448, 10.1146/annurev.food.080708.100836 (2010).
6. Fernández-Pan, I., Ignacio, J. & Caballero, M. In *Biopolymers – New Materials for Sustainable Films and Coatings* (ed. Plackett, D.) 233–254 (John Wiley & Sons, Ltd, 2011).
7. Ribeiro, C., Vicente, A. A., Teixeira, J. A. & Miranda, C.Optimization of edible coating composition to retard strawberry fruit senescence. *Postharvest Biology and Technology* **44**, 63–70, http://dx.doi.org/10.1016/ j.postharvbio. 2006. 11.015 (2007).
8. Guilbert, S., Gontard, N. & Cuq, B. Technology and applications of edible protective films. *Packaging Technology and Science* **8**, 339–346 (1995).
9. Mannheim, C. H. & Soffer, T. Permeability of different wax coatings and their effect on citrus fruit quality. *Journal of Agricultural and Food Chemistry* **44**, 919–923 (1996).
10. Arnon, H., Zaitsev, Y., Porat, R. & Poverenov, E. Effects of carboxymethyl cellulose and chitosan bilayer edible coating on postharvest quality of citrus fruit. *Postharvest Biology and Technology* **87**, 21–26, 10.1016/j.postharvbio. 2013.08.007 (2014).
11. Arnon, H., Granit, R., Porat, R. & Poverenov, E. Development of polysaccharides-based edible coatings for citrus fruits: A layer-by-layer approach. *Food Chemistry* **166**, 465–472, 10.1016/j.foodchem.2014.06.061 (2015).
12. Bai, J., Hagenmaier, R. D. & Baldwin, E. A. Coating selection for 'Delicious' and other apples. *Postharvest Biology and Technology***28**, 381–390, 10.1016/S0925-5214(02)00201-6 (2003).
13. Hagenmaier, R. D. & Shaw, P. E. Gas Permeability of Fruit Coating Waxes. *Journal of the American Society for Horticultural Science***117**, 105–109 (1992).
14. Tanada-Palmu, P. S. & Grosso, C. R. F. Effect of edible wheat gluten-based films and coatings on refrigerated strawberry (Fragaria ananassa) quality. *Postharvest Biology and Technology* **36**, 199–208, http://dx.doi.org/ 10.1016/ j.postharv bio.2004.12.003(2005).
15. Negri, A. S. *et al.* Comparative analysis of fruit aroma patterns in the domesticated wild strawberries "Profumata di Tortona" (F. moschata) and "Regina delle Valli" (F. vesca). *Frontiers in Plant Science* **6**, 56, 10.3389/fpls.2015.00056 (2015).
16. Omenetto, F. G. & Kaplan, D. L. New Opportunities for an Ancient Material. *Science* **329**, 528–531, 10.1126/science.1188936 (2010).
17. Tao, H., Kaplan, D. L. & Omenetto, F. G. Silk Materials–A Road to Sustainable High Technology. *Advanced Materials* **24**, 2824–2837, 10.1002/adma.201104477 (2012).
18. Jin, H. J. & Kaplan, D. L. Mechanism of silk processing in insects and spiders. *Nature* **424**, 1057–1061, 10.1038/nature01809 (2003).
19. Hwang, S.-W. *et al.* A Physically Transient Form of Silicon Electronics. *Science* **337**, 1640–1644, 10.1126/science.1226325 (2012).
20. Kim, D.-H. *et al.* Epidermal Electronics. *Science* **333**, 838–843, 10.1126/ science. 1206157 (2011).

21. Kim, D.-H. *et al.* Dissolvable films of silk fibroin for ultrathin conformal bio-integrated electronics. *Nature Materials* **9**, 511–517, 10.1038/nmat2745 (2010).
22. Tao, H. *et al.* Silk-Based Conformal, Adhesive, Edible Food Sensors. *Advanced Materials* **24**, 1067–1072, 10.1002/adma.201103814 (2012).
23. Tao, H. *et al.* Silk-based resorbable electronic devices for remotely controlled therapy and *in vivo* infection abatement. *Proceedings of the National Academy of Sciences of the United States of America* **111**, 17385–17389, 10.1073/pnas.1407743111 (2014).
24. Kim, S. *et al.* All-water-based electron-beam lithography using silk as a resist. *Nature Nanotechnology* **9**, 306–310, 10.1038/nnano.2014.47 (2014).
25. Hardy, J. G. & Scheibel, T. R. Production and processing of spider silk proteins. *Journal of Polymer Science, Part A: Polymer Chemistry* **47**, 3957–3963, 10.1002/pola.23484 (2009).
26. Hu, X. *et al.* Regulation of Silk Material Structure by Temperature-Controlled Water Vapor Annealing. *Biomacromolecules* **12**, 1686–1696, 10.1021/bm200062a (2011).
27. Altman, G. H. *et al.* Silk-based biomaterials. *Biomaterials* **24**, 401–416, 10.1016/s0142-9612(02)00353-8 (2003).
28. Lawrence, B. D., Pan, Z., Liu, A., Kaplan, D. L. & Rosenblatt, M. I.Human corneal limbal epithelial cell response to varying silk film geometric topography in vitro. *Acta Biomaterialia* **8**, 3732–3743 (2012).
29. Lawrence, B. D., Marchant, J. K., Pindrus, M. A., Omenetto, F. G. & Kaplan, D. L. Silk film biomaterials for cornea tissue engineering. *Biomaterials* **30**, 1299–1308 (2009).
30. Lawrence, B. D. *et al.* Effect of Hydration on Silk Film Material Properties. *Macromolecular Bioscience* **10**, 393–403, 10.1002/mabi.200900294 (2010).
31. Pritchard, E. M., Dennis, P. B., Omenetto, F., Naik, R. R. & Kaplan, D. L. Review physical and chemical aspects of stabilization of compounds in silk. *Biopolymers* **97**, 479–498, 10.1002/bip.22026 (2012).
32. Rockwood, D. *et al.* Materials fabrication from Bombyx mori silk fibroin. *Nature Protocols* **6**, 1612–1631, 10.1038/nprot.2011.379 (2011).
33. Janjarasskul, T. & Krochta, J. M. In *Annual Review of Food Science and Technology, Vol. 1 Annual Review of Food Science and Technology* (eds M. P. Doyle & T. R. Klaenhammer) 415–448 (Annual Reviews, 2010).
34. Pritchard, E. M. *et al.* Encapsulation of oil in silk fibroin biomaterials. *Journal of Applied Polymer Science* **131**, 10.1002/app.39990 (2014).
35. Matsumoto, A., Lindsay, A., Abedian, B. & Kaplan, D. L. Silk Fibroin Solution Properties Related to Assembly and Structure. *Macromolecular Bioscience* **8**, 1006–1018, 10.1002/mabi.200800020 (2008).
36. World_Health_Organization. Guidelines for drinking-water quality - Volume 1: Recommendations. **668** (2008).
37. Hu, X., Kaplan, D. & Cebe, P. Dynamic protein-water relationships during βD-sheet formation. *Macromolecules* **41**, 3939–3948, 10.1021/ma071551d (2008).
38. ASTM. in *E96 / E96M-14* Vol. **E96**, (ASTM International, West Conshohocken, PA, 2014).
39. Broumand, A., Emam-Djomeh, Z., Hamedi, M. & Razavi, S. H.Antimicrobial, water vapour permeability, mechanical and thermal properties of casein based Zataraia multiflora Boiss. Extract containing film. *LWT - Food Science and Technology* **44**, 2316–2323, http://dx.doi.org/10.1016/j.lwt.2011.07.002 (2011).
40. Mauer, L. J., Smith, D. E. & Labuza, T. P. Water vapor permeability, mechanical, and structural properties of edible βD-casein films. *International Dairy Journal* **10**, 353–358, http://dx.doi.org/10.1016/S0958-6946(00)00061-3 (2000).
41. Valentin, J. E. *et al.* Oxygen diffusivity of biologic and synthetic scaffold materials for tissue engineering. *Journal of Biomedical Materials Research Part A* **91A**, 1010–1017, 10.1002/jbm.a.32328 (2009).

42. Androjna, C., Gatica, J. E., Belovich, J. M. & Derwin, K. A. Oxygen Diffusion through Natural Extracellular Matrices: Implications for Estimating "Critical Thickness" Values in Tendon Tissue Engineering. *Tissue Engineering Part A* **14**, 559–569, 10.1089/tea.2006.0361 (2008).
43. Velickova, E., Winkelhausen, E., Kuzmanova, S., Alves, V. D. & Moldão-Martins, M. Impact of chitosan-beeswax edible coatings on the quality of fresh strawberries (Fragaria ananassa cv Camarosa) under commercial storage conditions. *LWT - Food Science and Technology* **52**, 80–92, http://dx.doi.org/10.1016/j.lwt.2013.02.004 (2013).

Index